W9-BMV-371

The Night Sky in March

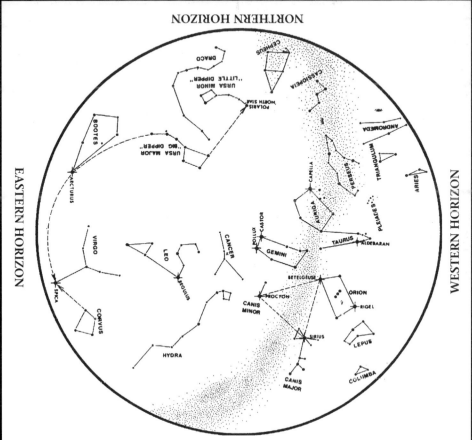

Latitude of chart is 34°N, but it is practical throughout the continental United States.

To use: Hold chart vertically and turn it so the direction you are facing shows at the bottom.

Chart time (Local Standard):

10 P.M. First of month

9 P.M. Middle of month

8 P.M. Last of month

Star Chart from *Griffith Observer*, Griffith Observatory, Los Angeles

ALPHA

The Night Sky in June

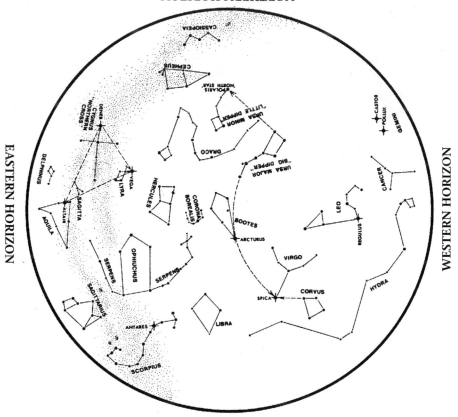

NORTHERN HORIZON

EASTERN HORIZON

WESTERN HORIZON

SOUTHERN HORIZON

Latitude of chart is 34°N, but it is practical throughout the continental United States.

To use: Hold chart vertically and turn it so the direction you are facing shows at the bottom.

Chart time (Local Standard):

10 P.M. First of month

9 P.M. Middle of month

8 P.M. Last of month

Star Chart from *Griffith Observer*, Griffith Observatory, Los Angeles

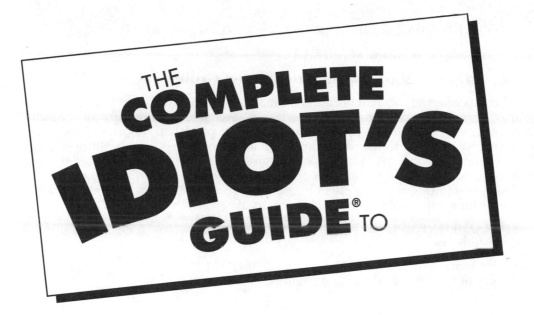

THE COMPLETE IDIOT'S GUIDE® TO

Astronomy

Second Edition

by Christopher De Pree and Alan Axelrod

ALPHA

A Pearson Education Company

To my girls, Julia, Claire, and Madeleine (CGD)

For my stars, Anita and Ian (AA)

Publisher
Marie Butler-Knight

Product Manager
Phil Kitchel

Managing Editor
Jennifer Chisholm

Acquisitions Editor
Mike Sanders

Development Editor
Amy Gordon

Production Editor
Billy Fields

Copy Editor
Amy Lepore

Illustrator
Brian Moyer

Cover Designers
Mike Freeland
Kevin Spear

Book Designers
Scott Cook and Amy Adams of DesignLab

Indexer
Lisa Wilson

Layout/Proofreading
Angela Calvert
Svetlana Dominguez
Mary Hunt
Gloria Schurick

Contents at a Glance

Appendixes

Contents

6 You and Your Telescope 81

Do I Really Need a Telescope?...82
Science Aside, What Will It Cost?..87
Decisions, Decisions 89
Refractors: Virtues and Vices 89
Reflectors: Newton's Favorite 90
Rich-Field Telescopes: Increasing in Popularity 90
Schmidt-Cassegrain: High-Performance Hybrid...................90
Maksutov-Cassegrain: New Market Leader 91
Dobsonians: More for Your Money? 92
The Go-To Revolution 93
I've Bought My Telescope, Now What? 94
Grab a Piece of Sky 94
Become an Astrophotographer 95
Light Pollution and What to Do About It..............................96
Finding What You're Looking For...97
Learning to See 98
Low-Light Adjustment 98
Don't Look Too Hard 99

7 Over the Rainbow 101

Making Waves 102
Anatomy of a Wave..102
New Wave..104
Big News from Little Places 104
Full Spectrum...105
The Long and the Short of It 106
What Makes Color?...107
Heavenly Scoop 108
Atmospheric Ceilings and Skylights 109
The Black-Body Spectrum..110
Watch Your Head, Here Comes an Equation........................111
Read Any Good Spectral Lines Lately? 112

8 Seeing in the Dark 117

Dark Doesn't Mean You Can't See 118
A Telephone Man Tunes In 118
Anatomy of a Radio Telescope 121
Bigger Is Better: The Green Bank Telescope 121
Interference Can Be a Good Thing 123
What Radio Astronomers "See" 124

Part 4: To the Stars   235

16 Our Star   237

17 Of Giants and Dwarfs: Stepping Out into the Stars   251

Foreword

Astronomy is one of the oldest scientific disciplines. Observations of the sky by ancient civilizations provided important milestones. Solar and lunar eclipses were prominent events as were the discovery of comets and "guest stars," now recognized to be supernovae. These "guest stars" were observed by Chinese, Japanese, and Korean astronomers (or astrologers) for the last two millennia and possibly were sighted by the ancestors of the native Americans of the U.S. Southwest. The prime example of this was the Crab supernova in 1054, a drawing of which can be seen at the Chaco Culture National Historical Park in New Mexico.

Humans have had a fascination with astronomy for thousands of years. At the end of the twentieth century, public interest in astronomy is at an all-time high. Few scientific disciplines have so many active and successful amateurs. Many important discoveries are made by amateurs, including comets, minor planets, and supernovae.

Of course, Hollywood has also played a role in popularizing astronomy. A prominent recent example is the 1997 Warner Brothers film *Contact,* starring Jodie Foster. The film was made in 1995–1996, partly at the National Science Foundation's Very Large Array in New Mexico. Ironically, the main subject matter of the film is SETI (the Search for Extra-Terrestrial Intelligence), one of the very few areas of astronomical research in which the VLA plays no role.

Chris De Pree and Alan Axelrod present a comprehensive tour of the universe in *The Complete Idiot's Guide to Astronomy, Second Edition.* Readers will enjoy the historical approach, starting with the ancients, moving on to Copernicus and Galileo, and ending in the modern era with Neil Armstrong and others. This book provides an excellent guide not only for first-time observers, but also for experienced amateur astronomers.

Astronomical techniques, the solar system, stars, and the distant universe are described in a concise but thorough manner. The simple physical concepts underlying these phenomena are presented as they are required.

Finally, a few words about the senior author, Chris De Pree. Chris was a summer student at the Very Large Array a few years ago while he was a graduate student at the University of North Carolina, Chapel Hill. He later moved to the VLA for two years, where he completed his UNC Ph.D., working on radio observations of compact HII regions. He received his doctorate in 1996 and then moved to Decatur, Georgia, to join the faculty of Agnes Scott College as (not surprisingly) a professor of astronomy.

Astronomy at Agnes Scott has begun a new and vital era, and readers of *The Complete Idiot's Guide to Astronomy, Second Edition* are in for a treat that is informative and exhilarating as well as challenging.

W. Miller Goss, Ph.D.
Director, Very Large Array, Very Long Baseline Array
National Radio Astronomy Observatory of the National Science Foundation

Introduction

You are not alone.

Relax. That statement has nothing to do with the existence of extraterrestrial life—though we *will* get around to that, too, way out in Chapter 24, "Table for One." For the present, it applies only to our mutual interest in astronomy. For we (the authors) and you (the reader) have come together because we are the kind of people who look up at the sky a lot and have all kinds of questions about it. This habit hardly brands us as unique. Astronomy, the scientific study of matter in outer space, is among the most ancient of human studies. The very earliest scientific records we have—from Babylon, from Egypt, from China—all concern astronomy.

Recorded history spans about 5,500 years. The recorded history of astronomy starts at the beginning of that period. People have been sky watchers for a very, very long time.

And yet astronomy is also among the most modern of sciences. Although we possess the collected celestial observations of some 50 centuries, almost all that we know about the universe we have learned in the century just ended, and most of *that* knowledge has been gathered since the development of radio astronomy in the 1950s. In fact, the lifetime of any reader of this book, no matter how young, is filled with astronomical discoveries that merit being called milestones. Think it was a pretty big deal when Copernicus, in the early sixteenth century, proposed that the sun, not the earth, was at the heart of the solar system? Well, did you know that a Greek astronomer actually proposed the same idea nearly 2,000 years earlier? His pitch just wasn't as good.

Astronomy is an ancient science on the cutting edge. Great discoveries were made centuries ago. Great discoveries are being made today. And great leaps forward in astronomical knowledge have often followed leaps forward in technology: the invention of the telescope, the invention of the computer, the development of fast, cheap computers. So much is being learned every day that we've been asked to bring out a revised edition of this book, the first edition of which came out only two years ago. And even more recent discoveries will be on the table by the time you read this new edition.

Yet you don't have to be a government or university scientist with your eager fingers on millions of dollars' worth of equipment to make those discoveries. For if astronomy is both ancient and advanced, it is also universally accessible: up for grabs.

The sky belongs to anyone with eyes, a mind, imagination, a spark of curiosity, and the capacity for wonder. If you've also got a few dollars to spend, a good pair of binoculars or a telescope makes more of the sky available to you. (Even if you don't want to spend the money, chances are your local astronomy club will let you use members equipment if you come and join them for a cold night under the stars.) And if you have a PC and Internet connection available, you—yes, *you*—have access to much of the information that those millions of dollars in government equipment produce: images from the world's great telescopes and from a wealth of satellite probes, including the *Hubble Space Telescope* and the *Mars Global Surveyor*. This information is all free for the downloading. (See Appendix E, "Sources for Astronomers" for some starting points in your online searches.)

We are not alone. No science is more inclusive than astronomy.

Nor is astronomy strictly a spectator sport. You don't have to peek through a knot-hole and watch the game. You're welcome to step right up to the plate. Many new comets are discovered by astronomy buffs, backyard sky watchers, not Ph.D. scientists in a domed observatory. Most meteor observations are the work of amateurs. You can even get in on such seemingly esoteric fields as radio astronomy and the search for extraterrestrial intelligence (see Chapter 7, "Over the Rainbow" for both).

But most important are the discoveries you can make for yourself: like *really* seeing the surface of the moon, or looking at the rings of Saturn for the first time through your own telescope, or observing the phases of Venus, or suddenly realizing that the fuzzy patch of light you're looking at is not just Messier Object 31, but Andromeda, a whole galaxy as vast as our own. Those photons that left Andromeda millions of years ago are landing on *your* retina.

We'd enjoy nothing more than to help you get started on your journey. Here's a map.

How This Book Is Organized

Part 1, "Finding Our Place in Space," orients you in the evening sky and presents a brief history of astronomy.

Part 2, "Now You See It (Now You Don't)," explains how telescopes work, offers advice on choosing a telescope of your own, and provides pointers to help you get the most from your telescope. You'll also find an explanation of the electromagnetic spectrum (of which visible light is only one part) and how astronomers use radio telescopes and other instruments to "see" the invisible portions of that spectrum. Finally, we'll take you into the cosmos aboard a host of manned and unmanned probes, satellites, and space-borne observatories.

Part 3, "A Walk Around the Block," begins with a visit to our nearest neighbor, the moon, and then ventures out into the rest of the solar system. You'll find here a discussion of the birth and development of the solar system and a close look at the planets and their moons, as well as such objects as asteroids and comets.

Part 4, "To the Stars," begins with our own sun, taking it apart, showing how it works, and providing instructions for safely viewing it both day to day and during an eclipse. From our sun, we venture beyond the solar system to the other stars and learn how to observe them meaningfully. The last three chapters in this section discuss the birth and evolution of stars, ending with their collapse as neutron stars and black holes.

Part 5, "Way Out of This World," pulls back from individual stars to take in entire galaxies, beginning with our own Milky Way. We learn how astronomers observe, measure, classify, and study galaxies and how those galaxies are all rushing away from us at incredible speed. The section ends with the so-called active galaxies, which emit unimaginably huge quantities of energy and can tell us much about the origin and fate of the universe.

Part 6, "The Big Questions," asks how the universe was born (and offers the Big Bang theory by way of an answer); asks whether the existence of extraterrestrial life and even civilizations is possible, probable, or perhaps inevitable; and, finally, asks if (and how) the universe will end.

At the back of the book, you'll find a series of appendixes that defines key terms, lists upcoming eclipses, catalogs the constellations, provides the classic Messier Catalog of deep-space objects that amateurs can readily observe, and lists sources of additional information, including great astronomy Web sites.

Extras

In addition to the main text and illustrations of *The Complete Idiot's Guide to Astronomy, Second Edition*, you'll also find other types of useful information, including definitions of key terms, important statistics and scientific principles, amazing facts, and special subjects of interest to sky watchers. Look for these features:

Astro Byte

Here are some startling astronomical facts and amazing trivia. Strange—but true!

Close Encounter

In these boxes, you'll find discussions elaborating on important events, projects, issues, or persons in astronomy.

Star Words

These boxes define some key terms used in astronomy.

Astronomer's Notebook

This feature highlights important statistics, scientific laws and principles, measurements, and mathematical formulas.

Trademarks

All terms mentioned in this book that are known to be or are suspected of being trademarks or service marks have been appropriately capitalized. Alpha Books and Pearson Education Inc., cannot attest to the accuracy of this information. Use of a term in this book should not be regarded as affecting the validity of any trademark or service mark.

Part 1
Finding Our Place in Space

We know this isn't your first night out and that you've certainly looked up at the sky before. Maybe you can find the Big Dipper and even Orion—or at least his Belt—but, for the most part, all the stars look pretty much the same to you, and you can't tell a star from a planet.

The first chapter of this part gets you started with the constellations. The second chapter introduces ancient astronomy. The third looks at the motions of the solar system and why planets behave differently from stars. The last chapter in this part presents the work of the great astronomers of the Renaissance.

Chapter 1

Naked Sky, Naked Eye: Finding Your Way in the Dark

In This Chapter

➤ What you can see with your naked eye

➤ The celestial sphere

➤ Orienting yourself among the stars

➤ Celestial coordinates and altazimuth coordinates

➤ Identifying constellations: how and why

Want to make a movie on an extraterrestrial theme? Hollywood has been using space as a backdrop for quite some time, and it's especially big box office these days. Audiences are thrilled by special effects: blazing comets, flaming meteors, brightly banded planets, strange, dark moons. Just be prepared to spend upward of $100 million to make your film. Those special effects don't come cheap, and today's moviegoers are spoiled by one dazzling spectacle after another. Whatever did people do for excitement before *2001: A Space Odyssey, Star Wars, Star Trek, Independence Day,* and *Contact?*

They looked at the sky.

This chapter will tell you what they saw.

Sun Days

We've become jaded—a bit spoiled—by the increasingly elaborate and costly special effects in today's sci-fi flicks, but none of us these days is nearly as spoiled as the sky most of us look at.

Imagine yourself as one of your ancestors, say ten thousand years ago. Your reality consists of a few tools, household utensils, perhaps buildings (the city-states were beginning to appear along the Tigris) and, of course, all that nature has to offer: trees, hills, plants, rivers, streams—and the sky.

The sky is the biggest, greatest, most spectacular object you know. During the day, the sky is crossed by a brightly glowing disk from which all light and warmth emanate. Announced in the predawn hours by a pink glow on the eastern horizon, the great disk rises, then arcs across the sky, deepening toward twilight into a ruddy hue before slipping below the horizon to the west. Without electric power, your working hours are dictated by the presence of the sun's light.

Flat Earth, Big Bowl

As the sun's glow fades and your eyes become accustomed to the night, the sky gradually fills with stars. Thousands of them shimmer blue, silvery white, some gold, some reddish, seemingly set into a great dark bowl, the *celestial sphere,* overarching the flat earth on which you stand.

Thousands of stars in the night sky?

Maybe that number has brought you back through a starlit ten thousand years and into the incandescent lamp light of your living room or kitchen or bedroom or wherever you are reading this: "*I've* never seen thousands of stars!" you protest.

We said earlier that, from many locations, our sky is spoiled. The sad fact is that, these days, fewer and fewer of us can see anything like the three thousand or so stars that *should* be visible to the naked eye on a clear evening. Ten thousand years ago, the night sky was not lit up with the *light pollution* of so many sources of artificial illumination. Unless you sail far out to sea or travel to the high, dry desert of the Southwest, you might go through your entire life without *really* seeing the night sky, at least not the way our ancestors saw it.

Star Words

The **celestial sphere** is an imaginary sphere that we picture surrounding the earth upon which the stars are fixed. Some ancient cultures believed such a sphere (or bowl) really existed. Today, however, astronomers use the concept as a way to map the location of stars relative to observers on Earth.

Star Words

Light pollution is the result of photons of light that goes up instead of down. Light that goes **down** (from fixtures) illuminates the ground. Light that goes **up** makes the stars harder to see. Contact your local astronomy club to find ways to combat light pollution.

Man in the Moon

Even in our smog- and light-polluted skies, however, the Moon shines bright and clear. Unlike the Sun, which appears uniform, the surface of the Moon has details we can see, even without a telescope. Even now, some three decades after human beings walked, skipped, and jumped on the Moon and even hit a golf ball across the lunar surface, the Moon holds wonder. Bathed in its silver glow, we may feel a connection with our ancestors of 10 millennia ago. Like them, we see in the lunar blotches the face of the "Man in the Moon."

Neil Armstrong took this picture of fellow astronaut "Buzz" Aldrin about to join him on the surface of the Moon, July 20, 1969.

(Image from arttoday.com*)*

If the face of the Moon presented a puzzle to our ancestors, they were also fascinated by the way the Moon apparently changed shape. One night, the Moon might be invisible (a new moon); then, night by night, it would appear to grow (wax), becoming a crescent; and, by one week later, be a quarter moon (which is a *half* moon in shape). Through the following week, the Moon would continue to wax, entering its *gibbous* phase, in which more than half of the lunar disk was seen. Finally, two weeks after the new moon, all of the lunar disk would be visible: The full moon would rise majestically at sunset. Then, through the next two weeks, the Moon would appear to shrink (wane) night after night, passing back through the gibbous, quarter, and crescent phases, until it became again the all-but-invisible new moon.

Star Words

Gibbous is a word from Middle English that means "bulging"—an apt description of the Moon's shape between its quarter phase and full phase.

Close Encounter

For untold generations, people have discerned a human face in the crater-scarred markings of the Moon. The *Man* in the Moon is sometimes interpreted as an old woman cooking. Among Native Americans, the face or faces in the Moon have been described (for example) as a frog charged with protecting the Moon from a bear who would otherwise swallow it. An ancient Scandinavian folktale speaks of Hjuki and Bill, perhaps the original Jack and Jill, who, carrying a pail of water, tumbled down a hill as they ran from their cruel father. They were rescued by the embrace of the Moon. For Scandinavian kids, the "Man in the Moon" is the image of Hjuki and Bill, complete with pail.

The cycle takes a little more than 29 days, a month, give or take, and it should be no surprise that the word "month" derived from the word "moon." In fact, just as our ancestors learned to tell the time of day from the position of the Sun, so they measured what we call weeks and months by the lunar phases. The lunar calendar is of particular importance in many world religions, including Judaism and Islam. For those who came before us, the sky was more than something to marvel at. It could also be used to guide and coordinate human activity. As we will see in Chapters 2 and 3, the ancients became remarkably adept at using the heavens as a great clock and calendar.

The phases of the Moon. The globe in the center is Earth. The inner circle shows how the sunlight illuminates the Moon as it orbits Earth. The outer circle shows how the Moon appears from Earth.

(Image from the authors' collection)

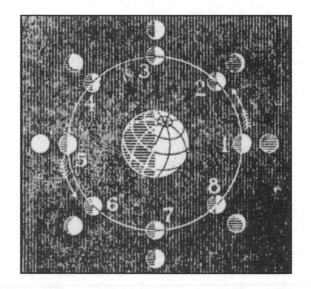

Lights and Wanderers

Ten thousand years ago, family time at night was not occupied with primetime sitcoms followed by the news and *David Letterman*. Our ancestors were not glued to television screens, but presumably to the free show above, the celestial sphere. Early cultures noticed that the bowl above them rotated from east to west. They concluded that what they were seeing was the celestial sphere—which contained the stars—rotating, and not the individual stars. All the stars, they noticed, moved together, their positions relative to one another remaining unchanged. (That the stars "move" because of *Earth's* rotation was a concept that lay far in the future.)

The coordinated movement of the stars was in dramatic contrast to something else the ancient sky watchers noticed. While the vast majority of stars were clearly fixed in the rotating celestial sphere, a few—the ancients counted five—seemed to meander independently, yet regularly, across the celestial sphere. The Greeks called these five objects *planetes,* "wanderers," and, like nonconformists in an otherwise orderly society, the wanderers would eventually cause trouble. Their existence would bring the entire heavenly status quo into question and, ultimately, the whole celestial sphere would come crashing down.

Celestial Coordinates

But we're getting ahead of our story. In Chapter 4, "Astronomy Reborn: 1543–1687," you'll find out why we no longer believe that the celestial sphere represents reality; however, the notion of such a fixed structure holding the stars is still a useful model for us moderns. It helps us to communicate with others about the positions of the objects in the sky. We can orient our gaze into the heavens by thinking of the point of sky directly above the earth's North Pole as the north celestial pole, and the point below the South Pole as the south celestial pole. Just as the earth's equator lies midway between the North and South Poles, so the celestial

Star Words

Declination is the angular distance (distance expressed as an angle rather than in absolute units, such as feet or miles) north or south of the celestial equator. It is akin to lines of latitude on the earth.

Astronomer's Notebook

Declination is analogous to Earthly latitude. The declination of a star seen directly above the earth's equator would also be at the celestial equator—that is, 0 degrees. A star at the north celestial pole (that is, directly over the earth's North Pole) would be +90 degrees. At the south celestial pole, it would be -90 degrees. In the latitudes of the United States, stars directly overhead have declinations in the +30- to +40-degree ranges. The Bradley Observatory at Agnes Scott College is at a latitude of 33 deg, 45 min, 55.84 sec. That means that in Decatur, GA, the North Star (Polaris) is about 34 degrees above the northern horizon.

equator lies equidistant between the north and south celestial poles. Think of it this way: If you were standing at the North Pole, then the north celestial pole would be directly overhead. If you were standing at the equator, the north and south celestial poles would be on opposite horizons. And if you were standing at the South Pole, the south celestial pole would be directly overhead.

Astronomers have extended to the celestial sphere the same system of latitude and longitude that describes earthly coordinates. The lines of latitude, you may recall from geography, run parallel with the equator and measure angular distance north or south of the equator. On the celestial sphere, *declination* (dec) corresponds to latitude and measures the angular distance above or below the celestial equator. While earth-bound latitude is expressed in degrees north or south of the equator (Philadelphia, for instance, is 40 degrees north), celestial declination is expressed in degrees + (above) or – (below) the celestial equator. The star Betelgeuse, for example, is at a declination of +7 degrees, 24 minutes.

On a globe, the lines of longitude run vertically from pole to pole. They demarcate angular distance measured east and west of the so-called prime meridian (that is, 0 degrees), which by convention and history has been fixed at Greenwich Observatory, in Greenwich, England. On the celestial sphere, *right ascension* (R.A.) corresponds to longitude. While declination is measured in degrees, right ascension is measured in hours, minutes, and seconds, increasing from west to east, starting at 0. This zero point is taken to be the position of the sun in the sky at the moment of the vernal equinox (we'll discuss this in Chapter 3, "The Unexplained Motions of the Heavens"). Because the earth rotates once approximately every 24 hours, the same objects will return to their positions in the sky approximately 24 hours later. After 24 hours, the earth has rotated through 360 degrees, so that each hour of R.A. corresponds to 15 degrees on the sky.

Star Words

Right ascension is a coordinate for measuring the east-west position of objects in the sky.

If the celestial poles, the celestial equator, and declination are projections of earthly coordinates (the poles, the equator, and latitude), why not simply imagine R.A. as projections of lines of longitude?

There are good reasons why we don't. Think of it this way: The stars in the sky above your head in winter time are different than those in summer time. That is, in the winter we see the constellation Orion, for example, but in summer, Orion is gone, hidden in the glare of a much closer star, the sun. Well, although the stars above you are changing daily, your longitude (in Atlanta, for example) is not changing. So the coordinates of the stars cannot be fixed to the coordinates on the surface of the earth. As we'll see in later chapters, this difference comes from the fact that in addition to spinning on its axis, the earth is also orbiting the sun.

Measuring the Sky

The true value of the celestial coordinate system is that it gives the absolute coordinates of an object, so that two observers, anywhere on Earth, can direct their gaze to the exact same star. When you want to meet a friend in the big city, you don't tell her that you'll get together "somewhere downtown." You give precise coordinates: "Let's meet at the corner of State and Madison streets." Similarly, the right ascension and declination astronomers use tell them (and you) precisely where in the sky to look.

The celestial coordinate system can be confusing for the beginning sky watcher and is of little practical value to an observer armed with nothing but the naked eye. However, it can help the novice locate the North Star, and to know approximately where to look for planets.

There is a simpler way to measure the location of an object in the sky as observed from your location at a particular time. It involves two angles. You can use angles to divide up the horizon by thinking of yourself as standing at the center of a circle. A circle may be divided into 360 degrees (and a degree may be subdivided into 60 minutes, and a minute sliced into 60 seconds). Once you decide which direction is 0 degrees (the convention is to take due north as 0 degrees), you can measure, in degrees, precisely how far an object is from that point. Now that you have taken care of your horizontal direction, you can fix your vertical point of view by imagining an upright half circle extending from horizon to horizon. Divide this circle into 180 degrees, with the 90-degree point directly overhead. Astronomers call this overhead point the *zenith*.

Altitude and azimuth are the coordinates that, together, make up the altazimuth coordinate system, and, for most people, they are quite a bit easier to use than celestial coordinates. An object's *altitude* is its angular distance above the horizon, and its compass direction, called *azimuth,* is measured in degrees increasing clockwise from due north. Thus east is at 90 degrees, south at 180 degrees, and west at 270 degrees.

Altazimuth coordinates, while perhaps more intuitive than the celestial coordinate system, do have a serious shortcoming. They are valid only for your location on Earth at a particular time of day or night. In contrast, the celestial coordinate system is universal because its coordinate system moves with the stars in the sky.

Star Words

Altazimuth coordinates are **altitude** (angular distance above the horizon) and **azimuth** (compass direction expressed in angular measure).

The Size of Things, or "I Am Crushing Your Head!"

In a television show called *Kids in the Hall,* there was a character who would look at people far away through one eye and pretend to crush their heads between his thumb and forefinger. If you try this trick yourself, you'll notice that people have to be at least five or so feet away for their heads to be small enough to crush. Their heads don't actually get smaller, of course, just the angular size of the head does. In fact, you can use this same trick (if sufficiently distant) to crush cars, or planes flying overhead. All because of the fact that as things get more distant, they appear smaller—their *angular size* is reduced.

The surface of the earth is real and solid. You can easily use absolute units such as feet and miles to measure the distance between objects. The celestial sphere, however, is an imaginary construct, and we do not know the distances between us and the objects. In fact, simply to locate objects in the sky, we don't need to know their distances from us. We get that information in other ways, which we will discuss in several chapters. Now, from our perspective on Earth, two stars may appear to be separated by the width of a finger held at arm's length when they are actually many trillions of miles distant from each other. You could try to fix the measurement between two stars with a ruler, but where would you hold the measuring stick? Put the ruler close to your eye, and two stars may be a quarter-inch apart. Put it at arm's length, and the distance between those same two stars may have grown to several inches.

Astronomers use angular size and *angular separation* to discuss the apparent size on the sky or apparent distance between two objects in the sky. For example, if two objects were on opposite horizons, they would be 180 degrees apart. If one were on the horizon and the other directly overhead, they would be 90 degrees apart. You get the picture. Well, a degree is made up of even smaller increments. One degree is made up of 60 minutes (or *arcminutes*), and a minute is divided into 60 seconds (*arcseconds*).

Let's establish a quick and dirty scale. The full moon has an angular size of half a degree, or 30 arcminutes, or 1,800 arcseconds (these are all equivalent). The "smallest" celestial object the human eye can resolve is about 1 arcminute across. The largest lunar craters are about 2 arcminutes across, and separating objects that are 1–2 arcseconds apart is impossible (at least at optical wavelengths) from all but the best sites on Earth. This difficulty is due to atmospheric turbulence and is a limitation of current ground-based optical observing. Now that you know the full moon is about half a degree across, you can use its diameter to gauge other angular sizes.

Star Words

Angular size and **angular separation** are size and distance expressed as angles on the sky rather than as absolute units (such as feet or miles). Since many of these measurements are less than a full degree, we point out that a degree is made up of 60 **arcminutes** and an arcminute of 60 **arcseconds**.

To estimate angles greater than a half-degree, you can make use of your hand. Look at the sky. Hold your hand upright at arm's length, arm fully extended outward, the back of the hand facing you, your thumb and index finger fully and stiffly extended, your middle finger and ring finger folded in, and your pinky also fully extended. The distance from the tip of your thumb to the tip of your index finger is about 20 degrees (depending on the length of your fingers!). From the tip of your index finger to the tip of your pinky is 15 degrees; and the gap between the base of your index finger and the base of your pinky is about 10 degrees.

Celestial Portraits

Well, now that you're standing there with your arm outstretched and your head full of angles, what can you do with this wealth of information?

We now have some rough tools for measuring separations and sizes in the sky, but we still need a way to *anchor* our altazimuth measurements, which, remember, are relative to where we happen to be standing on Earth. We need the celestial equivalent of landmarks.

Fortunately for us, our ancestors had vivid imaginations.

Human brains are natural pattern makers. We have all seen elephants and lions masquerading as clouds in the sky. Present the mind with the spectacle of 3,000 randomly placed points of light against a sable sky, and, before you know it, it will start "seeing" some pretty incredible pictures. The *constellations*—arbitrary formations of stars that are perceived as figures or designs—are such pictures, many of them inspired by mythological heroes, whose images (in the western world) the Greeks created by connecting the dots.

By the second century C.E., Ptolemy (whom we'll meet in Chapter 3) listed 48 constellations in his *Almagest,* a compendium of astronomical knowledge. Centuries later, during the late Renaissance, more constellations were added, and a total of 88 are recognized today. We cannot say that the constellations were really discovered, because they do not exist except in the minds of those who see them. Grouping stars into constellations is an arbitrary act of the imagination and to present-day astronomers are a convenience. In much the same way that states are divided into counties, the night sky is divided into constellations. The stars thus grouped have no physical relationship to one another and, in fact, are many, many trillions of miles apart. Nor do they necessarily lie in the same plane with respect to the earth; some are much farther from us than others. But, remember, we simply imagine that they are embedded in the celestial sphere as a convenience.

If the constellations are outmoded figments of the imagination, why bother with them?

The answer is that they are convenient (not to mention poetic) celestial landmarks. We all use landmarks to navigate on land. "Take a right at the

Astronomer's Notebook

Of the 88 constellations, 28 are in the northern sky and 48 are in the southern sky. The remaining dozen lie along the ecliptic—a circle that describes the path that the sun takes in the course of a year against the background stars. This apparent motion is actually due to the earth moving around the sun. (We'll revisit the term ecliptic in Chapter 11). These 12 constellations are the zodiac, familiar to many as the basis of the pseudoscience (a body of lore masquerading as fact verified by observation) of astrology. All but the southernmost 18 of the 88 constellations are at least sometimes visible from part of the United States.

gas station," you might tell a friend. What's so special about that particular gas station? Nothing—until you invest it with significance as a landmark. Nor was there anything special about a group of physically unrelated stars—until they were invested with significance. Now these constellations can help us find our way in the sky and, unless you are using a telescope equipped with an equatorial mount, are more useful than either the celestial or altazimuth coordinate system.

Astro Byte

In an age when so many objects, of necessity, are referred to by rather cold catalog names (NGC 4258, W49A, K3-50A, to name a few), it is pleasing that we can still refer to some objects by their brightness within a given constellation. Cygnus X-1 is a famous x-ray source and black-hole candidate in the constellation of Cygnus, the swan.

Star Words

An **asterism** is an arbitrary grouping of stars within or associated with a constellation, which are perceived to have a recognizable shape (such as a Teapot or Orion's Belt) and, therefore, readily serve as celestial landmarks.

The Dippers First

Almost everybody knows the Big Dipper and maybe the Little Dipper, too. Actually, neither Dipper is a constellation, but are subsets of other constellations, Ursa Major and Ursa Minor, the big and little bears (official constellation names are in Latin). Such generally recognizable subgroups within constellations are called asterisms. The Big Dipper is not only bright, but it is easy to find in the northern sky in all seasons except fall, when it is low on the horizon. It might interest you to know that you'll find the Big Dipper between 11 and 14 hours R.A. and +50 to +60 degrees dec. Using your hand to estimate the Big Dipper's angular size, you'll see that it's about 25 degrees across the bowl to the end of the handle. But the really important thing is that its seven bright stars form a pattern that really does look like a dipper.

And that's what's so handy about *asterisms*. They are simpler, brighter, and more immediately recognizable than the larger, more complex constellations of which they are a subset. They will help you to find the constellations with which they are associated and generally help to orient you in the sky.

Seafarers and other wanderers have long used the Big Dipper as a navigational aid. If you trace an imaginary line between the two stars that mark the outer edge of the Big Dipper's bowl and extend that line beyond the top of the bowl, it points to Polaris, the North Star, about 25 degrees away. Polaris is very nearly at the north celestial pole (about 1 degree off), which means that it appears to move very little during the course of the night, and travelers have always used it as a compass. During the decades before the American Civil War, many Southern slaves escaped to the free North by following the North Star. For sky explorers, the combination of the Big Dipper and Polaris provide a major landmark useful in locating other constellations. One of those constellations—actually, it's another asterism—is the Little Dipper. Dimmer and smaller than the Big Dipper, it would be harder to find, except that Polaris, which we've just located, is at the tip of its handle. Like its big brother, this asterism consists of seven stars.

The Stars of Spring

Let's look at a few of the highlights of each season's sky.

With the arc of our Galaxy (the Milky Way) low and heading toward the western horizon, the spring sky offers fewer bright stars than any other season. This isn't necessarily a bad thing, because it makes identifying the three bright ones that much easier. Some 45 degrees south of the Big Dipper's bowl is the constellation Leo. If you can't quite pick out Leo, you might find it easier to identify the asterism called the Sickle, a kind of backwards question mark that forms Leo the Lion's mane. At the base of the Sickle is the bright star Regulus.

Arcturus, another bright star of spring, may be located by extending the curve of the Big Dipper's handle 35 degrees southward.

Yellow-orange in color, Arcturus is the brightest star of the constellation Boötes, the Charioteer.

Now extend the Big Dipper handle's curve beyond Arcturus, and you will find Spica (in Virgo), the third bright star of spring. In vivid contrast to the warm hue of Arcturus, Spica is electric blue. We'll find out in Chapter 17, "Of Giants and Dwarfs: Stepping Out into the Stars," that the color of a star actually tells us about its surface temperature. It can be quite a thrill looking at different stars in the sky and be able to "take their temperatures" simply from their colors.

Astronomer's Notebook

On the tear-out cards at the front and back of this book, you will find four "all-sky" star charts, which show the major constellations visible in the night sky during the four seasons. See Chapter 6 for tips on the best times and places to view the night sky and steps you can take to minimize the effects of light pollution.

Astro Byte

"Arc to Arcturus" is a handy mnemonic often taught to astronomy students to help them easily locate the star. Following the arc of the Big Dipper's handle leads to this bright star.

Summer Nights

Summer offers four bright stars. Three of them form a distinct right triangle called the summer triangle. Vega is at the triangle's biggest angle. Vega is also the brightest star in the constellation Lyra, the Lyre (or harp). South and east of Vega is the second brightest star in the triangle, Altair, which is in the constellation Aquila, the Eagle. Deneb is the third star of the summer triangle, and is in Cygnus, the Swan. Deneb is also part of the prominent asterism, the Northern Cross, and is the brightest star in that group.

Take a good long look at Deneb. Bright as it is—fourth brightest in the summer sky—it is one of the most distant stars visible to the naked eye, fifty times more distant than Vega and several hundred times farther than Alpha Centauri, our closest stellar companion.

If you are sufficiently far from sources of atmospheric and light pollution, and the night is clear and dry, you may notice that the Northern Cross lies within a kind of hazy band stretching across the sky. This band is the Milky Way, our own Galaxy, which we will explore in Chapter 21, "The Milky Way: Much More Than a Candy Bar," and whose haze is the light of some 100 billion or so stars.

Two other major summer constellations should not be missed. Sagittarius, the Archer, and Scorpius, the Scorpion, are found low in the southern sky about 30 degrees below the celestial equator. You can locate Scorpius by finding the fourth bright star of the summer sky, Antares, unmistakable for the red hue that gives it its name, which means "rival of Mars." If Scorpius is not below your horizon (and therefore out of sight), you should recognize its fishhook-shaped scorpion's tail.

One hour R.A. (15 degrees) east of Scorpius is Sagittarius. You may better recognize it by two asterisms within it: the Teapot, which looks as if it pours out on the tail of nearby Scorpius, and the Milk Dipper, called this because its dipper shape seems to dip into the milkiest part (thickest star cloud) of the Milky Way. As we will see, there is a reason that the Milky Way is thicker here, bulging slightly. Sagittarius is the direction of the center of our own Galaxy.

Fall Constellations

In the fall, the constellation Pegasus, winged horse of Greek mythology, is easy to locate. If you find it hard to imagine connecting the stars to trace out the horse, look for the highly recognizable asterism associated with Pegasus called the Great Square. At southern latitudes, by about 10 P.M. in early October, it should be directly above you. The four stars marking out its four corners aren't terribly bright, but the other stars in that area of the sky are fairly dim, so the figure should stand out clearly. The eastern side of Great Square also coincides with the 0 marking from which the hours of right ascension start, increasing to the east.

Some 20 degrees west and 5 degrees south of Markab, the star that marks the Great Square's southwest corner, is Enif, the brightest star in Pegasus. Its name means "the horse's mouth," and between Markab and Enif is the horse's neck. Look to the Great Square's northeast corner for the star Alpheratz, which is not part of Pegasus, but part of Andromeda, the Maiden in Chains.

If you trace a line from Alpheratz through Markab, continuing about 40 degrees southwest of Markab, you'll find the zodiacal constellation Capricornus, Capricorn, or the Sea Goat. Capricorn is distinguished by its brightest star, the brilliant Deneb Algiedi.

Return to the Great Square. About 20 degrees east of it, you'll find another zodiacal constellation, Aries, the Ram. This grouping is easy to identify, since it is marked by two fairly bright stars a mere 5 degrees apart.

Last to rise in the sky of fall is Perseus, slayer of snake-haired Medusa and other monsters of Greek mythology. About 45 degrees up in the northeast, it lies across the Milky Way and is marked by its brightest star, Mirfak.

Astro Byte

There is an abundance of stars that still have their Arabic names, a testament to the many contributions of Arab astronomers: Aldebaran, Mizar, Alcor, and so on.

Winter Skies

Winter nights, with the bright arc of the Milky Way overhead, offer more bright stars than are visible at any other season: Sirius, Capella, Rigel, Procyon, Aldebaran, Betelgeuse, Pollux, and Castor. Brightest and most readily recognizable of the winter constellations is Orion, the Hunter, which spans the celestial equator and sports the heavens' second most familiar asterism (after the Big Dipper): Orion's Belt, three closely spaced bright stars in a line 3 degrees long. The star Rigel, brightest in the Orion constellation, marks the hunter's foot, 10 degrees below and to the west of Orion's Belt. About the same distance and direction above the Belt is Betelgeuse, a reddish star, whose name is Arabic for "armpit of the giant." And that is precisely what Betelgeuse marks: Orion's armpit. If you look at the winter star chart on the tear-out card, you'll also see Bellatrix, which marks the shoulder of Orion's arm holding his shield, which is an arc of closely spaced, albeit dim stars. Suspended from Orion's Belt is a short sword, the middle "star" of which is actually a region where stars are being born. (We will discuss the Orion nebula and other regions like it in Chapter 12, "Solar System Family Snapshot.")

Saiph is Orion's eastern leg. About 15 degrees to the southeast of this star is Sirius, called the Dog Star, because it is in the constellation Canis Major, the Great Dog. Sirius is the brightest star in the heavens.

To the northeast of Orion you will readily see a pair of bright stars close together. These are Castor and Pollux, the Twins, which represent the two heads of the constellation Gemini. Moving in an arc to the northwest of Castor and Pollux, you should see another bright star, this one with a distinctly yellow-gold color. It is called Capella, which means "little she-goat," and the ancients thought the star was the color of a goat's eye. Capella is in the constellation Auriga, the Goatherd.

Return to Orion. Just to the northwest of his shield, you will find Taurus, the bull, which is marked by Aldebaran, a bright orange star that forms the constellation's bull's eye. Early sky watchers imagined Taurus eternally charging the shield of Orion, who stood eternally poised to strike the animal with his upraised club.

It is admittedly difficult to imagine the bull in Taurus, though you may at least be able to discern a V-shaped asterism called the Hyades, which is the bull's mouth. To the northwest of this feature are the Pleiades, or Seven Sisters, a strikingly beautiful cluster of seven stars that are part of an open cluster.

Who Cares?

Enjoy the constellations. The pleasures of getting to know them can occupy a lifetime, and it's a lot of fun pointing them out to your friends, as well as to sons and daughters. You will also find familiarity with them useful for quickly navigating the heavens. But you won't be hearing a lot more about the constellations in this book. Recognizing them as the products of human fantasy and not the design of the universe, modern astronomy has little use for them.

The Least You Need to Know

➤ For the ancients, even without telescopes, the night sky was a source of great fascination, which we can share.

➤ To view the sky meaningfully, you need a system of orienting yourself and identifying certain key features. Celestial coordinates and altazimuth coordinates offer two such systems. While the celestial coordinate system is how professional astronomers designate location, altazimuth coordinates are used on telescopes that have an altazimuth mount.

➤ Astronomers use angular size and angular distance to describe the apparent sizes and separations of objects in the sky.

➤ Constellations are imaginative groupings of stars perceived as images, many of them influenced by Greek mythology; however, these groupings are arbitrary, reflecting human imagination rather than any actual relationships between those stars.

➤ Constellations are useful as celestial landmarks to help orient your observations.

Ancient Evenings: The First Watchers

One of the great attractions of astronomy is that it so new and yet so old. Astronomy asks many questions that push the envelope of human knowledge. What exactly are black holes? How did the universe begin and how will it end? How old is the universe? At the same time, it is the most ancient of sciences. The Babylonians, who lived in southeastern Mesopotamia between the Tigris and Euphrates rivers (present-day southern Iraq from Baghdad to the Persian Gulf), are the first people we know of who actively studied the stars and planets. As early as 3000 B.C.E., they seem to have identified constellations and, sometime later, developed a calendar tied to the recurrence of certain astronomical events (they didn't have NCAA basketball tournaments back then to let them know it was springtime).

Astronomy was only one of the Babylonian areas of knowledge basic to civilization. From ancient Babylonia came the first system of writing, cuneiform; the earliest known body of law, the Code of Hammurabi; the potter's wheel; the sailboat; the seed plow; and even the form of government known as the city-state. And whenever people sought to bring order and understanding to their world, astronomy was part of the effort.

If you are reading this book, it's a safe bet that you're interested in astronomy. You're not alone today, and you haven't been alone for at least 5,000 years and probably a lot longer. This chapter provides a glimpse of astronomy's ancient roots.

Star Words

A **conjunction** occurs when two celestial bodies appear to come close to one another on the celestial sphere.

A distinctly Western conception of the monster the ancient Chinese believed menaced the sun during an eclipse.

(From Century Magazine, 1885)

A Dragon Eats the Sun: Ancient Chinese Astronomy

As we said, the ancient Babylonians began making systematic observations of the heavens by 3000 B.C.E., and the Chinese weren't far behind. Records exist that show they had observed a grouping of bright planets (called a *conjunction*) that occurred around 2500 B.C.E., and, sometime before this, had arrived at the concept of a 365-day year, based on what appeared to be the sun's annual journey across the background stars.

Why the Emperor Executed Hsi and Ho

Like human beings everywhere throughout history, the Chinese in ancient times were a self-centered people. In fact, the Chinese word for their own country means "Middle Kingdom." Their belief was that the objects in the heavens had been put there for the benefit of humankind in general and for the emperor in particular.

Perhaps for this reason, they felt particularly threatened when, occasionally, something seemed to take a bite right out of the sun, then nibble away, gradually and ominously darkening the sky and the earth below.

The Chinese reasoned that a great dragon was attacking the sun, trying to consume it, and that since it was a beast, it might be susceptible to fear. So, in the midst of an *eclipse,* people would gather to shout, strike gongs, and generally make as much noise as possible—the more noise the better, since the beast was very big and was certainly very far away. Eventually the noise appeared to always scare off the dragon.

Because it was important to assemble as many people to make as much noise as possible, it was of inestimable value to get advance warning of an eclipse. With infinite patience, generations of Chinese astronomers observed solar eclipses and discovered something they called the Saros, a cycle in which sun, moon, and earth are aligned in a particular way every 18 years, 11.3 days—more or less. Armed with a knowledge of the Saros, the Chinese were able to predict eclipses—usually.

We know of this because in 2136 B.C.E. there was an unpredicted eclipse, which caught the noisemakers unawares. It was only by great good fortune that the sun wasn't consumed entirely. The Imperial Court astronomers Hsi and Ho weren't so fortunate. They were executed for having fallen down on the job. (The royal astronomer position may have been particularly difficult to fill after the "departure" of Hsi and Ho.)

Time, Space, Harmony

More than 4,000 years later, the fate of Hsi and Ho is still regrettable, but not nearly as important as the fact that we know about their fate at all. The Chinese *made records* of their astronomical observations and, indeed, along with the Babylonians were some of the earliest people to do so. Some oracle bones (animal bones used to foretell the future) from the Bronze Age Shang dynasty (about 1800 B.C.E.) bear the early Chinese ideogram character for

Star Words

An **eclipse** is an astronomical event in which one body passes in front of another, so that the light from the occulted (shadowed) body is blocked. For example, the sun, moon, and earth can align so that the moon blocks the light from the sun. This alignment produces a solar eclipse.

Astro Byte

The ancient Chinese also observed the nearly 12-year cycle of the planet Jupiter. In fact, the Chinese zodiac (and its 12 animals) is directly related to the yearly change in the position of the planet Jupiter. In the Chinese system, you and your father may be born 36 years apart, and both fall within the "Year of the Horse."

"pillar." Scholars believe that this ideogram is associated with a *gnomon,* a pillar or tower erected for the purpose of measuring the sun's shadow in order to determine, among other things, the dates of the solstices.

Writings from the Zhou dynasty, in the seventh century B.C.E., reveal that a special tower was built to measure the sun's shadow. During the Han era (C.E. 25–220), the town of Yang-chhêng was judged to be the center of the world, probably because the principal gnomon was installed there (or the gnomon may have been installed there because it was considered the center of the world). By C.E. 725, many smaller gnomons—what might be called field stations—were set up along a single line of longitude extending some 2,200 miles from the principal gnomon at Yang-chhêng. With this system, the Chinese could calculate calendars with considerable precision. In subsequent eras, even more elaborate gnomon towers—observatories, really—were built, including that of the astronomer Guo shou jing at Gao cheng zhen in Henan province, in C.E. 1276.

Why this passion to measure the heavens and the passage of time? Living in harmony with nature has always been important in Chinese philosophies, and, in terms of practical politics, exact knowledge of the heavens aided rulers in establishing and maintaining their absolute authority.

Star Words

A **gnomon** is any object designed to project a shadow used as an indicator. The upright part of a sundial is a gnomon.

Babylon Revisited

Back in Babylon, by the sixth century B.C.E., astronomers were tabulating in advance the intervals between moonrise and moonset and between sunrise and sunset, as well as the daily shift of the sun with respect to the background stars. They also predicted eclipses and made elaborate attempts to explain planetary movement.

The Venus Tablet

Some time between 1792 and 1750 B.C.E., during the reign of Hammurabi, the Babylonian king who gave the world its first code of laws, the so-called Venus Tablet was inscribed, devoted to interpreting the behavior of that planet. Babylonian astronomers believed that the movements and positions of planets with respect to the constellations could influence the fate of kings and nations. This interest in the positions of planets as a portent of the future was one early motivation for careful study of the heavens.

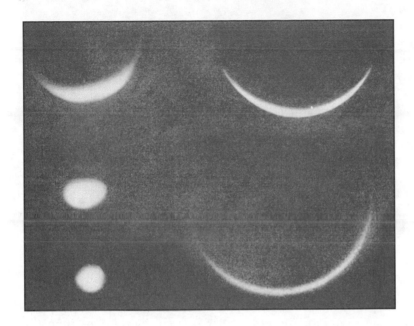

It is easy to imagine why, of all the planets, Venus captured the attention of the Babylonians. If you see a bright, steadily shining object in the west at or before sunset, or in the eastern sky at or before sunrise, it is almost certainly Venus—the brightest celestial object after the Sun and the Moon. Like the Moon, Venus has distinct phases, seen here.

(Image from arttoday.com)

Close Encounter

Using binoculars or a telescope, you may be able to see Venus in its phases, from thin crescent to full. During much of the year, the planet is bright enough to see even in daylight, always quite near the sun. Venus is closer to the sun than we are, and that fact keeps it close to the sun in the sky (never more than 47 degrees, or about a quarter of the sky from horizon to horizon). The best times to observe Venus are at twilight, just before the sky becomes dark, or just before dawn. The planet will be full when it is on the far side of the sun from us and crescent when it is on the same side of the sun as we are. With a telescope on a dark night, you may be able to observe the "ashen light" phenomenon. When the planet is at quarter phase or less, a faint glow makes the unilluminated face of Venus visible.

Draftsmen of the Constellations?

Babylonian astronomers systematically used a set of 30 stars as reference points to record the celestial position of various phenomena. Then, in 625 B.C.E., the neighboring Chaldeans invaded and captured Babylon. Although they ransacked the

Babylonian capital, the Chaldeans did not ignore the astronomy of their conquered foes. Indeed, they seem to have gone well beyond merely identifying 30 reference stars and actually recorded some of the first constellation patterns.

Astronomer's Notebook

The Great Pyramid is the oldest and biggest of three Fourth Dynasty (ca. 2575–ca. 2465 B.C.E.) pyramids built near Giza, Egypt. The length of each side at the base averages 755¾ feet, and it was originally 481⅖ feet high. Some 2,300,000 blocks of stone, each weighing about 2½ tons, went into building the Great Pyramid.

Star Words

The **circumpolar stars** are those stars near the celestial north pole, that is, close to Polaris, the North Star. Depending upon your location on the earth, these stars might be directly overhead (if you live at the North Pole) or at some angle to the horizon.

Egypt Looks

While the Chinese, the Babylonians, and the Chaldeans used astronomical observations to help them rule and regulate the living, the ancient Egyptians used the observations and measurements they made to help the dead find their place in the afterlife. Actually, Egyptian astronomers worked for both the living and the dead. They created a calendar to help predict the annual flooding of the Nile—essential information for a people whose entire agriculture was subject to the whims of that river. To create their calendar, Egyptian sky watchers concentrated in particular on the rising of the star Sirius (which they called Sothis). Working from these data (measuring the time from one rising of Sirius to the next), Egyptian astronomers were able to determine that a year was 365.25 days long.

Celestial Pyramids

To think of the Egyptians is to think of the pyramids, the great tombs of the pharaohs. Prayers carved into the walls of pyramid chambers make reference to the stars and to the pharaoh's ascent into the sky among them. Texts inscribed on the monuments at Saqqara, Egypt, tell of the pharaoh joining the *circumpolar stars,* which neither rise nor set, and therefore live eternally. These texts also tell of the pharaoh's journey to the constellation Orion—identified by the Egyptians with Osiris, the eternally resurrected god.

Replete as the pyramids are with such astronomical texts, it is little wonder that many archaeologists and others have speculated about the astronomical significance of the pyramid structures themselves. Certainly the Great Pyramid, by far the largest of the 80 or so known pyramids along the Nile's west bank, is celestially aligned. Internal shafts or ducts point to the star Thuban, which in ancient Egyptian times was the

North Star (see the discussion of *precession* in Chapter 3, "The Unexplained Motions of the Heavens"). Other shafts point to Orion's Belt at certain times of the year, as if to indicate the afterlife destiny of the pharaoh, toward the deathless North Star (which does not rise or set) on the one hand, and toward the constellation associated with the eternally reborn god Osiris on the other.

The Universe-in-a-Box

The pyramids were not observatories, although it is tempting to think of them as such. The astronomical alignments they demonstrate were symbolic or magical rather than practical. Indeed, except for their very accurate calendar, the Egyptians seem to have made little of what we would call scientific use of their many astronomical observations. The ancient Egyptians drew images of constellations (an Egyptian star map was discovered by one of Napoleon's generals in 1798 when the French army campaigned in Egypt) and made accurate measurements of stellar positions. However, they also reached the fantastic conclusion that the universe was a rectangular box, running north and south, its ceiling flat, supported by four pillars at the *cardinal points*—due north, south, east, and west. Joining the pillars together was a mountain chain, along which the celestial river Ur-nes ran, carrying boats bearing the Sun, the Moon, and other heavenly deities. In fact, many early cultures developed an accurate astronomical calendar side by side with a fanciful mythology.

Stonehenge and the New World

The ancient peoples of the Far East and the Middle East had no monopoly on the stars. The endlessly fascinating Stonehenge, built in stages between about 2800 B.C.E. and 1550 B.C.E. on the Salisbury Plain in Wiltshire, England, seems almost certainly to have been designed as a kind of astronomical observatory or, as some scholars have argued, a computer of astronomical phenomena. Various features of Stonehenge are aligned on the positions of the sun and moon where each rise and set on particular days known as *solstices* (the days of longest and briefest daylight, which begin the summer and winter, respectively). Thus it is widely agreed that Stonehenge was at least in part used for the keeping of a calendar.

Elsewhere in England and on the continent other circular stone monuments, akin to Stonehenge, are to be found. These have also been studied for the relation they bear to astronomical phenomena.

Star Words

The **summer solstice** occurs on June 22, is the longest day in the Northern Hemisphere, and marks the beginning of summer. The **winter solstice** occurs on December 22 and, again in the Northern Hemisphere, is the shortest day and the start of winter. The solstices are fully explained in Chapter 3. Many of our winter festivals (Chanukah, Christmas) have their roots in ancient celebrations of the winter solstice.

23

The New World, too, has its celestially oriented ancient structures. Around C.E. 900, at Cahokia, in southern Illinois, a Native American people known to us as the Mound Builders erected more than a hundred earthen mounds, the layout of which seem to mirror a concept of a cosmic plan. Farther south, the Maya of Mexico built magnificent stepped pyramids, like the one at Chichén Itzá in the Yucatan, clearly oriented toward the sunset at the winter solstice, as if to mark the annual "death" of the sun.

The Maya (and later, the Aztecs) used celestial observation to formulate a calendar as accurate as that which the Chinese had developed. A host of North American, Mesoamerican, and South American Indian monuments reveal careful orientation toward specific astronomical events. For example, the great *kiva* (a subterranean ceremonial chamber perhaps dating from C.E. 700 to 1050) at Chaco Canyon, New Mexico, built by the Anasazi, ancestors of the Zuni, Hopi, and Navajo, is oriented to mark the sunrise on the day of the summer solstice. Mayan and Aztec structures found throughout Mexico, such as the Hall of Columns at Alta Vista, dating to about C.E. 700, seem to be oriented to mark sunrise on the days of the *equinoxes* (the days of equal night and day, which mark the beginning of spring and fall). Monte Albán, built by the Zapotec of ancient Oaxaca, Mexico, as early as the eighth century B.C.E., seems to be oriented to the sun at zenith and to the rising of the star we call Capella.

Star Words

The **autumnal equinox** occurs on September 23, and the **vernal equinox** on March 21, marking the beginning of fall and spring, respectively. On these dates, day and night are of equal duration. The sun's apparent course in the sky (the ecliptic) intersects the celestial equator at these points. We'll discuss the equinoxes in more detail in Chapter 3.

Grecian Formula

We could go on with speculation about the astronomy of the Far East, Near East, and New World (and those interested should read E. C. Krupp's *Echoes of the Ancient Skies: The Astronomy of Lost Civilizations,* New York, 1983), but it would be just that: fascinating speculation. For these early astronomers left few written records, and those they did leave either note only their observations or link such observations to religion and mythology. These earliest records do not show any effort to use astronomical observation to explain the physical realities of the world and the cosmos. For these first attempts, we must turn to the Greeks.

Anaximander Puts Earth in Space

The word *philosopher* means "lover of wisdom," which accurately describes the passion of the Greek philosophers. These were not idle thinkers eating grapes in secluded gardens. These were men who observed the world around them and wondered how the elements that made up the earth worked together and how human beings fit into the resulting grand scheme. For some eight centuries, Greek philosophers confronted some of the most fundamental questions in the natural sciences. What is the smallest division of matter? What are we and the world made of? How big are the earth and the universe? Beginning with Thales, the first of the important Greek philosophers (born about 624 B.C.E.), and culminating with Ptolemy (who died about C.E. 180 and whom we'll meet in the next chapter), a series of Greek philosophers thought most intensely about the sky and the wonders it presented.

Thales' junior colleague and student, Anaximander (610–546/545 B.C.E.), is often called the founder of astronomy. He might even more accurately be called the father of a particular branch of astronomy, *cosmology,* which deals with the structure and origin of the universe. Anaximander theorized that the world and everything in it were derived from an imperceptible substance he called the *apeiron* (unlimited), which was separated into various contrasting qualities and eventually differentiated into all matter, including the earth. Importantly, Anaximander rejected what was then the prevailing notion that the earth was suspended from or supported by something in the heavens. He held that the earth floated freely in space at the center of the universe. Without reason to move anywhere, the earth, shaped like a short cylinder (we'd call it a soup can today), floated motionless. As for the stars, they were fiery jets, and the sun a chariot wheel whose rim was hollow and filled with fire.

Anaximenes Says Stars Burn

Anaximenes of Miletus (active around 545 B.C.E.) theorized that what he called *aer* (air or vapor) was the most basic form of matter and was also the substance that formed the life spirit of animals, the soul of humankind, and the divine essence of the gods. He also believed that, when rarified, *aer* turned to fire, and he held that the sun, moon, and stars were collections of rarified *aer,* masses of fire, which, he believed, were set into a great crystal hemisphere.

Pythagoras Calls Earth a Globe

Anaximander and Anaximenes may no longer be household names, but a lot of us remember Pythagoras (ca. 580–ca. 500 B.C.E.) from high school geometry. We all heard about the man who is credited with the Pythagorean theorem ("The sum of the squares of the sides of a right triangle is equal to the square of the hypotenuse," or $A^2 + B^2 = C^2$). He also taught that the earth was a globe—not a cylinder and certainly not flat—and that it was fixed within a sphere that held the stars. The planets and the sun moved against this starry background.

Anaxagoras Explains Eclipses

Anaxagoras (ca. 500–ca. 428 B.C.E.) believed that the earth was flat, but speculated that the sun was a large, red-hot body and that the moon was much like the earth, complete with mountains and ravines. Most important, Anaxagoras theorized that solar eclipses were caused by the passage of the moon between the sun and the earth. His was the first explanation of an eclipse that didn't involve the supernatural and certainly didn't summon up any dragons.

Aristarchus Sets the Sun in the Middle and Us in Motion

Living in a technology-driven society, we've become accustomed to thinking of linear progress in science, a movement from point A, which takes us to point B, then to point C, and so on. We don't think that steps backward can ever occur. If this were the way knowledge actually was built, the model of a *geocentric* (or earth-centered) universe would have died during the second century B.C.E.

Star Words

Geocentric means earth–centered, and the geocentric model of the universe is one in which the earth is believed to be at the center of the universe. **Heliocentric** means sun–centered and accurately describes our solar system, in which the planets and other bodies orbit the sun.

Far in advance of his peers, Aristarchus of Samos (ca. 310–230 B.C.E.) proposed that the earth is not at the center of the universe or the solar system, but that it orbits the sun while also rotating. This theory sounds completely reasonable to our modern ears, but it did not sit well with the Greek philosophical establishment, nor with common sense. Why, one might ask, if the earth is orbiting the sun, and spinning on its axis do we not all go flying into space as we would if the earth were a large merry-go-round? Without a theory of gravity (which keeps everything stuck to the surface of the earth as it spins), there was no good answer to this valid question. One philosopher, Cleanthes the Stoic, went so far as to declare that Aristarchus should be punished for impiety. Maybe if he had been punished, becoming a martyr to his idea, the *heliocentric* (sun-centered) solar system would have caught on much sooner than it did. But it didn't. The geocentric model of Aristotle and others held sway for millennia.

Eratosthenes Sizes Up the Earth

Anaxagoras's explanation of eclipses was a bold exercise in the use of science to understand a phenomenon well beyond everyday experience. One other such exercise came from Eratosthenes of Cyrene (ca. 276–ca. 194 B.C.E.). A careful observer,

Eratosthenes noted that at the town of Syene (present-day Aswan, Egypt), southeast of Alexandria, the rays of the sun are precisely vertical at noon during the summer solstice. That is, a vertical stick in the ground would cast no shadow. He further noted that, at Alexandria, at exactly the same date and time, sunlight falls at an angle of 7.5 degrees from the vertical.

Close Encounter

Aristarchus also tried to estimate the size of the sun and moon and their distance from the earth. His geometry was off, so the values he derived proved inaccurate. However, he did fix a reasonably accurate value for the *solar year* (see Chapter 3 for a definition and discussion of the term) and is honored by a lunar crater named for him. The peak in the center of the Aristarchus crater is the brightest feature on the face of the Moon.

As we'll see in the next chapter, small differences and apparently inconsequential discrepancies often have profound implications in astronomy. Eratosthenes instinctively understood the importance of details. Assuming—correctly—that the sun is very far from the earth, he reasoned that its rays are essentially parallel when they strike the earth. Eratosthenes believed (as did Aristotle, whom we'll meet in the next chapter) that the earth was a sphere. He further reasoned that the angle of the shadow cast in Alexandria (7.5 degrees) was equal to the difference in *latitude* (see Chapter 1, "Naked Sky, Naked Eye: Finding Your Way in the Dark") between the two cities. How did he figure that? Think of it this way: Imagine poking a stick vertically into the earth at the equator and at the North Pole, and imagine the sun is directly over the stick at the equator. The stick at the equator will have no shadow, and the stick at the North Pole will cast a shadow at an angle of 90 degrees from the stick. Now, move the stick from the North Pole to a latitude of 45 degrees. The shadow will now fall at an angle of 45 degrees from the stick. As the stick that was at the North Pole gets closer and closer to the equator (where the other stick is), the angle of its shadow will get smaller and smaller until it is beside the other stick at the equator, casting no shadow. Noting that a complete circle has 360 degrees, and 7.5 degrees is approximately ¹⁄₅₀ of 360 degrees, Eratosthenes figured that the two cities were separated by ¹⁄₅₀ of the earth's circumference, and that the circumference of the earth must simply be fifty times the distance between Alexandria and Syene.

The distance from Syene to Alexandria, as measured in Eratosthenes's time, was 5,000 *stadia*. He apparently paid someone (perhaps a hungry grad student) to pace out the distance between the two cities. So he calculated that the circumference of the earth was 250,000 *stadia*. Assuming that the *stadion* is equivalent to 521.4 feet, Eratosthenes calculation of the earth's circumference comes out to about 23,990 miles and the diameter to about 7,580 miles. These figures are within 4 percent of what we know today as the earth's circumference—24,887.64 miles—and its diameter, 7,926 miles. And he figured that out with only a few sticks—and one long hike.

Close Encounter

Did you ever wonder why "clockwise" is the way it is and not the other way? On a sunny Saturday morning, put a stick vertically in the ground, as Eratosthenes did so many years ago. Now, as the sun comes up, mark where the shadow of the stick points at 1 hour intervals—10 A.M., 11 A.M., 12 P.M., 1 P.M., 2 P.M. You should notice a few interesting things. Stand to the south of your stick and look at it. Where is the mark for 12 P.M. (noon)? Is it at about the same place as noon on your wall clock? How about the marks for 1 P.M. and 2 P.M.? The sun's daily motion across the southern sky determines the "clockwise" motion of the shadow. The shadow of the gnomon was the "hour hand" on one of the oldest clocks—the sundial—and you can see the echo of this ancient time-keeping technique in the orientation of the numbers, and the "clockwise" sweep of the hands on your wristwatch.

Eratosthenes made other important contributions to early astronomy. He accurately measured the tilt of the earth's axis with respect to the plane of the solar system, and compiled an accurate and impressive star catalog and a calendar that included leap years.

We may consider Eratosthenes the first astronomer in the modern sense of the word. He used careful observations and mathematics to venture beyond a simple interpretation of what his senses told him. This combination of observation and interpretation is the essence of what astronomers (and all scientists) do. It is a cruel irony that Eratosthenes lost his eyesight in old age. Deprived of his ability to observe, he committed suicide by starvation.

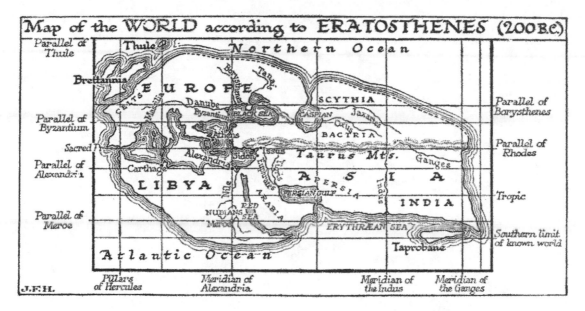

Eratosthenes drew a fanciful map of the world whose circumference he had estimated so precisely.

(Image from arttoday.com*)*

The Least You Need to Know

➤ The Babylonians were the first known astronomers, making observations as early as 3000 B.C.E.

➤ The Chinese used astronomical observation to aid in predicting such events as eclipses and to create and reinforce the sense of divine order in the universe.

➤ Egyptian astronomy was in large part intended to help the dead find their way in the afterlife.

➤ From our perspective, the Greeks are the most important of the ancient astronomers because they used their careful observations to analyze, measure, and explain physical reality.

➤ With his keen eye for observation and his genius for using observation to draw conclusions beyond what could be directly observed, the Greek philosopher Eratosthenes may be considered the first astronomer in the modern sense of the word.

Chapter 3

The Unexplained Motions of the Heavens

In This Chapter

➤ How does celestial movement mark time?

➤ The discrepancy between the solar and sidereal day—and what it means

➤ The sidereal month versus the synodic month

➤ Understanding the tropical year and the sidereal year

➤ The reason for seasons

➤ The solar system according to Ptolemy

The next time you're outside doing yard work in the sun, put a stick in the ground, call it a gnomon, and watch the motion of its shadow. Believe it or not, you have made a simple *sundial,* which was one of the earliest ways that human beings kept track of time. In fact, keeping time was one of the two major reasons that early civilizations kept a close watch on the skies. The other reason, of course, was to use the motions of the planets through the constellations to predict the future for the benefit of kings and queens and empires. Well, the first practice (keeping track of time) has continued to this day. The U.S. Naval Observatory is charged with being the time-keeper for the nation, using technology a bit more advanced than a stick in the ground. The second practice (predicting the future) is also alive and well, but astronomers have turned those duties over to The Psychic Network™—at least for the time being.

In the days before movies, television, video games, and the Internet, the starry sky (untouched by city lights and automobile exhaust) was truly the greatest show on Earth. Generations of sky watchers looked and imagined and sought to explain. Common sense told many of these early watchers that they were on a kind of platform overarched by a rotating bowl or sphere that held the stars. We have seen that in various cultures, other explanations surfaced from time to time.

It doesn't matter right now whether these explanations were right or wrong (well, many of them were wrong). What matters is that the explanations were, to many astronomers, *unsatisfying.* None of the explanations could account for everything that happened in the sky. For example, if the stars were all fixed in this overarching bowl, how did the planets break free to wander among the stars? And they didn't wander randomly. The planets were only found in certain regions of the sky, close to the great circle on the sky called the ecliptic. Why was that? The sky is filled with thousands of bright points of light that move, and none of the ancient explanations adequately explained all of these movements.

In this chapter, we examine the sky in motion and how early astronomers came to be the keepers of the clock.

Time on Our Hands

Why did the ancients concern themselves about things moving in the sky when they were stuck here down on Earth? Chalk it up in part to human curiosity. But their interest also had even more basic motives.

You're walking down the street, and a passerby asks you for the time. What do you do?

You look at your watch and tell him the time.

But what if you don't have a watch?

If you still want to be helpful, you might estimate the time, and you might even do this by noting the position of the sun in the sky.

The ancients had no wrist watches, and, for them, time—a dimension so critical to human activity—was measured by the movement of objects in the sky, chiefly the sun and the moon. What, then, could be more important than observing and explaining the movement of these bodies?

What Really Happens in a Day?

We define a day as a period of 24 hours—but not just *any* 24 hours. Usually, by a "day," most people, and certainly ancient cultures would have meant the period from one sunrise to the next.

But how long is a day, really? What do we mean when we say a day? It turns out that there are two different kinds of days: a day as measured by the rising of the sun (a *solar day*), and a day as measured by the rising of a star (a *sidereal day*). Let's think about this a bit.

Even casual observation of the night sky reveals that the position of the stars relative to the sun are not identical from one night to the next. Astute ancient skywatchers noticed that the celestial sphere shifted just a little each night over the course of days, weeks, and months. In fact, the aggregate result of this slight daily shift is the well-known fact that the constellations of summer and winter, for example, are different. We see Orion in the winter, and Leo in the spring.

Astronomers call the conventional 24-hour day (the time from one sunrise [or sunset] to the next) a solar day. They call a day that is measured by the span from star rise to star rise a sidereal day. The sidereal day is almost 4 minutes (3.9 minutes) shorter than the solar day. We'll see why in a moment.

It works like this: Relative to the rotating and orbiting Earth, we may imagine that the sun is essentially at rest at the center of the solar system. However, from any spot on Earth it looks as if the *sun* is rising, traveling across the sky, and setting. A solar day is not really the 24 hours from *sun*rise to *sun*rise, but the 24 hours it takes the *earth* to rotate one full turn.

While spinning on its axis, the earth is also orbiting the sun, with one complete revolution defining a year (we'll get into the details of years in a moment). Remember that a circle consists of 360 degrees; therefore, one complete circuit around the sun is a journey of 360 degrees. It takes the earth one year to make a complete circuit around the sun. And for the moment, let's say that a year consists of 365 days. To find out how far the earth travels through its orbit in one day, divide 360 degrees by 365 days. The result is about 1 degree.

A *sidereal day* is defined as the time between risings of a particular star (all of which are very distant relative to the distance to the sun). Since the earth is

Star Words

A **solar day** is measured from sunup to sunup (or noon to noon, or sunset to sunset). A **sidereal day** is measured from star rise to star rise, for example, the time between the star Betelgeuse rising above the horizon one day and the next. A solar day is 3.9 minutes longer than a sidereal day.

Astro Byte

You can check out for yourself the difference between the solar day and the sidereal day by noting, over the course of a month, what constellations are directly overhead at 9 P.M. You should see that the constellations exhibit a slow drift. In what direction are they drifting? East or west? This drift is the cumulative effect of the 3.9 minute difference between the sidereal day and the solar day.

constantly moving ahead in its orbit around the sun, we will see the same star rise again on the next day slightly sooner than we see the sun rise—precisely 3.9 minutes sooner each day. In one solar day, then, the earth has advanced in its orbit almost a degree, and it takes the earth 3.9 minutes to rotate through this one-degree angle and bring about another sunrise.

A Month of Moons

And then there is the moon. As we saw in Chapter 1, "Naked Sky, Naked Eye: Finding Your Way in the Dark," the words "moon" and "month" are closely related, and with good reason. The moon takes about a month (29½ days) to cycle through all its phases.

➤ The invisible (or almost invisible) new moon

➤ The waxing crescent (fully visible about four days after the new moon)

➤ The first quarter (the half moon, a week after the new)

➤ The waxing gibbous (75 percent of the moon visible, 10 days after new moon)

➤ The full moon (two weeks after new moon)

➤ The waning gibbous (75 percent, 18 days old)

➤ The third quarter (half a moon at 22 days old)

➤ The waning crescent (a sliver by the 26th day)

➤ New moon at day 29

What's going on here? Doubtless, many of our ancestors believed the moon changed shape, was consumed and reborn. The Greeks, however, surmised that the moon had no light of its own, but reflected the light of the sun—and therein lies the explanation for the phases of the moon.

Close Encounter

Wondering what the phase of the moon is right now? Too lazy to go outside and look for yourself? Then check out www.tycho.usno.navy./mil/vphase.html. This cool Web site allows you to see the moon phase for any day you choose.

The full disc of the moon is always present, but we see what we call the full moon only when the sun and the moon are located on opposite sides of the earth. When the moon comes between the sun and the earth, the side of the moon away from the earth is illuminated, so we see only its shadowed face as what we call the new moon. In the periods between these two phases (new and full), the sun's light reveals to us varying portions of the moon, depending upon the relative position of the earth, moon, and sun.

There is another observation that must have been made early on, which probably puzzled early sky watchers. The moon's face is irregular, marked with what we now know to be craters (holes and depressions left by meteor impacts) and other features (which we will investigate in Chapter 10, "The Moon: Our Closest Neighbor"). Even naked eye observations make it abundantly clear that the moon always presents this same, identifiable face to us, and that we never see its other side—a fact that has given rise to innumerable myths.

For generations, the far side of the moon, or the "dark side," has been the subject of wild speculation, including stories of mysterious civilizations hidden there. Humankind didn't get so much as a glimpse of it until a Soviet space probe radioed images back to Earth in 1959. As it turns out, the "far side" of the moon hid no mysterious civilizations, but did look rather different from the near side, with more craters and fewer large grey areas (seas). Those differences support certain theories of how the moon formed, as we'll see in Chapter 10.

Astro Byte

Calling the face of the moon that we never see the "dark side" is a misnomer, since at new moon, the side of the moon that we do not see is fully illuminated by the sun and not dark at all. (We just don't see it.) "Far side" is a far better term.

The Moon as you might see it through a telescope.

(Image from arttoday.com*)*

Close Encounter

Want a peek at the far side? A phenomenon called *libration,* a kind of swaying (or wobbling) motion to which the moon is subject, means that you can occasionally catch the smallest glimpse of the far side. Libration can reveal 59 percent of the lunar surface—though, of course, never more than 50 percent at once, since the swaying of part of the moon toward us means that the part opposite it must sway away from us. After you become an experienced lunar observer, look at the extreme northern and southern regions of the moon. With the help of a good lunar map, see if you can find features there that you never saw before and that later seem to disappear.

The explanation of why we never see the far side requires understanding a few more of the solar system's timing mechanisms. Like the earth, the moon rotates as well as orbits. It rotates once on its axis in 27.3 days, which is exactly the amount of time it takes for the moon to make one complete orbit around the earth. Synchronized in this way, the rotating and orbiting moon presents only one face to the earth at all times.

But hold the phone! Twenty-seven-and-one-third days to orbit the earth? Why, then, does it take *29.5 days* for the moon to cycle through all of its phases? It seems as if the solar system is playing games with time again.

Close Encounter

Hold your hand out at arm's length with the back of your hand facing you. Imagine that your head is the earth, and the back of your hand (facing you) is the face of the moon. Now, move your arm in an arc from right to left, keeping the back of your hand facing you. This exercise should help you to appreciate that the moon (your hand) must rotate to keep the same side facing the earth (your head).

True enough. But given what we just discussed, the difference between a *sidereal month* (the 27.3 days it takes the moon to complete one revolution around the earth) and the *synodic month* (the 29.5 days required to cycle through the lunar phases) should now be easy to understand. The difference is explained by the same principle that accounts for the difference between the solar day and the sidereal day. Because the earth's position relative to the sun—the source of light that reveals the moon to us—changes as the earth travels in its orbit, the moon must actually complete slightly more than a full orbit around the earth to complete its cycle through all the phases.

That is, the moon will have traveled through 360 degrees of its orbit around the earth after 27.3 days, but due to the earth's motion, the moon has to continue a little farther in its orbit before it will be full (or new) again.

A few minutes here, a few days there. The progress of science, of coming to a greater understanding of reality, is often measured in small differences and discrepancies. These slight differences are akin to the apparently inconsequential "accounting error" that reveals to the careful auditor some vast scheme of financial manipulation.

Another Wrinkle in Time

Measured from equinox to equinox, a year is 365.242 days long. This length of time is called a *tropical year.* Yet the time it actually takes the earth to complete one circuit around the sun is twenty minutes longer: 365.256 days—a *sidereal year.*

Well, here we go again! More wrinkles in how we measure time. The slight discrepancy between the sidereal year and the tropical year is due to the fact that the earth's rotational axis slowly changes direction over long periods of time. Astronomers call this phenomenon *precession.* If you spin an old-fashioned top, you will notice that the toy spins rapidly on its own tilted axis *and* that, as the top starts to slow down, the axis itself slowly revolves around the vertical, the handle of the top tracing out a circle. The earth, subject to gravitational pull from sun and moon, behaves much like a toy top, spinning *rapidly* on its axis, even as that axis *gradually* rotates.

Star Words

A **sidereal month** is the period of 27.3 days it takes the moon to orbit once around the earth. A **synodic month** is the 29.5 days the moon requires to cycle through its phases, from new moon to new moon or full moon to full moon.

Star Words

A **tropical year** is measured from equinox to equinox, a span of 365.242 days. A **sidereal year** is the time it takes the earth to complete one circuit around the sun: 365.256 days, about 20 minutes longer than a tropical year.

How gradually does the earth's axis precess? This is astronomy we're talking about, and things take a long time. A complete cycle of precession takes about 26,000 years. In practical terms, this means that, whereas Polaris is the pole star today—the star almost directly above the North Pole—a star called Thuban was the pole star in 3000 B.C.E., and Vega will be the pole star in C.E. 14000. Moreover, if we linked our calendars to the sidereal year instead of the tropical year, February would be a mid-summer month in the northern hemisphere some 13,000 years from now. The effect on the celestial sphere is that points in the sky are not exactly fixed, but ever so slowly drift. The "zero mark" of right ascension (called the vernal equinox) is where the *ecliptic* and the *celestial equator* cross on the celestial sphere. That point in the sky is currently in the constellation Virgo. But it won't always be. The vernal equinox will drift through the constellations as the earth slowly precesses.

We don't think that our distant ancestors were aware of the causes of precession, but they did have a good deal of trouble coming up with accurate calendars, and making calendars was often a subject of intense debate. Hipparchos, in the second century B.C.E., was the first astronomer to explain the difference in length between the sidereal year and the tropical year as being due to the "precession of the equinoxes."

Close Encounter

The slow drift of the earth makes it important to know the reference time, or epoch of your coordinates. Since the coordinates slowly drift, astronomers have to know whether they are talking about where objects were in the coordinate system as it was in 1950, or in the year 2000, or are exactly at this moment. To observe a source in the sky, of course, you want to know the coordinates at that particular moment. However, in order to compare their results, most astronomers convert their coordinates to what are called J2000 coordinates—the positions of sources in the sky in the year 2000. Not that there is anything special about the year 2000, it just serves (for now) as a common reference time to compare coordinates.

To Everything a Season

Both authors of this book remember learning how to drive a car with a stick shift. One of us (the bearded one) was once in line at a New Jersey Turnpike toll booth, waiting to pay. The car behind gently nudged him each time the line of cars inched forward. Not once, not twice, but several times. Finally, enough was enough. He got out of his car to give the other driver a piece of his mind. That fellow, however, rolled down his window first: "Why do you keep rolling back into me?"

That's when this author realized that he needed more practice coordinating the clutch with the shift. But until that rude awakening, it seemed to him that the other person's car was moving, and that he was standing still.

So it was for thousands of years with the people on Earth. As the earth orbits the sun, it appears (to an earthbound observer) that the sun gradually moves across the sky, relative to the position of the stars, in the course of a year. Early astronomers charted what was apparently the sun's path across the celestial sphere and found that it described a great circle inclined at 23½ degrees relative to the celestial equator. This apparent path of the sun is called (as we have seen) the *ecliptic*. Like the odd pieces of time that are apparently lost or gained in the course of a day or a month, this curious angle can actually tell us a lot about how the solar system works, and why summer follows spring each year.

Star Words

The **ecliptic** traces the apparent path of the sun against the background stars of the celestial sphere. This "great circle" is inclined at 23½ degrees relative to the celestial equator, which is the projection of the earth's equator onto the celestial sphere.

We now know that the *sun* doesn't travel across the celestial sphere in a course that is inclined at 23½ degrees with respect to the celestial equator. In fact, as we all know, the *earth* orbits the sun while also rotating around its own north-south axis. The earth's rotational axis, however, is tilted at 23½ degrees relative to an axis perpendicular to its orbital plane. Think of it this way: The earth is not standing up straight in the plane of the solar system, but is tipped over on its side by an angle of 23½ degrees. It is this inclination in the earth's axis that makes it appear that the *sun* is traveling an inclined course across the sky. If the earth were not "tipped," then the sun would move directly along the celestial equator.

The earth's inclination has a profound effect on us all. When the sun appears to be at its northernmost point above the celestial equator, we have the *summer solstice* (June 21). As the earth rotates on this date, locations north of the earth's equator enjoy the longest day of the year, because these locations spend the greatest portion of their time exposed to the sunlight. In the southern hemisphere, this day is the shortest of them all, because that portion of the earth is tilted away from the sun. Six months after the summer solstice comes the *winter solstice* (on December 21). On this day, the situations of the northern and southern hemispheres are reversed: The sun is low in the sky in the northern hemisphere, making it the shortest day of the year, and the southern hemisphere has the sun high in the sky.

Between the two solstices come the *equinoxes*. On these dates— September 21 for the autumnal equinox, and March 21 for the vernal equinox—day and night are of equal duration; the sun's apparent course (one great circle on the celestial sphere) intersects the celestial equator (another great circle) at these points, as it passes into the northern hemisphere of the celestial sphere.

In the Northern Hemisphere, the summer solstice marks the beginning of summer, the autumnal equinox the beginning of fall, the winter solstice is the first day of winter, and the vernal equinox is the start of spring.

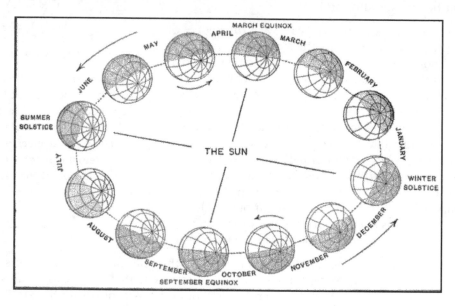

The seasons as a function of the earth's orbit around the sun.

(Image from the authors' collection)

The equinoxes are more than just a matter of marking time. Because of the earth's tilt, the summer sun is high in the sky and the days are long, which means the weather tends to be warm. In the winter, the sun is low in the sky, the days are short, and the weather is cold. As the height of the sun in the sky changes, the area of the earth's surface over which the sun's energy is distributed also changes: a larger area in winter (when the sun is low in the sky) and a smaller area in summer (when the sun is more overhead). The effect is that the earth absorbs less energy per unit area in the winter (cooler temperatures), and more energy per unit area in the summer (warmer temperatures). The combination of this change in energy absorbed and the length of the day due to the position of the sun creates what we know as seasons.

Close Encounter

In ancient times, sky watchers paid special attention to the 12 constellations grouped along the ecliptic, the apparent path of the sun. These constellations are known as the zodiac and are, accordingly, associated with certain seasons. For example, the earth squarely faces Gemini at about the time of the winter solstice, Virgo at the vernal equinox, Sagittarius and Capricorn at the summer solstice, and Pisces during the autumnal equinox. Our ancestors believed that the constellations of the zodiac exerted a special influence on events and human traits and character relative to the time of year and other factors. Of particular importance to early astrologer-astronomers was the interplay between the positions of planets and the constellations. Thus the pseudoscience of astrology came into being as an attempt to correlate seasonal, celestial, and earthly events. The precession of the earth has shifted the zodiac since Babylonian times. Your "true" astrological sign is (generally) shifted one sign earlier in the year. And you thought you were a Capricorn!

The Sun Goes Dark, the Moon Becomes Blood

The ancients had another, far more dramatic, celestial irregularity to contend with. On rare occasions, the moon would gradually dim and turn a deep red for a time, only to reemerge in its full glory after a short time. On even rarer occasions, daylight would fade as a great shadowy disk stole across the sun.

We will discuss lunar eclipses in Chapter 10 and solar eclipses in Chapter 16, "Our Star." As we saw in the previous chapter, eclipses were events that could cause great fear, and governments and rulers put tremendous pressure on astronomers (or soothsayers or astrologers or whatever they called their official sky watchers at the time) to come up with dependable ways of predicting when eclipses would occur. Here was yet another set of celestial events that certainly weren't random, yet, without a complete understanding of how the various parts of the solar system were put together, they were hard to predict accurately.

Aristotle Lays Down the Law

Fortunately for the hard-pressed astronomers of yore, total eclipses of the sun are relatively rare events. More immediately, they had *nightly* occurrences they couldn't explain.

They could watch the sky all night and see the stars glide predictably across the sky, as if affixed to a moving celestial sphere. Likewise, the moon made its traversal with perfect regularity. But there were (so far as the ancients could see) five heavenly bodies that didn't behave with this regularity. Mercury, Venus, Mars, Jupiter, and Saturn looked pretty much like stars, but, in contrast to stars, they *wandered* across the sky and so were called by the Greeks *planetes,* or "wanderers."

While the planets always remain near the ecliptic, they have strange motions. Relative to the stars, planets seem to slow down and speed up, moving (from an earthbound observer's perspective) generally eastward (*prograde* motion), but sometimes westward (*retrograde* motion) with respect to the background stars.

How did the ancients explain this strange observation?

The great Greek philosopher Aristotle (384–322 B.C.E.), tutor of Alexander the Great, formulated a picture of the solar system that put the earth at its center with all the other heavenly bodies orbiting around it. The orbits were described as perfect circles.

Now, Aristotle was a smart man. He provided observational evidence that the earth was spherical rather than flat. Moreover, his descriptions of the orbits of the moon and the sun accorded very well with what people actually observed. As far as the stars went, the celestial sphere idea explained them well enough.

The cosmos according to Aristotle, as published in Peter Apian's Cosmographia *of 1524.*

(Image from Rice University)

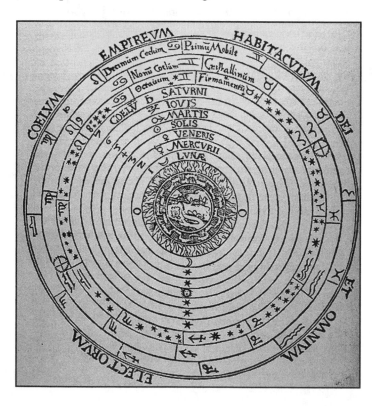

The problem was the planets. Aristotle's simple geocentric (earth-centered) model did not explain why the planets varied in brightness, why they moved at varying speeds, and why their motion was sometimes prograde and sometimes retrograde.

Ptolemy's Picture

Why didn't the world's astronomers just toss out Aristotle's geocentric model of the solar system?

There are at least three reasons.

First, and most important, the geocentric picture appeals to common sense. On a given day, as we watch the sun and the stars rise and set, we do not feel like we are in motion. There are, for example, no great winds whipping at us as one might expect if the earth were tearing through space.

Second, human beings are egocentric creatures. We tend to see ourselves as being at the center of things. Extend this egocentric tendency into the heavens, and you have an explanation for our species' geocentric tendency. If, individually, we feel ourselves at the center of things, we also feel this collectively, as a planet.

The final reason?

Because Aristotle was Aristotle.

His opinions were regarded with awe for centuries. Few dared—few even thought—to question his teachings.

Instead, generations of European astronomers wrestled with Aristotle's model in an effort to show how it was actually correct. The most impressive of these wrestling matches was fought by Claudius Ptolemaeus, a Greek astronomer better known as Ptolemy, active in Alexandria during C.E. 127–145. Drawing on the work of the Greek astronomer Hipparchus (who died after 127 B.C.E.), Ptolemy developed the geocentric model into an impressively complex system of "deferents," large circular orbits centered on the earth, and "epicycles," small circles whose centers traveled around the circumferences of the deferents. Some 80 deferents and epicycles came into play, along with several other highly complex geometric arrangements, which allowed Ptolemy to account for many of the perplexing planetary motions actually observed.

Close Encounter

Almost nothing is known about the life of Ptolemy, but his astronomical work was collected in a single great volume, at first called (in Greek) *He mathematike syntaxis* (*The Mathematical Collection*). The book was later known an *Ho megas astronomos* (*The Great Astronomer*), but perhaps best known by what Arab astronomers called it in the ninth century, *Almagest*, essentially, *The Greatest Book*.

It is *Almagest* that has come down to us today. Divided into thirteen books, it is nothing less than a synthesis of all of Greek astronomy, a star catalog started by Hipparchus and expanded by Ptolemy into an identification of 1,022 stars. In addition, *Almagest* presents Ptolemy's geocentric model of the solar system, which was generally accepted as an accurate predictive model until Copernicus offered a heliocentric alternative in 1543.

Night Falls

Even if today's astronomers knew nothing of the solar system, they would likely reject the Ptolemaic model on the grounds of its unnecessary complexity. And for all of its intricacy, the Ptolemaic model was not a particularly accurate predictor of astronomical phenomenon. Indeed, the errors in the model became more glaring when better data on planetary positions (from Tycho Brahe and others) became available.

A love of elegant simplicity also characterized the classical world, but Ptolemy's era was already falling away from classical elegance and toward the cobwebbed mysteries that so appealed to people in the Dark Ages, when complexity and obscurity, not simplicity and clarity, were taken as the hallmarks of truth. Besides, Ptolemy's model, while highly imperfect, agreed pretty well with actual observation; it kept Aristotle safely on his pedestal, and it let humankind stay right where the Church said that God had intended: at the center of everything.

There were others who came after Aristotle and before Ptolemy, most notably Aristarchus of Samos (310–230 B.C.E.), who actually proposed that the earth and the planets orbit the sun (see Chapter 2, "Ancient Evenings: The First Watchers"). But Aristotle, Ptolemy, and common sense drowned out such voices that, for some thirteen centuries, few wanted to hear. For the light of classical learning had been dimmed, and the spirit of scientific inquiry muffled (at least in the West) in a long age of orthodoxy and obedience.

The Least You Need to Know

➤ Astronomy has always been associated with the keeping of time. The observed repeating motions of the sun, moon, and stars were first used to divide time into now familiar intervals: days, months, and years. Unfortunately, none of these intervals is evenly divisible into another.

➤ The earth's rotational axis is ever so slowly precessing, tracing out a circle on the sky. It takes the earth a long time (26,000 years) to precess once.

➤ Seasonal temperature variations and the difference in the length of days on the earth in the summer and winter are due to the 23½-degree tilt of the earth on its axis. The earth is *not* closer to the sun in summer and farther away in winter—in fact, the opposite is true.

➤ Equinoxes and solstices are special days in the astronomical year. The summer and winter solstices represent the longest and shortest days of the year, and the vernal and autumnal equinoxes are times when the day and night have equal length.

➤ Ptolemy made a highly successful—and complicated—attempt to account for all the complexities of movement among the celestial bodies. His geocentric model of the solar system was accepted until better observations of planetary positions in the sixteenth and seventeenth centuries made it clear that the model had to be significantly revised.

Astronomy Reborn: 1543–1687

The "Dark Ages" weren't dark everywhere, and astronomy didn't exactly wither and die after Ptolemy. Outside of Europe, there were some exciting discoveries being made. The Indian astronomer Aryabhātta (born ca. 476) held that the earth was a rotating globe, and he correctly explained the causes of eclipses. We have seen in Chapter 2, "Ancient Evenings: The First Watchers," that the Maya (and other Central American peoples) were actively observing the heavens around 1000 C.E.. The Maya in particular created a complex calendar, prepared planetary tables, and closely studied Venus, basing much of their system of timekeeping on its movements.

In Europe, however, astronomy was no longer so much studied as it was taught. Much Medieval learning discouraged direct observation in favor of poring over texts of recorded and accepted wisdom. Despite the work of Bishop Isidorus of Seville (570), who drew a sharp distinction between astronomy and astrology, medieval astronomy was mired in the superstition, which has its apologists and practitioners to this very day. But this chapter says little about astronomy's "Dark Ages" and turns, instead, to its rebirth in the Renaissance.

Arabian Nights

From our perspective just beyond the cusp of the millennium, it is easy to disparage Ptolemy for insisting that the earth stood at the center of the solar system. But we often forget that we live in a unique age, when images of the earth and other planets are routinely beamed from space. These stunning pictures of our cosmic neighborhood have become so familiar to us that commercial TV networks wouldn't think of elbowing aside this or that sitcom to show the images to the viewing public. Informed as we are with "the truth" about how the solar system works, we wonder how Ptolemy's complicated explanation could have been accepted for so long.

There is no doubt that his model of the solar system was wrong, but, wrong as he was, his book contained the heart and soul of classical astronomy and survived into an age that had turned its back on classical learning. During the early Middle Ages, Ptolemy's work remained unread in Europe, but his principal book found its way into the Arab world, and in 820 it was translated into Arabic as *Almagest* (roughly translatable as *The Greatest Book*). The circulation of Ptolemy's work renewed interest in astronomy throughout Arabia, with centers of learning being established in both Damascus and Baghdad.

Abu 'Abd Allah Muhammad Ibn Jabir Ibn Sinan Al-battani Al-harrani As-sabi', more conveniently known as al-Battani (ca. 858–929), became the most celebrated of the Arab astronomers, although it took many years before his major work, *On Stellar Motion,* was brought to Europe in Latin (about 1116) and in Spanish translation (in the thirteenth century). Al-Battani made important refinements to calculations of the length of the year and the seasons, as well as the annual precession of the equinoxes and the angle of the ecliptic. Moreover, he demonstrated that the Sun's *apogee* (its farthest point from Earth) is variable, and he refined Ptolemy's astronomical calculations by replacing geometry with sleek trigonometry.

Another Arab, Al-Sûfi (903–986), wrote a book translated as *Uranographia* (in essence, *Writings of the Celestial Muse*), in which he discussed the comparative brilliance of the stars. Like the scale of the Greek astronomer Hipparchos, the system of Al-Sûfi rated star brightness in orders of magnitude. Relative star brightness is still rated in terms of magnitude, and we will discuss this system in Chapter 17, "Of Giants and Dwarfs: Stepping Out into the Stars." Arab astronomers like Al-Sûfi also contributed star maps and catalogues, which were so influential that many of the star names in use today are of Arab origin (such as Betelgeuse, Aldebaran, and Algol), as are such basic astronomical terms as *azimuth* and *zenith*.

Heresy of a Polish Priest

In Europe, astronomy—as a truly observational science—did not revive until the Middle Ages had given way to the Renaissance. The German mathematician and astronomer Johann Müller (1436–1476) called himself Regiomontanus, after the Latinized form of Königsberg (King's Mountain), his birthplace. Enrolled at the University of Leipzig by the time he was 11, Regiomontanus assisted the Austrian mathematician Georg von Peuerbach in composing a work on Ptolemaic astronomy.

Close Encounter

Copernicus was born Mikolaj Kopernik on February 19, 1473, at Torun, and was educated at the finest universities in Europe, becoming master of the fields of mathematics, astronomy, medicine, and theology. In addition to his work in astronomy—effectively rebuilding the planetary system—Copernicus translated the Greek verses of Theophylactus, a seventh-century Byzantine poet, and laid the groundwork for currency reform in some Polish provinces. His serious rethinking of the Ptolemaic model of the planetary system began when church officials asked for his help with revising the outmoded and inaccurate Julian calendar. He demurred, saying that he could give no definitive answers because the motions of the moon and planets were insufficiently understood. But the problem gnawed at him, prompting him to delve into long-neglected Greek texts that argued for a heliocentric (sun-centered) universe.

Copernicus proceeded cautiously, privately circulating a brief summary of his heliocentric model, then gradually elaborated on his picture. Finally, Copernicus shared his manuscript with his student Georg Joachim Rhäticus, who took it to Germany for printing. But there he met opposition from Martin Luther and others, and Rhäticus quickly passed the manuscript to a printer in Leipzig, Andreas Osiander, who published the work. But Osiander took the precaution of writing a preface stating that the notion of a stationary sun around which Earth and other planets orbited was nothing more than a convenient way of simplifying planetary computations.

Despite the printer's caution, the Copernican idea eventually caught on with astronomers and triggered a revolution in thought about the solar system, the universe, and humankind's place in the cosmos.

Regiomontanus took his job seriously and, in 1461, journeyed to Rome to learn Greek and collect Greek manuscripts from refugee astronomers fleeing the Turks, so that he was able to read the most important texts, including the Greek translation of *Almagest*. In the meantime, his mentor Peuerbach had died and Regiomontanus completed the master's work in 1463. Three years later, he moved to Nürnberg, where a wealthy patron built him an observatory and gave him a printing press. Beginning in 1474, he used the press to publish *ephemerides*, celestial almanacs giving the daily positions of the heavenly bodies for periods of several years.

Star Words

Ephemerides are special almanacs that give the daily positions of various celestial objects for periods of several years.

The publications of Regiomontanus, which were issued until his death in 1476, did much to reintroduce to European astronomy the practice of scientific observation. He was so highly respected that Pope Sixtus IV summoned him to Rome to oversee revision of the notably inaccurate Julian calendar then in use.

Regiomontanus began this work on the calendar, but then died mysteriously—possibly from plague or from poison, perhaps administered by enemies resentful of his probing the cosmos too insistently. At that time, it could be dangerous to question accepted ideas, especially where the heavens were concerned. Nikolaus Krebs (1401–1464), known as Nicholas of Cusa, wrote a book called *De Docta Ignorantia* suggesting that the earth might not be the center of the universe. Fortunately for Nicholas (who was a cardinal of the Catholic church), few paid attention to the idea.

Another Nicholas (actually spelled Nicolaus)—Copernicus—was born in eastern Poland in 1473, almost a decade after Krebs's death. A brilliant youth, he studied at the universities of Kraków, Bologna, Padua, and Ferrara, learning just about everything that was then known in the fields of mathematics, astronomy, medicine, and theology.

Copernicus earned great renown as an astronomer and in 1514 was asked by the church for his opinion on the vexing question of calendar reform. The great Copernicus declared that he could not give an opinion, because the positions of the sun and moon were not understood with sufficient accuracy.

"More Pleasing to the Mind"

This uncertainty in the calendar bothered Copernicus deeply. Many astronomers thought that any errors in the Ptolemaic system might be due to the many small "typos" that had crept into the manuscript with centuries of copying by scribes. However, connections between east and west at this time meant that Copernicus was able to have a nearly pristine copy of Ptolemy's work *Almagest*. Any errors had to be errors in the model itself. Ptolemy's complex geocentric system of epicycles and deferents had allowed some degree of accuracy in predicting planetary motions and eclipses. But as more careful observations were made, the Ptolemaic system was proving to be more and more creaky.

Copernicus, as pictured in Charles F. Horne's Great Men and Famous Women *(1894).*

(Image from the authors' collection)

Copernicus questioned the Ptolemaic system on the very basis that a modern scientist might. The model had become too complicated, and scientists tend to seek simplicity (where possible) in their models of the universe. The printing presses that were firing up across Europe at the time made it possible for many more scholars to read good copies of ancient works. As Copernicus started reading Greek manuscripts that had been long neglected, he rediscovered Aristarchus's old idea of a heliocentric (sun-centered) universe. He concluded that putting the sun at (or near) the center of a solar system with planets in orbit around it created a model that was "more pleasing to the mind" than what Ptolemy had proposed and medieval Europe accepted for so many centuries.

But he did not rush to publish, only after much hesitation privately circulating a brief manuscript, *De Hypothesibus Motuum Coelestium a Se Constitutis Commentariolus* (*A Commentary on the Theories of the Motions of Heavenly Objects from Their Arrangements*) in 1514. He argued that all of the motion we see in the heavens is the result of the earth's daily rotation on its axis and yearly revolution around the sun, which is motionless at the center of the planetary system.

Sound familiar?

Aristarchus had suggested it almost 2,000 years earlier, but no one had listened. The earth, Copernicus explained, was central only to the orbit of the moon.

For almost two more decades he refined his thought before consenting, in 1536, to publish the full theory. But largely because of opposition from Martin Luther and other German religious reformers, *De Revolutionibus Orbium Caelestium* (*On the*

Revolutions of the Celestial Spheres) wasn't actually printed until 1543. As close as Copernicus's model came to representing the motion of planets in the solar system, it insisted on the perfection of circular orbits, so that it actually had no better predictive ability for planetary motions than the Ptolemaic model it replaced. For all its creakiness, the Ptolemaic model still predicted, for example, where Mercury would be on a particular night about as well as Copernicus's model did.

A Revolution of Revolutions

For centuries, astronomy had been a science in which errors of a few degrees in planetary position on the sky were acceptable. But to Copernicus, Kepler, and others who would follow, errors of that magnitude indicated that something was seriously wrong in our understanding of planetary motion. They were driven to discover their origin.

Now we turn to the details of Copernicus's model. The first of the six sections of *De Revolutionibus* sets out some mathematical principles and rearranges the planets in order from the sun: Mercury is closest to the sun, followed by Venus, Earth (with the Moon orbiting it), Mars, Jupiter, and Saturn. The second part applies the mathematical rules set out in the first to explain the apparent motions of the stars, planets, and sun. The third section describes Earth's motions mathematically and includes a discussion of precession of the equinoxes, attributing it correctly to the slow gyration of the earth's rotational axis. The last three parts of the book are devoted to the motions of the moon and the planets other than Earth.

While Copernicus's purpose in reordering the planetary system may have been relatively modest, it soon became apparent that the new theory required the most profound revision of thought.

First: The universe had to be a much bigger place than previously imagined. The stars always appeared in the same positions with the same apparent brightness. But if the earth really were in orbit around the sun, the stars should display a small but noticeable periodic change in position and brightness. Why didn't they? Copernicus said that the starry celestial sphere had to be so distant from Earth that changes simply could not be detected. Building on this explanation, others theorized an *infinite* universe, in which the stars were not arranged on a celestial sphere, but were scattered throughout space.

The second required revision, while not as obvious, was even more basic.

Astronomer's Notebook

Properly speaking, Copernicus's solar system was *heliostatic* (a motionless sun) rather than *heliocentric* (a centered sun), because, in order to account more accurately for planetary motion, he did not place the sun at the precise center of the system, but offset it slightly. The perfection of shapes, and circles in particular, had been part of the astronomer's baggage since Greek times. It would take Kepler's brilliant and unconventional idea of elliptical, rather than circular, orbits eventually to reconcile all the apparent irregularities of planetary motion.

Why do things fall? Aristotle explained that bodies fell toward their "natural place," which, he said, was the center of the universe. That explanation worked as long as the earth was considered to be the center of the universe. But now that it wasn't the center, how could the behavior of falling bodies be explained? The answers would have to wait until the late seventeenth century, when Isaac Newton published his *Principia,* including a theory of universal gravitation.

Profound as were the astronomical and other scientific implications, the emotional shock of the Copernican universe was even greater. Suddenly, the earth, with humankind upon it, was no longer at the center of all creation, but was instead hurtling through space like a ball on a string. Why did we stay on the ball? What was the string? These were all unanswered questions that must have been very unsettling for those who thought about them.

The Man with the Golden Nose

The astronomer Tycho Brahe (1546–1601) was born in Denmark just three years after the publication of *De Revolutionibus.* As a youth, he studied law, but was so impressed by astronomers' ability to predict a total solar eclipse on August 21, 1560, that he began to study astronomy on his own. In August 1563, he made his first recorded astronomical observation, a conjunction (a coming together in the sky) of Jupiter and Saturn, and discovered that the existing ephemerides were highly inaccurate. From this point on, Tycho decided to devote his life to careful astronomical observation.

On November 11, 1572, he recorded the appearance of a new star (a nova, an object explained in Chapter 19, "Black Holes: One-Way Tickets to Eternity"), brighter than Venus, in the constellation Cassiopeia. The publication of his observations (*De Nova Stella,* 1573) made him famous. King Frederick II of Denmark gave him land and financed the construction of an observatory Tycho called Uraniborg (after Urania, the muse of astronomy). Here Tycho not only attracted scholars from all over the world, but designed innovative astronomical instruments and made meticulous astronomical observations—the most accurate possible before the invention of the telescope.

Astro Byte

Not only did Tycho Brahe chart the position of some 777 stars, prove that the nova of 1572 was a star, and carry out a refined and comprehensive study of the solar system, he also invented a pressure-operated indoor toilet—unheard of in Renaissance Europe.

Tycho Brahe in his observatory, as seen in an early print.

(Image from arttoday.com*)*

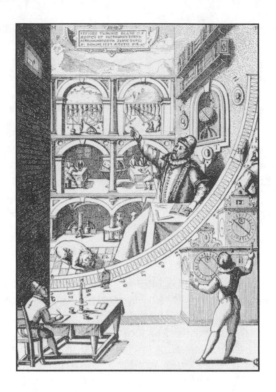

Kepler Makes Sense of It

When King Frederick died in 1588, Tycho lost his most understanding and indulgent patron. Frederick's son Christian IV was less interested in astronomy than his father had been, and, to Christian, Tycho was an unreasonably demanding protégé, who repeatedly sought more money. At last, the astronomer left Denmark and ultimately settled in Prague in 1599 as Imperial Mathematician of the Holy Roman Empire. In the Czech city, he was joined by a persistent younger German astronomer, Johannes Kepler (1571–1630), who, after writing several flattering letters to Tycho, became his student and disciple.

Tycho Brahe, a colorful character, who lost part of his nose in a duel (he replaced it with a golden prosthesis), died ingloriously in 1601, apparently from a burst bladder after drinking too much at a dinner party. After a bit of a struggle, Kepler got a hold of the mass of complex observational data Tycho had accumulated.

While Tycho had been a brilliant observer, he was not a particularly good theoretician. Kepler, a sickly child, had grown into a frail adult with the mind of a brilliant theorist—though with very little aptitude for close observation, since he also suffered from poor eyesight. Indeed, Tycho and Kepler were the original odd couple, who argued incessantly; yet their skills were perfectly complementary. And when Tycho died, instead of using his instruments to make new observations, Kepler dived into

Tycho's data, seeking in its precise observations of planetary positions a unifying principle that would explain the motions of the planets without resorting to epicycles. It was clear, especially with Tycho's data, that the Copernican system of planets moving around the sun in perfectly circular orbits was not going to be sufficient. Kepler sought the missing piece of information that would harmonize the heliostatic solar system of Copernicus with the mountain of data that was Tycho's planetary observations.

Johannes Kepler.

(Image from arttoday.com*)*

Three Laws

Kepler had predicted that with Tycho's data, he would solve the problem of planetary motion in a matter of days. After almost eight years of study, trial, and error, Kepler had a stroke of genius. He concluded that the planets must orbit the sun not in perfect circles, but in elliptical orbits (an *ellipse* is a flattened circle). He wrote to a friend: "I have the answer … The orbit of the planet is a perfect ellipse."

Kepler was able to state the fundamentals of planetary motion in three basic laws. That planets move in elliptical orbits, with the sun at one focus of the ellipse, is known as Kepler's First Law. It resolved the discrepancies in observed planetary motion that both Ptolemy and Copernicus had failed to explain adequately. Both of those great minds had been convinced that in a perfect universe, the orbits of planets *had* to be circular.

Star Words

An **ellipse** is an oblong circle drawn around two foci instead of a single center point.

Close Encounter

The nature of an ellipse and Kepler's Second Law are both difficult to explain, but easy to visualize. You can draw an ellipse with the aid of a couple of tacks, string, a piece of cardboard, and a pencil. Push the two tacks into a piece of cardboard and loop the string around them. The tacks are the foci of the ellipse. The closer together the foci, the less elliptical (more perfectly circular) the ellipse. To draw an ellipse, simply place the pencil inside the loop of the string, and pull it taut—now, pull the pencil around the tacks all the way. Depending on the separation of the tacks, you will draw different looking ellipses.

To illustrate Kepler's Second Law, imagine the sun as being located at the focus nearest *a* (that is, the tack *f* nearest *a*). The pencil point is a planet, and the portion of the string connecting the pencil point to the tack nearest *a* is the imaginary line connecting the sun to a planet.

If you would accurately model the orbit of the planet around the sun, you would have to be able to draw the ellipse such that the pencil would sweep out equal areas in equal intervals of time. This would not be an easy trick, because you would have to draw faster when the pencil/planet was closest to the tack/sun (the position called *parahelion*) than when it was farthest from the sun (*aphelion*). To make an area swept near *parahelion* equal to an area swept at *aphelion* you have to cover more of the circumference of the ellipse nearer the sun–tack than you have to cover farther from the sun–tack.

Another apparent attribute of elliptical orbits determined Kepler's Second Law. It states that an imaginary line connecting the sun to any planet would sweep out equal areas of the ellipse in equal intervals of time. (See the preceding "Close Encounter" sidebar for an explanation of focus and for a demonstration of Kepler's Second Law.) The Second Law explained the variation in speed with which planets travel. They will move faster when they are closer to the sun. Kepler did not say *why* this was so, just that it was apparently so. Later minds would confront that *why*.

The first two of Kepler's laws were published in 1609. The third did not appear until ten years later and is slightly more complex. It states that the square of a planet's *orbital period* (the time needed to complete one orbit around the sun) is proportional to the cube of its *semi-major axis* (see the following figure and the preceding "Close Encounter" sidebar for an illustration and explanation of the semi-major axis). Since the planets' orbits, while elliptical, are very nearly circular, the semi-major axis can be considered to be a planet's average distance from the sun.

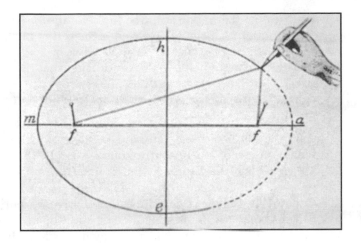

Drawing an ellipse to illustrate Kepler's Second Law. See the "Close Encounter" sidebar on the previous page for an explanation.

(Image from the authors' collection)

Galileo's Eye

While Kepler was theorizing from Copernicus's data, an Italian astronomer, Galileo Galilei (1565–1642), directed his gaze skyward, amplifying his eyesight with the aid of a new Dutch invention (which we'll meet in the next chapter), the telescope.

Through this instrument, Galileo explored the imperfect surface of the moon, covered as it was with craters, "seas," and features that looked very much like the surface of the earth. Galileo was even more surprised to find that the surface of the sun was blemished. These "sunspots" (which we'll investigate in Chapter 16, "Our Star"), he noted, changed position from day to day. From this fact, Galileo did not conclude that the spots changed, but that the sun was rotating, making a complete revolution about once each month.

Close Encounter

Popular legend has it that Galileo dutifully repeated aloud what the Inquisitors ordered him to say, that the earth does not move, only to mutter under his breath, as he marched off to house arrest, "Still, it moves." House arrest did not prevent Galileo from continuing to work and even from corresponding with other scientists. "Though my soul may set in darkness, it will rise in perfect light," he wrote. "I have loved the stars too fondly to be fearful of the night."

His telescope also revealed for the first time that moons orbited Jupiter—another observation that strongly supported the notion that the earth was not the center of all things. He observed that the planet Venus cycled through phases, much like the moon, and that the size of the planet varied with its phase. From this, he concluded that Venus must orbit not the earth, but the sun.

Galileo published these many independent experimental proofs of a heliocentric solar system in 1610. Six years later, the Catholic Church judged the work heretical and banned them, as well as the work of Copernicus. Galileo defied the ban and, in 1632, published a comparison of the Ptolemaic and Copernican models written as a kind of three-way discussion. He was so bold as to write in Italian, instead of learned Latin, which meant that common folk (at least those who were literate) were being invited to read a theory that challenged the teaching of the Church.

Galileo was silenced by the Holy Inquisition in 1633, forced to recant his heresy under threat of death, and placed under house arrest for the rest of his life.

Star Words

The **orbital period** of an object is the time required to complete one full orbit. The orbital period of the earth around the sun, for example, is about 365 days. If the period (P) is given in Earth years, and the distance (a) in astronomical units (or the average distance between the earth and the sun), then Kepler's Third Law is simply $P^2 = a^3$.

Galileo Galilei.

(Image from arttoday.com*)*

Holding It All Together

One of us knew a young man who owned a pocket watch, which he would habitually twirl, holding the end of the chain and allowing it to orbit around the focus of his finger. One day, the chain broke, sending the costly timepiece flying off into space—and against a wall, with predictably catastrophic results.

Why don't the planets suffer the same fate? Despite his brilliant explanation of planetary motion, Kepler had not explained *how* the planets orbited the sun without flying off into space, and *why* they traveled in ellipses.

The answer came in the late seventeenth century when an Englishman, Isaac Newton (1642–1727), one of the most brilliant mathematicians who ever lived, formulated three laws of motion and the law of universal gravitation in a great work, *Philosophiae Naturalis Principia Mathematica* (*Mathematical Principles of Natural Philosophy*), better known simply as *Principia*.

Star Words

An **Astronomical Unit** (A.U.) is equal to the mean distance between the earth and the sun. It is a conveniently large unit for measuring the large distances in the solar system.

Newton's Three Laws of Motion

Newton's first law of motion states that, unless acted upon by some external force, a body at rest remains at rest and a moving object continues to move forever in a straight line and at a constant speed. This property is known as *inertia*. The measure of an object's inertia is its *mass* (in effect, the amount of matter the object contains). The more massive an object, the greater its inertia.

Newton's first law explains why the planets move in nearly circular orbits—essentially because an external force (gravity) acts on each planet. Without gravity, the planets would all fly off in straight lines, like so many pocket watches.

Newton's second law states that the acceleration of an object is directly proportional to the force applied to the object and inversely proportional to the mass of the object. Pull two objects with the same force, and the more massive object will accelerate more slowly than the less massive one. We all know this intuitively. Your subcompact car's engine would have a much harder time accelerating than an 18-wheel truck!

Star Words

Inertia is the tendency of a body in motion to remain in motion in a straight line and at a constant velocity unless acted on by an external force; it is also the tendency of a body at rest to remain at rest unless acted upon by an external force. **Mass** can be thought of as the amount of matter an object contains or as a measure of an object's inertia. You have experienced the inertia of your body when rounding the corner in a car. Your body wants to keep moving in a straight line, but the car door (fortunately) makes you move in a curved path.

Newton's third law of motion states that forces do not act in isolation. If object A exerts a force on object B, object B exerts an *equal but opposite* force on object A. A hammer, for example, exerts a force on the nail, driving it into the wall. The nail exerts an equal and opposite force on the hammer, stopping its motion.

Weighty Matters

Throw a ball up into the air, and you will observe that it travels in a familiar curved (parabolic) trajectory: first up, up, up, leveling off, then down, down, down. Common sense tells us that the force of gravity pulls the ball back to Earth.

Newton's brilliance was in postulating not only that there is a force, gravity, that pulls the ball (or apple) back to the earth, but that such a force applies to everything in the universe that has mass. The gravitational force due to the mass of the earth also pulls on the moon, holding in its orbit, and pulls on each of us, keeping us in contact with the ground. Finally there was an answer to those who thought the earth could not be spinning and orbiting the sun. What was there to keep us firmly footed on the earth? Newton had the answer: the force of gravity.

Astro Byte

Newton's laws of motion were the result of brilliant insight and considerable hard work. When asked later how he had unraveled the mathematics behind the mysteries of planetary motion, he responded, "By thinking of them without ceasing."

Sir Isaac Newton.

(Image from arttoday.com*)*

It's Not Just a Good Idea ...

In subsequent chapters discussing the planets, we will look more precisely at how mathematics describes planetary motion. For now, consider the following general points.

In *Principia,* Newton proposed that the force of gravity exerted by objects upon one another is proportional to the mass of the two objects, and weakens as the square of the distance between those objects.

Specifically, he postulated that the gravitational force between two objects is directly proportional to the product of their masses (mass of object A times mass of object B). So two objects, one very massive and the other with very little mass, will "feel" the same mutual attraction. In addition, he claimed that the force between two objects will decrease in proportion to the square of the distance. This "inverse-square law" means that the force of gravity mutually exerted by two objects, say 10 units of distance apart, is 100 times (10^2) weaker than that exerted by objects only 1 unit apart—yet this force never reaches zero. The most distant galaxies in the universe exert a gravitational pull on one another. These relations between mass, distance, and force comprise what we call Newton's Law of Universal Gravitation.

Consider the solar system, with the planets moving in elliptical orbits around the sun. Newton's *Principia* explained not only what holds the planets in their elliptical orbits (an "inverse-square" force called gravity), but also predicted that the planets themselves (massive Jupiter in particular) would have a small but measurable effect on each other's orbits.

Close Encounter

Newton's mathematical prowess was legendary, and he had both the brilliance and the determination to solve problems that had stymied others for decades. His friend, the great astronomer Edmund Halley, and others had been thinking for some time about a way to account for the elliptical orbits of planets, and thought that if the force of gravity decreased with the square of the distance, that the elliptical orbits of Kepler could be explained. But Halley was unable to prove this hunch mathematically. One day, Halley asked Newton what would be the trajectory of a planet moving in a gravitational force field that decreased in strength with the square of the distance from the sun. Newton instantly responded: "An ellipse."

Halley asked to see the mathematical proof, and a few months later, Newton sent him a letter in which he derived Kepler's three laws of planetary motion from the basis of a gravitational attraction that decreased with the square of the distance between the sun and the planets. With Halley's encouragement, Newton began his masterwork, *Principia,* in which he put forth his famous Laws of Motion and the Law of Universal Gravitation.

Like any good scientific theory, Newton's laws not only explained what was already observed (the motion of the planets), but was able to make testable predictions. The orbit of Saturn, for example, was known to deviate slightly from what one would expect if it were simply in orbit around the sun (with no other planets present). The mass of Jupiter has a small, but measurable, effect on its orbital path. Newton noted with a sense of humor that the effect of Jupiter on Saturn's orbit made so much sense (according to his theory) that "astronomers are puzzled with it."

For the first time, a scientist had claimed that the rules of motion on the earth were no different from the rules of motion in the heavens. The moon was just a big apple, much farther away, falling to the earth in its own way. The planets orbit the sun following the same rules as a baseball thrown up into the air, and the pocket watch of the earth is held in its orbit by a chain called gravity.

Did Newton bring the celestial sphere down to Earth, or elevate us all to the status of planets? Whatever you think, we have never looked at the solar system or the universe in the same way since.

The Least You Need to Know

➤ While Europe labored through the Middle Ages, Arabian astronomers, inspired by a translation of Ptolemy, were busy making remarkable observations.

➤ Seeking an explanation of the motion of the planets accurate enough to enable him to revise inaccurate European calendars, Copernicus (following the lead of Aristarchus before him) put the sun at the center of the solar system, with the planets, including Earth, in orbit around it.

➤ Tycho Brahe made extraordinary astronomical observations, but it was his student, Johannes Kepler, who reduced these data to three basic laws of planetary motion. These laws, based on elliptical planetary orbits, brought the heliocentric model of the solar system into close agreement with observations.

➤ Galileo used the newly invented telescope to provide experimental proofs of the Copernican idea.

➤ With the sun-centered universe in place, Isaac Newton put forth in *Principia* the three laws of motion and the law of universal gravitation, which accounted for the forces behind planetary motion. For the first time, someone proposed that the laws governing the motion of objects here on Earth and in the heavens were one and the same.

Part 2

Now You See It (Now You Don't)

This part begins with two chapters on telescopes and includes advice on choosing a good amateur instrument and getting the most out of it once you've made your purchase.

But professional astronomers don't limit their observations to the visible spectrum. Modern astronomy can tune into radio, infrared, ultraviolet, and even high-energy gamma radiation. Chapter 7 orients you within the electromagnetic spectrum, and Chapter 8 explores the range of the modern invisible astronomies.

The last chapter in this part is a compact history of space exploration by manned as well as robotic craft, including such wonders as the Hubble Space Telescope, Mars Observer, *and the forthcoming* International Space Station.

Chapter 5

The Art of Collecting Light (with a Telescope)

In This Chapter

➤ Overcoming the tyranny of vast distance in the universe

➤ Light as energy that conveys information

➤ A brief look at the spectrum and waves

➤ An introduction to the telescope

➤ Refractor versus reflector telescope design

➤ Why a big telescope is a better telescope

➤ Resolution and atmospheric interference

➤ The *Hubble Space Telescope* and other cutting-edge projects

---

You have every reason and right to look at the night sky as the greatest free show in the universe. Most amateur stargazers—in fact, most professional astronomers—are strongly attracted by the great beauty of the sky. But the sky is more than beautiful. Celestial objects are full of information just waiting to be interpreted—information like: how distant the stars and galaxies are, how large they are, and whether they are moving toward or away from us.

How does the information reach us?

It travels to us in the form of electromagnetic radiation, a small portion of which is visible light. We begin this chapter by defining light and then talk about how we can most effectively collect it to see more of the solar system and the universe.

Slice of Light

The universe is ruled by the tyranny of distance. That is, the universe is so vast, that we are able to see many things that we will never be able to visit. Light is able to travel at extraordinary speeds (about 984,000,000 feet, or 300,000,000 meters, every second), but the light that we now see from many objects in the sky left those sources thousands, millions, or even billions of years ago. It is possible, for example, to see the Andromeda galaxy, even with the naked eye, but will we ever travel there?

Well, Andromeda is about two million light-years away, and a *light-year* is the distance light travels in one year—about 9,461,000,000,000,000 meters (some 6 trillion miles). Now, light can travel that far every year, so to get the distance to Andromeda, you multiply the velocity of light (6 trillion miles in a year) by the amount of time it took the light to get here (2 million years), and you get a lot of miles—approximately a 1 with 19 zeroes after it. Another way to think about these unbelievable distances: If you could travel at the speed of light (an impossibility, according to Einstein's theory of relativity, as we will see in later chapters), it would *still* take you two million years to reach Andromeda.

But we can't travel at anywhere near the speed of light. Right now, the fastest rockets are capable of doing 30,000 miles per hour (48,000 km/h). Maybe—someday—technology will enable us at least to approach the speed of light, but that still means a trip of two million years up and two million back. All of recorded history consumes no more than 5,500 years, and civilization, perhaps 10,000 years.

Why not go faster than the speed of light? In Chapter 7, "Over the Rainbow," we'll see that, according to our understanding of space and time, the speed of light is an absolute speed limit, which cannot be exceeded.

So revel in the fact that, on a clear night, you are able to gaze at the Andromeda galaxy, an object so distant that no human being will likely ever visit it.

Star Words

A **light-year** is the distance light travels in one year. In the vastness of space beyond the solar system, astronomers use the light-year as a basic unit of distance.

Star Words

Electromagnetic radiation transfers energy and information from one place to another, even in the vacuum of space. The energy is carried in the form of rapidly fluctuating electric and magnetic fields and includes visible light in addition to the similar but less familiar radio, infrared, ultra-violet, x-ray, and gamma ray radiation.

The Andromeda galaxy, as photographed by the Yerkes Observatory, 1930s.

(Image from arttoday.com*)*

Space ships may be severely limited as to how fast they can travel, but as we've said, the information conveyed by *electromagnetic radiation* can travel at the speed of light. The information from Andromeda, it is true, is not exactly recent news by the time we get it. In fact, the photons that we are receiving from Andromeda left that galaxy long before *Homo sapiens* walked the earth. But everything we know about Andromeda and almost all other celestial bodies (aside from the few solar system objects we have visited with probes or landers), we know by analyzing their electromagnetic radiation: radio, infrared, and ultraviolet radiation, as well as x-rays and gamma rays and what we call light.

The Whole Spectrum

Electromagnetic radiation travels though the vacuum of space in waves, which we shall also examine in some detail in Chapter 7. A wave—think of a water wave—is not a physical object, but a pattern of up-and-down or back-and-forth motion created by a disturbance. Waves are familiar to anyone who has thrown a rock in a pond of still water or watched raindrops striking a puddle. The wave pattern in the water, a series of concentric circles, radiates from the source of the energy, the impact of the rock or the rain drop. If anything happens to be floating on the surface of the water— say a leaf—the waves will transfer some of the energy of the splash to the leaf and cause it to oscillate up and down. The important thing to remember about waves is that they convey both energy and information. Even if we didn't actually see the rock or the raindrop hit the water, we would be able to surmise from the action of the waves that something had disturbed the surface of the water at a particular point.

A familiar example of waves propagated by the energy of a tossed stone.

(Image from the authors' collection)

Star Words

Wavelength is the distance between two adjacent wave crests (high points) or troughs (low points). This distance is usually measured in meters or multiples thereof. Water waves may have wavelengths of a few meters, radio waves of a few centimeters, and the wavelengths of optical light are very short (~0.0000005 meters). **Frequency** is the number of wave crests that pass a given point per unit of time. By convention, frequency is measured in hertz. 1 Hz is equivalent to one crest-to-crest cycle per second and named in honor of the nineteenth-century German physicist Heinrich Rudolf Hertz.

The type of energy and information created and conveyed by electromagnetic radiation is more complex than that created and conveyed by the waves generated by a splash in the water. We will wait until Chapter 7 for a little lesson in wave anatomy, but do take a moment now to make sure that you understand two properties of waves: *wavelength* and *frequency*. Wavelength is the distance between two adjacent wave crests (high points) or troughs (low points), measured in meters. Frequency is the number of wave crests that pass a given point per unit of time (and has units of 1/second).

We think of the light from our reading lamp as very different from the x-rays our dentist uses to diagnose an ailing tooth, but both are types of electromagnetic waves, and the only difference between them is their wavelengths. Frequency and wavelength of a wave are inversely proportional to one another, meaning that if one of them gets bigger, the other one must get smaller. The particular wavelength produced by a given energy source (a star's photosphere, a planetary atmosphere) determines whether the electromagnetic radiation produced by that source is detected at radio, infrared, visible, ultraviolet, x-ray, or gamma ray wavelengths.

The waves that produce what we perceive as visible light have wavelengths of between 400 and 700 nanometers (a nanometer is 0.000000001 meter, or 1×10^{-9} m) and frequencies of somewhat less than 10^{15} Hz. Light waves, like the other forms of electromagnetic radiation, are produced by the change in the energy state of an atom or molecule. These waves, in turn, transmit energy from one place in the universe to another. The special nerves in the retinas of our eyes, the emulsion on photographic film, and the pixels of a CCD (Charge Coupled Device) electronic detector are all stimulated (energized) by the energy transmitted by waves of what we call visible light. That is why we "see."

The outer layers of a star consist of extremely hot gas. This gas is radiating away some fraction of the huge amounts of energy that a star generates in its core through nuclear fusion (see Chapter 16, "Our Star"). That energy is emitted at some level in all portions of the *electromagnetic spectrum,* so that when we look at a distant or nearby star (the sun) with our eyes, we are receiving a small portion of that energy.

Star Words

The **electromagnetic spectrum** *is* the complete range of electromagnetic radiation, from radio waves to gamma waves and everything in between, including visible light.

Buckets of Light

Of course, the fraction of the emitted energy we receive from a very distant star—or even a whole galaxy, like far-off Andromeda—is very small, having been diminished by the square of the distance (but never reaching zero). Imagine a sphere centered on a distant star. As the sphere becomes larger and larger (that is, as we get farther and farther from the star), the same amount of energy will pass through ever larger spheres. Your eye (or your telescope) can be thought of as a very tiny fraction of the sphere centered on that distant star. You are collecting as much light from the distant source as falls into your "light bucket." If your eye is a tiny "bucket," then a 4-inch amateur telescope is a slightly larger bucket, and the *Hubble Space Telescope* is an even larger bucket. The larger the bucket, the more light you can "collect." And if we collect more light in our bucket, we get more information.

One early question among astronomers (and others) was, "How can we build a better bucket than the two little ones we have in our head?"

The answer came in the early seventeenth century.

The Telescope Is Born

In 1608, lens makers in the Netherlands discovered that if they mounted one lens at either end of a tube and adjusted the distance between the lenses, the lens that you put to your eye would magnify an image focused by the lens at the far end of the tube. In effect, the lens at the far end of the tube gathered and concentrated (focused) more light energy than the eye could do on its own. The lens near the eye enlarged to various degrees that concentrated image. This world-changing invention was dubbed a telescope. The word *telescope* comes from Greek roots meaning "far-seeing." Optical telescopes are arrangements of lenses and/or mirrors designed to gather visible light efficiently enough to enhance observation of distant objects and phenomena.

Many, perhaps most, inventions take time to gain acceptance. Typically, there is a lapse of more than a few years between the invention and its practical application. Not so with the telescope. By 1609, within a year after the first telescopes appeared, the Italian astronomer Galileo Galilei demonstrated their significance in military matters (seeing a distant naval foe), and was soon using them to explore the heavens. The largest of his instruments was quite small, with only modest magnifying power, but, as we've seen in the preceding chapter, Galileo was able to use this tool to describe the valleys and mountains on the Moon, to observe the phases of Venus, and to identify the four largest moons of Jupiter.

Refraction ...

Galileo's instrument, like all of the earliest telescopes, was a *refracting telescope,* which uses a glass lens to focus the incoming light. For all practical purposes, astronomical objects are so far away from us that we can consider that light rays come to us parallel to one another—that is, unfocused. Refraction is the bending of these parallel rays.

Star Words

A **refracting telescope** or **refractor** creates its image by refracting (bending) light rays in glass lenses. The point at which those incoming parallel bent rays converge is called the focus, and the distance from the cross-sectional center of the lens to the focus is the focal length of the lens.

The convex (bowed outward) piece of glass we call a lens bends the incoming rays such that they all converge at a point called the *focus,* which is behind the lens directly along its axis. The distance from the cross-sectional center of the lens to the focus is called the *focal length* of the lens. Positioned behind the focus is the eyepiece lens, which magnifies the focused image for the viewer's eye.

Modern refracting telescopes consist of more than two simple lenses. At both ends of the telescope tube, compound (multiple) lenses are used, consisting of assemblies of individual lenses (called elements) designed to correct for various distortions simple lenses produce. For example, the exact degree to which light bends or refracts in a piece of glass depends on its wavelength. Since light consists of many different wavelengths, a single lens will produce a distortion called "chromatic aberration." The compound eyepiece of many modern telescopes also corrects the image, which a simple eyepiece would see upside down and reversed left to right.

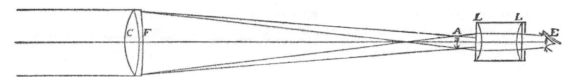

Diagram of a refracting telescope. CF *represents the objective lens and* LL *the eyepiece. The observer's eye is identified by* E.

(Image from arttoday.com*)*

... or Reflection?

The refracting telescope was one of humankind's great inventions, rendered even greater by the presence of a genius like Galileo to use it. However, the limitations of the refracting telescope soon became apparent:

➤ Even the most exquisitely crafted lens produces distortion, which can be corrected only by the introduction of other lenses, which, in turn, introduce their own distortion and loss of brightness, since a little of the energy is absorbed in all that glass. The chief distortion is chromatic aberration.

➤ Excellent lenses are expensive to produce, and this was even more true in the days when all lenses were painstakingly ground by hand. Lenses are particularly difficult to produce because both sides have to be precision crafted and polished. For mirrored surfaces, like those found in reflecting telescopes, only a single side must be polished.

➤ Generally, the larger the lens, the greater the magnification and the brighter the image; however, large lenses get heavier faster than large mirrors. Lenses have volume, and the potential for imperfections (such as bubbles in the glass) is higher in a large lens. All of this means that large lenses are much more difficult and expensive to produce than small ones.

Astro Byte

Newton gets credit for inventing the reflecting telescope, but another Englishman, John Gregory, actually beat him to it, with a design created in 1663. It was, however, the Newtonian reflector that caught on. The French lens maker Guillaume Cassegrain introduced another variation on the reflector design in 1672. In his design, there is a primary and a secondary mirror, and the focal point of the primary mirror is located behind the primary mirror surface.

Recognizing the deficiencies of the refracting telescope, Isaac Newton developed a new design, the reflecting telescope, in 1668.

Instead of the convex lens of a refractor, the reflector uses a concave mirror (shaped like a shallow bowl) to gather, reflect, and focus incoming light. The hollow side of your breakfast spoon is a concave mirror (the other side is a convex one). This curvature means that the focal point is in front of the mirror—between the mirror and the object being viewed. Newton recognized that this was at best inconvenient—your own head could block what you are looking at—so he introduced a secondary mirror to deflect the light path at a 90-degree angle to an eyepiece mounted on the side of the telescope.

Diagram of a Newtonian reflector. Light enters at the left and is focused by the primary mirror (M) at the back of the telescope. The focused image is sent by a secondary mirror (G) through the eyepiece (LL). The observer's eye is labeled E.

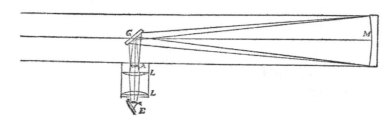

(Image from arttoday.com)

Refracting telescope design continued to develop throughout the eighteenth and nineteenth centuries, culminating in the 40-inch (that's the diameter of the principal lens) instrument at Yerkes Observatory in Williams Bay, Wisconsin, installed in 1897.

Star Words

Nebula is a term with several applications in astronomy, but it is used most generally to describe any fuzzy patch seen in the sky. Nebulae (plural) are often (though not always) vast clouds of dust and gas.

But due to the limitations just mentioned, the biggest, most powerful telescopes have all been reflectors. In the eighteenth century, the great British astronomer Sir William Herschel persuaded the king to finance an instrument with a 47-inch (1.2-meter) mirror. With this telescope, Herschel had a big enough light bucket to explore galaxies beyond our own Milky Way (though he did not know that's what they were). By the middle of the nineteenth century, William Parsons, third Earl of Rosse, explored new *nebulae* (fuzzy patches of light in the sky, some of which are galaxies) and star clusters with a 73-inch (1.85-meter) instrument constructed in 1845. It ranked as the largest telescope in the world well into the twentieth century, until the 100-inch reflector was installed at the Mount Wilson Observatory (near Pasadena, California) early in the century.

Sir William Herschel (1738–1822), who catalogued more than 800 double stars and 2,500 nebulae—mostly using the mammoth telescope he convinced the king of England to finance. He is depicted on a plaque designed by the English ceramicist Josiah Wedgwood.

(Image from arttoday.com)

Variations on an Optical Theme

While the two major types of optical telescopes are the refractor and the reflector, it is also useful to be aware of the basic variations in reflector design, especially when you think about choosing a telescope for yourself (see the next chapter). We have already seen that the simplest reflector (prime focus) focuses its image at the front of the telescope, introducing the possibility that the observer may block the image. The Newtonian focus instrument, as mentioned, overcomes this problem by introducing a secondary mirror to direct the focus to an eyepiece at the upper side of the instrument. This remains a popular arrangement for small reflecting telescopes used by amateur astronomers. This arrangement is unwieldy, however, for a large telescope. Imagine trying to get to the "top" of a telescope 6 feet long, perched on a 6-foot pedestal.

Some larger reflecting telescopes employ a Cassegrain focus. The image from the primary mirror is reflected to a secondary mirror, which again reflects the light rays down through an aperture (hole) in the primary mirror to an eyepiece at the back of the telescope.

Finally, a coudé-focus (*coudé* is French for "bent") reflector sends light rays from the primary mirror to a secondary mirror, much like a Cassegrain. However, instead of focusing the light behind the primary mirror, another mirror is employed to direct the light away from the telescope, through an aperture and into a separate room, called the coudé-focus room. Here astronomers can house special imaging equipment that might be too heavy or cumbersome to actually mount to the barrel of the telescope.

Reflecting telescopes have their problems as well. The presence of a secondary mirror (or a detector, in the case of a prime-focus reflector) means that some fraction of the incoming light is necessarily blocked. Although reflectors do not experience "chromatic aberration" (since light does not have to pass through glass), their spherical shape does introduce spherical aberration, light being focused at different distances when reflecting from a spherical mirror. If not corrected, this aberration will produce blurred images. One common solution to spherical aberration is to use a very thin "correcting" lens at the top of the telescope. This type of telescope, which we will discuss more in the next chapter, is called a Schmidt-Cassegrain, and is a popular design for high-end amateur telescopes.

Size Matters

Throughout the nineteenth and well into the twentieth century, astronomers and others interested in science and the sky avidly followed news about every new telescope that was built, each one larger than the last. In 1948, the Hale telescope at Mount Palomar, California, was dedicated. Its 200-inch (5-meter) mirror was the largest in the world. It was designed flexibly to be used as a prime-focus instrument (with the astronomer actually ensconced in a cage at the front end of the telescope), a Cassegrain-focus instrument (with the observer perched on an adjustable platform at the back of the telescope), or a coudé-focus instrument.

The Hale telescope was the largest in the world until 1974, when the Soviets completed a 74-ton, 236-inch (6-meter) mirror, which was installed at the Special Astrophysical Observatory in Zelenchukskaya in the Caucasus Mountains. In 1992, the first of two Keck telescopes, operated jointly by the California Institute of Technology and the University of California, became operational at Mauna Kea, Hawaii. A second Keck telescope was completed in 1996. Each of these instruments combines thirty-six 71-inch (1.8-meter) mirrors into the equivalent of a 393-inch (10-meter) reflector. Not only do these telescopes now have the distinction of being the largest telescopes on Earth, they are also among the highest (of those based on Earth), nestled on an extinct volcano 2.4 miles above sea level.

Astro Byte

In theory, the 6-meter Russian telescope in the Caucasus can detect the light from a single candle at a distance of 14,400 miles. The presence of the earth's atmosphere and other real-world factors do not permit the practical achievement of this theoretical potential, however.

The Power to Gather Light

Why this passion for size?

As we mentioned before, the bigger the bucket, the more light you can collect, so the more information you can gather. The observed brightness of an object is directly proportional to the area (yes, area; not diameter) of the primary mirror. Thus a 78-inch (2-meter) diameter mirror yields an image *4 times* brighter than a 39-inch (1-meter) mirror, because area is proportional to diameter squared, and the square of 2 (2 times 2) is 4. A 197-inch (5-meter) mirror would yield images *25 times* brighter (5 times 5) than a 1-meter mirror, and a 393-inch (10-meter) mirror would yield an image 100 times brighter than a 1-meter mirror.

Now, things that are farther away are always going to be more faint. It should be obvious that a 100-watt light bulb will appear more faint if it is 1 mile away versus 1 foot away. Thus, a telescope that can see more faint objects is able to see things that are farther away. So, in general, the bigger the telescope, the more distant are the objects that can be viewed. As we'll see near the end of this book, being able to see very distant (faint) objects is important to answering some fundamental questions about the ultimate fate of the universe.

The Power to Resolve an Image

Collecting more light is only one advantage of a larger mirror. Large telescopes have greater *resolving power*—that is, the ability to form distinct and separate images of objects that are close together. Low resolution produces a blur. High resolution produces a sharp image.

Twinkle, Twinkle

Theoretically, the giant Hale telescope at Mount Palomar is capable of a spectacular angular resolution of a .02" (or 20 milliarcseconds); that would be its resolution in the absence of complicating factors like the earth's atmosphere. In actual practice,

Astronomer's Notebook

Astronomers speak of the angular resolution of a telescope, which is a measurement of the smallest angle separating two objects that are resolvable as two objects. Generally, the earth's major telescopes, located at the best sites, can resolve objects separated by as little as 1" (that is, 1 arcsecond, which is 1/60 of 1 arcminute, which, in turn, is 1/60 of 1 degree). The theoretical resolution of these telescopes is much higher than this value, but turbulence in the earth's atmosphere means that, except for exceptional nights, this is the best that an Earth-based optical telescope can do. No matter how big the telescope, conventional telescopes cannot have resolutions higher than this value unless they employ adaptive optics.

Star Words

Resolving power is the ability of a telescope to render distinct, individual images of objects that are close together.

it has a resolution of about 1". The source of this limit is related to the reason that stars twinkle. The earth's turbulent atmosphere stands between the telescope's gigantic primary mirror and the stars, smearing the image just as it sometimes causes starlight viewed with the naked eye to shimmer and twinkle. If you took a still photograph of a twinkling star through a large telescope, you would see not a pinpoint image, but one that had been smeared over a minute circle of about 1" (1 arcsecond). This smeary circle is called the seeing disk, and astronomers call the effect of atmospheric turbulence *seeing*. When weather fronts are moving in (even if the skies appear clear), or have just moved out, the seeing can be particularly bad.

Star Words

Seeing is the degradation of telescopic images as a result of atmospheric turbulence. "Good seeing" means conditions that are minimally impacted by the effect of turbulence.

High, dry locations generally have the best seeing. To achieve resolutions better than about 1" from the surface of the earth is possible, but it requires a few tricks. Adaptive optics, for example, are being increasingly employed on new research telescopes. This method allows a mirror in the optical path to be slightly distorted in real time (by a series of actuators) in order to compensate for the blurring effects of the atmosphere. Of course, much higher resolutions are possible at other wavelengths. As we will see, radio interferometers regularly provide images with resolutions better than 0.001" (or 1 milliarcsecond).

Computer Assist

Beginning in the late nineteenth century, most serious telescope viewing was done photographically. Astronomers (despite the popular cartoon image) didn't peer through their telescopes in search of new and exciting information, but studied photographic plates instead. Photographic methods allowed astronomers to make longer observations, seeing many more faint details than could ever be distinguished with visual observing. In recent years, chemical-based photography has increasingly yielded to digital photography, which records images not on film but on CCDs (charge-coupled devices), in principle the same device at the focal plane of your camcorder lens.

CCDs are much more sensitive than photographic film, which means they can record fainter objects in briefer exposure times; moreover, the image produced is digital and can be directly transferred to a computer. Remember the sound of old-fashioned 12-inch, vinyl LP records? Even the best of them had a hiss audible during quiet musical passages, and the worst served up more snap, crackle, and pop than a popular breakfast cereal. CDs, recorded digitally, changed all that by electronically filtering out the nonmusical noise found at high frequencies. Analogous digital computer techniques can be used to filter out the "visual noise" in an image to improve its quality. The disadvantage of current CCDs is that they are relatively small. That is, CCD chips are much smaller than a photographic plate, so that only relatively small areas of the sky can be focused on a single CCD chip.

Close Encounter

This might be a good time to take a break and turn to your computer. Log on to the World Wide Web and point your browser to the site of the Space Telescope Science Institute at www.stsci.edu/public.html and peruse some of the wonderful images from the *Hubble Space Telescope* (which we will discuss in just a moment).

You may notice the somewhat strange geometry of the images. They are sort of L-shaped. This results from the fact that, for wide-field imaging, the HSI focuses the image on a set of four CCD chips, in order to "catch" more of the sky than would fall on a single small chip.

Fun House Mirrors

The greatest limitation of ground-based observations is that Earth's atmosphere gets in the way. The turbulence present in the upper atmosphere means that the best resolution attainable with a traditional telescope from the surface of Earth is about 1 arcsecond, or $\frac{1}{1800}$ the size of the Moon. Now that might seem like a pretty sharp picture, but for the largest telescopes on the surface of Earth, it is only a fraction of the theoretical resolution, the resolution that a telescope *should* have, based on its size. It was thought to be a shame, for example, that the 10-m diameter Beck telescope, while it could collect more light, would have no better resolution than a 1-m diameter telescope.

A new technology has been developed to get around this limitation. Dubbed adaptive optics, it allows astronomers to counteract the distortions introduced by the atmosphere with distortions of their own. The distortions are made to another reflective surface inserted into the optical path, the path that light follows through the telescope. The idea is that if the distortions can be removed quickly enough, then large telescopes would have both of the advantages that they should have, namely more sensitivity and more resolution. This technology has produced stunning results recently on the Keck Telescope and the Gemini North Telescope located on Mauna Kea, Hawaii. What does this all mean? As the technology is perfected, ground-based telescopes will be able to make images as sharp as those made from space—in a more easily maintained and upgradeable package.

This technology is very dependent on fast computers and rapidly movable motors that can make tiny, precise adjustments to the surfaces of small mirrors.

Image made with the Gemini North Adaptive Optics System showing the bow shock a star generates as it plows through a cloud of gas in our own galaxy.

(Photo courtesy of Gemini Observatory, National Science Foundation and the University of Hawaii Adaptive Optics Group)

Observatory in Space: the Hubble Space Telescope

There are other ways to escape the seeing caused by the earth's atmosphere: You can get above and away from the atmosphere. In fact, for observing in some portions of the electromagnetic spectrum, it is absolutely required to get above the earth's atmosphere. That is just what NASA, in conjunction with the European Space Agency, did with the *Hubble Space Telescope*. High above the earth's atmosphere, the HST regularly achieves its theoretical resolution.

Astro Byte

At $3 billion, the *Hubble Space Telescope* is one of the most expensive scientific instruments ever made.

The HST was deployed from the cargo bay of the space shuttle *Discovery* in 1990. The telescope is equipped with a 94-inch (2.4-meter) reflecting telescope, capable of 10 times the angular resolution of the best Earth-based telescopes and approximately 30 times more sensitive to light, not because it is bigger than telescopes on the earth, but because it is above the earth's atmosphere. Unfortunately, due to a manufacturing flaw, the curvature of the 2.4-meter mirror was off by literally less than a hair (it was too flat by $\frac{1}{50}$ of the width of a human hair), which changed its focal length. The telescope still focused light, but not where it needed to, in the plane of the various detectors. Astronauts aboard the shuttle *Endeavour* rendezvoused with the HST in space in 1993 and made repairs—primarily installing a system of small corrective mirrors. HST then began to transmit the spectacular images that scientists had hoped for and the world marveled at.

Subsequent repair missions have installed the short-lived but productive infrared camera (NICMOS) and other instrumentation. A final servicing mission is planned for 2003, after which HST will be replaced by the Next Generation Space Telescope (NGST) near 2010.

Close Encounter

As we have said, yet another way to achieve better resolution is to observe at another wavelength. The VLA or Very Large Array was dedicated in 1980, and has been operating continuously since then. It is located near the town of Socorro, New Mexico, and consists of twenty-seven 25-meter antennas that are connected to form an *interferometer* that acts as a single telescope. In the largest of its variable configurations and at the highest frequency, the VLA has a spatial resolution of about 0.03", or three one hundredths of an arcsecond. That resolution is as good as the HST—even though the VLA is an Earth-based telescope. How can this be? We'll talk more about radio astronomy in Chapter 8, "Seeing in the Dark."

The Least You Need to Know

➤ Light is a form of electromagnetic radiation. Radiation carries energy and conveys information.

➤ Objects in space produce or reflect the various forms of electromagnetic radiation (including radio, infrared, visible light, ultraviolet, x-rays, and gamma rays); this radiation is what we see with our eyes or detect with special instruments.

➤ The two basic optical telescope types are the lens-based refractors and the mirror-based reflectors.

➤ The two main functions of telescopes are to collect light, and resolve nearby objects. Larger telescopes (barring the effects of the earth's atmosphere) are better able to perform both these functions.

➤ New technologies, such as adaptive optics, allow ground-based telescopes to achieve much sharper images while maintaining the convenience of being on the ground.

Chapter 6

You and Your Telescope

In This Chapter

➤ Should you buy a telescope or binoculars?

➤ How much should you spend?

➤ Value versus junk: what to look for, what to avoid

➤ Refractor, reflector, Cassegrain, Dobsonian?

➤ Navigating your telescope by computer

➤ Tips for enjoying your telescope

At nearly $3 billion for the Chandra X-Ray Observatory, astronomy can be a dauntingly expensive pursuit. Fortunately, you don't have to spend quite that much to get started. In fact, you don't really *have* to spend anything. A lot of observation can be done with the naked eye, and many local communities have active amateur astronomers who would be happy to let you gaze at the heavens through their telescopes. Some veteran amateur astronomers even warn newcomers that they will be disappointed with a telescope unless they first obtain some star charts and guidebooks and make an effort to learn the major constellations, perceive differences in brightness, and learn to explain the phases of the moon. "Learn to use your eyes before you buy a telescope," they say.

There's some real value in this advice. You need at least a little working knowledge of the sky before you can locate much of anything with a telescope. In addition, the type of telescope you buy will depend in part on the type of observing that you want to do, and you won't know that until you have a little experience. So our first piece of advice is to be patient: Don't run out to a sale at your local Mega-Lo-Mart and buy a telescope just yet.

But let's face it—part of the fun of astronomy is making faint objects look brighter and distant objects look closer. To many, a big part of the fun of astronomy is its *tools*.

In Chapter 5, "The Art of Collecting Light (with a Telescope)," we reviewed the history and basic principles of the optical telescope. Now let's get some hands-on experience with one.

Do I Really Need a Telescope?

Few experiences with the night sky are more instantly rewarding than your first look at the moon, a nebula, or a planet through a telescope. Saturn, in particular, can look almost too perfect. One of us taught students (while in graduate school) who refused to believe that the planet that they were looking at through the telescope was real. This student insisted that Saturn was a sticker on the telescope lens. However, it is also true that such an experience can be singularly disappointing if that shiny new telescope you bought at the mall turns out to be a piece of wobbly, hard-to-use junk. If you are willing to invest in a good telescope (we'll talk about the magnitude of the investment in just a moment), and if you are willing to invest the time to learn how to use it, a telescope can be a wonderful thing to have.

But will you use it?

If you are an urban dweller who never escapes the streetlights of the city and are hemmed in by tall buildings, you may be better advised to spend your money elsewhere. Then again, owning a sufficiently portable telescope gives you a good excuse to pack up every once in a while and head for the country, where the skies are darker and the seeing is better.

You might consider an alternative to both the naked eye and the telescope: a good pair of binoculars. For hand-held viewing, 7× magnification is comfortable for most people. If you have steady hands, 10× may work well for you, and if you have the hands of a (successful) brain surgeon, even 12× may work. Remember, the greater the magnification, the harder it will be to hold the binoculars steady because objects will wobble farther in your ever smaller field of view.

Magnification means less than you think when it comes to viewing stars. While pointing a telescope or binoculars at the night sky will make stars that are too

Astronomer's Notebook

We'll talk about light pollution later in this chapter, but let's allow that, for a beginner, it may not all be bad. By blotting out many of the fainter stars, urban lighting certainly simplifies the night sky, making it less dazzlingly beautiful, it is true, but also less dazzling period. For beginners, the simplified urban sky may, in fact, be a good place to start viewing the sky. To many a city dweller, the first view of a dark night sky filled with several thousand stars can be overwhelming, although "go-to" controls (also discussed later in this chapter) now make navigating the night sky much simpler.

Star Words

Aperture is the diameter of the objective lens (that is, the primary or "big" lens) or primary mirror of a telescope and the main lenses of binoculars.

faint to see with the naked eye visible, all stars are so incredibly far away (the closest beyond our sun, Alpha Centauri, is about four light-years away) that a given star at higher magnification will still be nothing more than a point of light. Magnification is also largely wasted if what you look at is too dim to see well. Get binoculars with the largest *aperture* (the diameter of the *objective,* or main lens) you can afford. An aperture of 50 millimeters is a good choice. Couple this with a 7× magnification, and you have a 7 × 50 pair of binoculars—a good all-around choice for handheld viewing.

If you want to successfully use binoculars with a magnification of more than 10× or 12×, you will need to mount them on a camera tripod equipped with a binocular adaptor clamp or a specially designed binocular tripod; otherwise, the sky will be a blur.

Binoculars have the advantage of being very portable, and whole guidebooks have been written about observing the sky with them (for example, *Exploring the Night Sky with Binoculars,* by Patrick Moore [3rd ed., Cambridge University Press, 1996]). However, at anywhere from $200 to $1,000 and more, binoculars with high quality optics are not cheap; if you're thinking about buying a pair of big, expensive binoculars, there are other possibilities you may want to consider.

Astronomer's Notebook

The magnification of your binoculars (or telescope) is calculated by a simple ratio of the objective (primary lens) focal length to the eyepiece focal length (see Chapter 5). If your objective has a focal length of 1,000 mm and your eyepiece has a focal length of 100 mm, then the ratio is 1,000 mm/ 100 mm, yielding a magnification of 10 times. There is a useful upper limit to magnification. The rule of thumb is that you get 60× magnification for each inch of aperture. Thus a telescope with a 4-inch objective lens (in the case of a refractor) or primary mirror (in a reflector) can usefully magnify up to 4 × 60, or 240×.

Astronomer's Notebook

Specially designed binocular mounts, created to hold the binoculars in place and allow precision aiming, are available from companies that specialize in equipment for amateur astronomers. Very good mounts range in price from about $150 to almost $600. The latter sum, however, is enough to buy you an entry-level Maksutov-Cassegrain telescope. Indeed, a decent pair of binoculars costs $200 to $400—considerably less than the cost of the best mounts. But keep in mind, the stability of your image comes down to the stability of your mount.

Before we leave the subject of binoculars, here are a few words of shopping advice—much of which applies to telescopes as well:

➤ Examine and feel the binoculars. They should strike you as a well-crafted precision instrument.

➤ Test the focusing mechanism. It should be smooth and offer steady resistance.

➤ Look for antireflection coating on all lenses. This thin coating will make the lenses appear blue, yellow, magenta, or purple when held at an angle to the light.

➤ Look through the binoculars. Try focusing on a point of light (a distant bulb, for example). It should be absolutely sharp, at least until the point of light gets very near the edge of the field of view.

➤ Focus on a vertical straight line such as the corner of a building. Even with very good binoculars, the straight line will bend at the extreme edges of your field of view. However, if the line remains bent a third of the way from the edge, the quality of the optics is poor.

➤ The twin barrels of binoculars must be perfectly parallel with one another. If they aren't, you will see a double image. Your eyes will work hard to compensate and fuse that double image, but ideally, there should be no double image to fuse.

Close Encounter

The "field of view" is the piece of the sky you can see through your telescope or binoculars. You can easily determine the field of view of a telescope-eyepiece combination if you know the apparent field of the eyepiece (this will be listed with the specifications of the eyepiece) and the magnification. Let's say you have a magnification of 10×. The field of view is equal to the field of your eyepiece divided by the magnification. If your eyepiece has an apparent field of 45 degrees ($^1/_4$ of the sky from horizon to horizon) with 10× magnification, your telescope will have a 4.5-degree field of view. If the magnification were 100× with the same apparent field of the eyepiece, the field of view would be 0.45 degrees. What should be clear is that as the magnification is increased, the field of view (or how much of the sky you see) decreases.

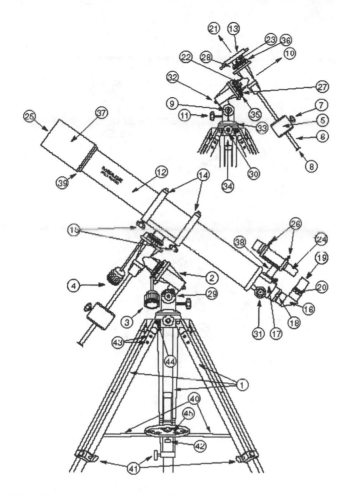

A typical amateur refracting telescope, Meade Model 395.

(Image from Meade Instruments Corporation)

1. Tripod legs	17. Focuser	33. Azimuth base
2. Equatorial mount	18. Focuser thumbscrew	34. Azimuth shaft bolt
3. R.A. flexible cable control	19. Eyepiece	35. R.A. worm block assembly
4. Dec. flexible cable control	20. Diagonal thumbscrew	36. Dec. worm block assembly
5. Counterweight	21. Declination axis	37. Dew shield
6. Counterweight shaft	22. R.A. lock	38. Viewfinder bracket
7. Counterweight lock	23. Dec. lock	39. Objective lens cell
8. Safety washer/knob	24. 6 × 30 viewfinder	40. Leg brace supports
9. Latitude lock	25. Telescope front dust cover	41. Tripod leg lock knobs
10. Polar axis	26. Viewfinder bracket thumbscrews	42. Accessory shelf central mounting knob
11. Latitude adjustment knob	27. R.A. setting circle	43. Tripod leg Phillips-head fastener screws
12. Optical tube assembly	28. Dec. setting circle	44. Tripod leg bolt 1/2" nut
13. Optical tube saddle plate	29. Latitude scale	45. Accessory shelf
14. Cradle rings	30. Azimuth lock	
15. Cradle ring lock knobs	31. Focus knobs	
16. Diagonal mirror	32. Polar shaft acorn cap nut	

A typical amateur reflector, Meade Model 4500.

(Image from Meade Instruments Corporation)

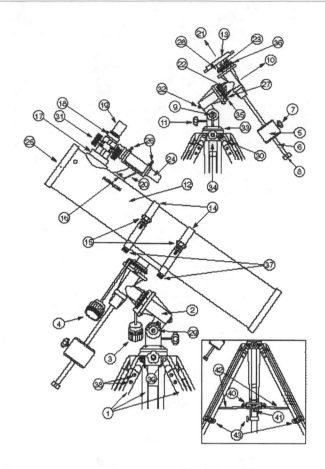

1. Tripod legs
2. Equatorial mount
3. R.A. flexible cable control
4. Dec. flexible cable control
5. Counterweight
6. Counterweight shaft
7. Counterweight lock
8. Safety washer/thumbscrew
9. Latitude lock
10. Polar axis
11. Latitude adjustment knob
12. Optical tube assembly
13. Optical tube saddle plate
14. Cradle rings
15. Cradle ring lock knobs
16. Viewfinder bracket mounting bolts

17. Focuser
18. Focuser thumbscrew
19. Eyepiece
20. Viewfinder bracket
21. Declination axis
22. R.A. lock
23. Dec. lock
24. 6 × 30 viewfinder
25. Telescope front dust cover
26. Viewfinder bracket thumbscrews
27. R.A. setting circle
28. Dec. setting circle
29. Latitude dial
30. Azimuth lock
31. Focus knobs
32. Polar shaft acorn cap nut

33. Azimuth base
34. Azimuth shaft bolt
35. R.A. worm block assembly
36. Dec. worm block assembly
37. Cradle ring attachment knobs
38. Tripod leg Phillips-head fastener screws
39. Tripod-to-mount attachment points
40. Accessory shelf
41. Accessory shelf central mounting knob
42. Tripod leg brace supports
43. Tripod leg lock knobs

Science Aside, What Will It Cost?

Amateur telescopes come in a wide range of prices, from a low of under $200 to a high of $5,000 and more, *much* more. There really is no upper limit. Someone out there will be happy to take as much money as you'd care to spend. In fact, the Beck Telescope (a Cassegrain-focus telescope with a 30" diameter primary mirror) located in Bradley Observatory at Agnes Scott College was originally owned privately by one Henry Gibson, who had it housed in a dome near his house. He sold the telescope to the College in 1947 for $15,000—a lot of money back then! The beginner need not invest in the four-digit range; however, spending less than about $300 on a new telescope (except in the case of certain rich-field instruments, which we will get to shortly) is likely to result in disappointment.

If you've been hitting the malls and looking at telescopes in department stores, camera stores, hobby shops, and even some optical stores, you may be surprised that most of the instruments you've seen will not provide you with a satisfying observing experience. The market is full of telescopes in the $100 to sub-$300 range—and they sell! But they're mostly not worth their "bargain" prices.

That's not a subjective judgment. It's a cold, hard fact. Here are some typical attributes of cheap telescopes:

➤ **A cheap, wobbly mount.** If the view wiggles, you will have a very frustrating time looking at the sky, especially if there is the slightest breeze, or you bump the telescope trying to focus the image. We'll discuss mounts in just a few minutes, but be aware that a shoddy equatorial mount is inferior to a simpler altazimuth mount—if that mount is steady and well made. (See Chapter 1, "Naked Sky, Naked Eye: Finding Your Way in the Dark," for a definition of the equatorial and altazimuth coordinate systems.)

➤ **A telescope that trumpets its magnification but makes little or no mention of its aperture.** Aperture, the diameter of the telescope's objective lens (if it's a refractor) or primary mirror (if it's a reflector), determines how much light you will be able to collect. (Remember our light buckets from the preceding chapter?) After you get tired of looking at the moon, you'll find yourself increasingly interested in the dimmer objects of the night sky. Buy the largest aperture you can afford; aperture size and component quality are far more important than magnification numbers. A 2.75-inch (70 mm) aperture is a very good

Astronomer's Notebook

Consider contacting your local astronomy club to see if any of the members are "upgrading" and would like to sell you their old instrument. Amateur astronomers are very careful with their telescopes; it is almost as good as buying a used car from a mechanic!

minimum for a refractor, and a 4-inch (100 mm) aperture is a good minimum for a reflector. Excellent Maksutov-Cassegrain instruments start at 90 mm, but the entry-level Schmidt-Cassegrain is a 5-inch model.

➤ **Poor optical quality.** Stars should focus as bright, sharp points of light, not smears, blurs, or flares. Unfortunately, it is rarely possible to field test a telescope before you buy it, so purchase only an instrument that comes with a no-questions-asked return policy. Put the telescope through its paces by focusing on a reasonably bright star. You may want to find Altair, Betelgeuse or Arcturus, for example. On a night with good seeing, the star should focus to a clean point. Next, using the highest magnification, slightly unfocus the image by turning the focus knob first one way and then the other. With good optics that are properly collimated (the optical elements made perfectly parallel with one another), the out-of-focus image will look the same regardless of which way you turn the knob. If the two out-of-focus images are significantly different, the optical collimation is poor.

➤ **Small eyepiece.** The modern standard for an eyepiece barrel diameter is 1.25". Two-inch diameter eyepieces are common on larger, more expensive telescopes (14" diameter mirrors and greater). A short-barrel eyepiece can restrict the field of view at low power and is generally a sign of an inferior telescope.

➤ **A junk finderscope.** The finderscope (or finder)—the small telescope attached to the side of the main telescope—is very important for locating the objects you wish to study, especially if your telescope lacks go-to capability. An inferior scope is of little use. Look for one with an aperture of at least 30 millimeters. Also make sure the bracket that mounts the finder is easily, firmly, and accurately adjustable. You'll need to align the finderscope with the main instrument frequently. You may want to also replace the factory finderscope with a "bulls-eye" on the sky, powered typically by a small red LED.

➤ **Obscure and/or skimpy instructions.** Not only is a clear and ample manual an important aid to using and enjoying a telescope, it is a sign of the quality of thought that has gone into making the instrument.

All of this said, how much, then, should you spend?

For a new telescope, you can expect to spend close to $400, although the Meade company now manufactures a good, but small (60 mm) $299 refractor and, for $349, a larger 70-mm model. Excellent small telescopes from Celestron are also available in this under-$300 range. For about $600, you can purchase an even better instrument.

While it is not likely that your $400-to-$600 telescope will be junk, there is no guarantee that it will thrill you. Make sure you purchase it on a returnable basis, and then be sure to put it through its paces before you irretrievably "file" your proof-of-purchase receipt. Also:

➤ Look for a telescope with a solid warranty. One year is standard for reliable instruments. Check out the manufacturer's repair policy, including prices for out-of-warranty repairs.

➤ Shop at stores that specialize in telescopes or at least carry a good selection of them. It is generally best to avoid department stores and toy stores, as well as camera stores and opticians who merely dabble in the merchandise. However, some high-quality camera stores have large telescope departments. Generally, look for stores that are authorized dealers for the major manufacturers.

➤ The best-known high-quality amateur telescope brands are Meade Instruments Corporation, Celestron, Orion, and Tele Vue. Edmund Scientific makes very good rich-field telescopes. Look through the pages of *Sky & Telescope Magazine* for information on excellent equipment from other smaller manufacturers.

Decisions, Decisions

Refractors, reflectors, or one of the newer variants: Which should you buy? Well, as you'll see, that depends on what you want to do with your telescope, how portable you want (or need) it to be, and how much you're willing to shell out.

Refractors: Virtues and Vices

Most astronomers agree that a good refractor is the instrument of choice for viewing the moon and the planets. Typically, the refractor's field is narrow, which enhances the contrast offered by good optics and brings out the details of such things as the lunar surface and planetary detail.

Refractors, however, are not the best choice for deep-sky work—looking at dim galaxies, for example. They are great for bright objects, but a refracting telescope with the same light-collecting ability of a decent reflecting telescope would be prohibitively expensive. See Chapter 5 "The Art of Collecting Life (with a Telescope)" for an explanation of the limitations of refractors.

Some of the cheapest, mass-market telescopes are refractors, but most of these will perform poorly. Most good refractors are long, heavy, and expensive—although the recently introduced Meade ETX-60AT and ETX-70AT are compact yet high-quality entry-level instruments. The disadvantage of expense is obvious, as is that of weight:

Astronomer's Notebook

What is the resolution of your telescope? Here's a good rule of thumb: To find the minimum separation (in arcseconds) of a double star (see Chapter 17, "Of Giants and Dwarfs: Stepping Out into the Stars") that can be resolved with a telescope, divide 12 by the aperture (in cm). Thus a telescope with a 4–inch (about 10 cm) diameter mirror could resolve stars that are separated by a mere $^{12}/_{10}$, or 1.2 arcseconds, depending on the seeing that night.

You'll be discouraged from taking the telescope with you on trips to the dark skies of the country. Length poses a less obvious problem. The longer the tube, the less inherently steady the telescope. A large refractor requires a very firm mount and tripod.

Reflectors: Newton's Favorite

Traditionally, the Newtonian reflector has been the most popular telescope with experienced amateurs, although, in recent years, affordable Schmidt-Cassegrain and Maksutov-Cassegrain instruments have found increasing favor. Generally, a reflector gives you more aperture—and thus more light—for your dollar than a refractor, and the reflector's mirror is not subject to chromatic aberration (the differences in the ways various colors, especially red and blue, are focused), which all but the most expensive refractor lenses suffer from. Although reflectors may be large, they are generally lighter than refractors; however, they tend not to be as robust, and unlike a good refracting telescope, they do require at least some minimal maintenance to re-align optical components occasionally.

Astro Byte

Price comparison: A high–quality 4-inch refractor from a leading manufacturer sells for $995. The same manufacturer sells a high-quality 10-inch reflector for about $900.

Rich-Field Telescopes: Increasing in Popularity

Worth investigating is a relatively new category of telescope. Ultra compact and reasonably priced, rich-field reflectors are typically handheld with a Newtonian focus. What they have in common is a short-tube design that offers low degrees of magnification but a bright, wide field of view (typically a few degrees). They range in price from about $250 to $400 and can weigh as little as 4 or 5 pounds. Highly portable and relatively rugged, these telescopes nevertheless have the disadvantage that, since they are handheld, they do not track with objects in the sky, and are only as steady as you are. The great advantages of these telescopes, besides price, are their portability and the brightness of the image they deliver.

Astronomer's Notebook

Rich-field telescopes are an exception to the under-$400-is-junk rule. A rich-field instrument in the $300 range can give very good value for the money. Just don't expect a high degree of magnification or stability.

Schmidt-Cassegrain: High-Performance Hybrid

Also called a catadioptric telescope, the Schmidt-Cassegrain design combines mirrors and lenses. Telescopes of this design are an increasingly popular choice for serious amateurs and introductory astronomy classes. The light passes through a corrector lens before it strikes the primary mirror, which reflects it to a secondary mirror. Since light bounces down the tube an extra time, the focal length of the telescope is effectively doubled, belying the very compact—wide but short and stubby—look of the instrument. A long effective focal length means that these telescopes can have a high magnification (remember that magnification is the ratio of objective focal length to eyepiece focal length) without a cumbersome long tube.

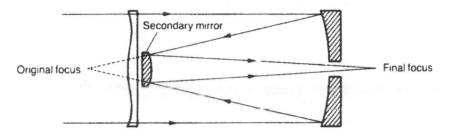

Diagram of a Schmidt-Cassegrain, or catadioptric, telescope. Light enters from the left and is focused by the primary mirror at the back of the telescope. Then it is refocused by a secondary mirror and sent out through an opening in the primary mirror to an eyepiece at the rear of the telescope.

(Image from the authors' collection)

Schmidt-Cassegrain telescopes are elegant instruments that offer some of the compactness of rich-field instruments but are much more powerful.

The catch?

These are usually more expensive amateur instruments, typically priced from $900 to much, much more, depending on aperture size and features. The portability of the Schmidt-Cassegrain design is a very big plus—not just because a compact telescope is easier to transport, but also because it is easier to keep a small scope stable during use.

Maksutov-Cassegrain: New Market Leader

Like the Schmidt-Cassegrain telescopes, the Maksutov-Cassegrain is a catadioptric design; however, these newer instruments optimize imaging performance by combining a special spherical meniscus (concave) lens with two mirrors. The secondary mirror multiplies the focal length of the telescope. The combined effect of the concave lens, the aspherical primary mirror, and the convex secondary mirror produces a telescope that is almost as well suited to lunar and planetary observation as a refractor, yet it has many of the reflector's advantages for deep-space viewing. These qualities are similar to the conventional Schmidt-Cassegrain design, but the Maksutov-Cassegrain variation tends to yield images of greater contrast than one gets from telescopes of the earlier design.

A 7-inch Maksutov is significantly more expensive than an 8-inch Schmidt-Cassegrain; however, Meade has marketed for some years now two extremely popular small Maksutov models, the ETX-90EC and ETX-125EC (90 mm and 125 mm, respectively), which trade aperture for price. The 90-mm model can be purchased for under $500, and the 125-mm model for less than $900.

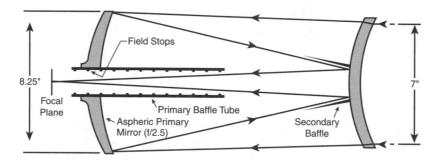

Diagram of a Maksutov-Cassegrain telescope. Light enters the concave lens at the right and is reflected by the aspheric primary mirror at the left, which sends it back to the spherical secondary mirror on the right. This, in turn, focuses the image on the focal plane on the left.

(Image from Meade Instruments Corporation)

Dobsonians: More for Your Money?

During the 1970s, an avid amateur astronomer named John Dobson began building large, standard Newtonian reflectors (10-inch mirrors were typical) and cutting costs by mounting them not on elaborate and expensive equatorial mounts but on inexpensive altazimuth mounts. Dollars were invested in optics and aperture—light-gathering ability—rather than in fancy mounting hardware and clock drives to aid in tracking objects. The result was a powerful reflecting telescope with a wide field of view.

Very nice Dobsonians can be purchased in the $300 to $1,000 range, or you could see if your local amateur astronomy club offers workshops in making your own telescope. Many astronomy club members make their own Dobsonians.

Is there a Dobsonian downside? Some users find the simple altazimuth mount—which lacks the ability to track objects—too limiting.

The Go-To Revolution

Books on amateur astronomy used to supply only two important pieces of advice about tripods and mounts.

First: Don't cheap out. Invest in something sturdy and steady.

Second: Choose between an altazimuth mount and an equatorial mount. The simple altazimuth mount is adjustable on two axes: up and down (altitude) and left and right (movement parallel to the horizon, or azimuth). There is nothing automatic about most altazimuth mounts. If you are trying to follow an object, you must continually adjust both the altitude and the azimuth. The alternative equatorial mount is aligned with the earth's rotational axis and, therefore, may be made to follow a celestial object by adjusting one axis only (to counteract the rotation of the earth).

These two pieces of advice used to be quite sufficient. In the late 1990s, however, popular manufactures started selling even some entry-level telescopes with *go-to computer controllers* that drive servo-motors built into the telescope mount. The handheld go-to controller stores a database of the locations of thousands of celestial objects. Select an object or punch it its coordinates, and (if properly trained and aligned) the telescope's servos will point the telescope at your target object.

In addition to servo motors for go-to capability, equatorial and altazimuth mounts typically include a clock drive that synchronizes the telescope with the earth's rotation so that a given object can be followed—"tracked"—without your having continually to re-aim the telescope.

Go-to capability can work on telescopes that have either altazimuth or equatorial mounts. For example, the go-to features on the Meade ETX 90EC telescope can be used in either equatorial or altazimuth mode. One just has to be careful that the computer has been informed of your choice (usually accomplished on the setup menu). The amazing thing is that go-to technology has become

Astronomer's Notebook

Some telescopes are sold complete with a tripod, pier support, or (in the case of Dobsonians) a simple altazimuth mount. Other instruments do not include a tripod. A tripod for a high-quality amateur telescope averages about $200. Buy a telescope tripod rather than a flimsy, low-end camera tripod. Not only will you have trouble keeping the telescope steady and pointed where you want it on an inadequate tripod, you also run a risk of the tripod falling—with most unfortunate consequences for the telescope.

Star Words

A **go-to computer controller** is a handheld "paddle" that stores location data on celestial objects. Select an object in the database or punch in right ascension and declination coordinates, and the controller will guide motors to point the telescope at the desired object or coordinates.

sufficiently affordable to be included in even entry-level telescopes. This feature has truly revolutionized amateur astronomy, greatly broadening its appeal. Keep in mind that the "go-to" hand paddle must typically be purchased as an accessory, and will cost several hundred dollars itself. If this capability is important to you, you should buy a telescope that can be updated at a later time.

I've Bought My Telescope, Now What?

Many subsequent chapters contain advice on observing various celestial objects, but for now, having bought your telescope (and having assembled it; typically, some assembly is required), what do you do with it?

Grab a Piece of Sky

In two words: *Use it.*

You don't need a plan, but many first-time sky watchers christen their new telescope by looking at the moon. A more original inaugural journey begins by marking out an interesting-looking piece of sky for yourself and studying it. Find what you can. Later, we'll talk about recording what you see.

IMPORTANT WARNING

Few truly interesting avocations are safer than astronomy, but it is critically important to be aware of one very serious danger.

Never use your telescope to look at the sun unless you have affixed to the front end of the telescope a specially designed solar filter. (Never use a solar filter that screws into the eyepiece. The heat of the sun's focused rays could easily damage such a filter, causing it to break without warning, suddenly sending unfiltered focused sunlight into your eye, causing blindness. Proper solar filters cover the front end of a telescope.)

Looking at the sun with an unfiltered telescope will burn your retina almost immediately.

Remember what you used to do to ants with a magnifying glass? Well, sunlight focused by your reflecting or refracting telescope (or by binoculars) will do the same thing to your retina. The resulting damage to your eyesight is permanent.

See Chapter 16, "Our Star," for instructions on viewing the sun safely.

Another good way to start is to go to your local library and check out *Astronomy* or *Sky & Telescope* magazine. Both of these periodicals (and their online equivalents) include a guide to the night sky in their center section every month, and you can check to see if there are any planets in the sky or which constellations are up. Also see Appendix E, "Sources for Astronomers," for recommended guidebooks.

An interesting second activity is to locate another piece of sky—one that looks almost empty—and try to find dim and distant objects. Test the limits of your new telescope and your own eyesight. Notice how many more stars you see with your finder telescope, which has a larger aperture than your eye, and then notice how many stars you see in your main telescope.

Close Encounter

A no-cost alternative to purchasing a telescope is to look at the night sky through the cardboard tube from a standard roll of toilet paper. No, this isn't a rather pathetic way of pretending that you have a real telescope. It is a genuine observing exercise: a method for estimating the number of stars visible in the sky.

Look through your toilet paper tube at five different parts of the sky and count how many stars you see in the field of view. Let's say that in your five separate fields you see 12, 15, 20, 11, and 10 stars. That is a total of 58. Now you have seen 58 stars in a small fraction of the sky (five toilet paper tubes' worth, to be exact). To estimate how many stars you would see in the entire night sky (with unaided eyesight), multiply this number by 15 (because we have assumed that the tube is 1.25" in diameter and 3.75" long). In our example, there are 58 times 15, or about 900, stars visible in the night sky in this example. In the country, your total number will be higher, and lower in the middle of a city.

Become an Astrophotographer

Once you get hooked on looking through a telescope, sooner or later you're going to want to start recording what you see. One very enjoyable activity is to make drawings, but many serious amateurs sooner or later turn to photography.

Astrophotography can be done with any good single-lens reflex (SLR) camera, the right kind of adapter to mate it with your telescope, and a sturdy tripod and mount with a tracking motor to compensate for the rotation of the earth during the long exposures are usually necessary.

As digital technology has greatly simplified and expanded the possibilities of finding objects in space, it has also simplified and expanded the field of professional as well as amateur astrophotography. In Chapter 5, we discussed how charge-coupled devices (CCDs) have largely replaced conventional photographic film for most astronomical imaging through major earth-based telescopes as well as such space-based instruments as the *Hubble Space Telescope.* Just as, in recent years, the cost of go-to technology has been greatly reduced, so now is digital imaging within the reach of serious amateurs. The operative word is "serious." Meade's Pictor 1616XTE CCD system costs more than $6,000, but the more "entry-level" Pictor 415XTE comes in at just under $2,000. And a very respectable camera from the Santa Barbara Instrument Group (SBIG) called the SBIG ST7 can be purchased (at the time of this writing) for under $3,000. It is likely that, over the years, the cost of CCD imaging will fall even further.

If you are interested in astrophotography, whether using conventional film or with CCD imaging, check out Michael A. Covington's excellent *Astrophotography for the Amateur* (Cambridge University Press, 1999) or Jeffrey R. Charles's *Practical Astrophotography* (Springer Verlag, 2000).

Light Pollution and What to Do About It

Light pollution is the obscuring of celestial objects by artificial light sources.

What do you do about it?

You avoid city lights, if you can. The recent trend toward those peach-colored, high-pressure, sodium-vapor and bluish metal-halide streetlights may make some people feel safer, but the lamps have also greatly increased light pollution, even in smaller urban areas.

Here are some ways to reduce the effects of light pollution:

➤ **Rise above the streetlights.** Set up your telescope on a hill or a safe roof. The cumulative effect of the streetlights will still blot out many of the less bright objects, especially near the horizon, but at least you won't be trying to look up *through* the nearest streetlights.

➤ **Study the sky in a direction away from light sources.** If your city's downtown area is east of your location, look west rather than east.

➤ **Get out of town.** Scout out some rural retreats away from the city lights but sufficiently clear of trees to allow reasonably unobstructed viewing. Local state parks are often a good option. It may be best to choose parks that offer overnight camping, since some public parks are open only from dawn to dusk. Don't trespass!

➤ **If you get very interested in observing, you can purchase filters for your telescope that will block out a good portion of the light pollution caused by streetlights.** Such filters are available from Meade Instruments, Orion, and other suppliers. Be aware, however, that these filters are most useful for astrophotography or digital imaging and are less effective if your primary imaging device is your own retina. Also, all filters block light, dimming the image you see; so small-aperture telescopes will suffer most from this side effect.

➤ **Write your local city government and encourage officials to install low-pressure, downward-facing sodium lamps.** These lights have a yellowish glow and are highly energy efficient. You can get many good ideas on how to reduce light pollution from the International Dark Sky Association (find more at www.darksky.org/).

Unless you live in a nest of searchlights, there should still be enough for a beginner to see.

Finding What You're Looking For

If your new telescope has go-to capability, all you need to do is follow the manufacturer's instructions for initially training the instrument and then use the go-to controller to point your telescope at whatever you wish to view. Bear in mind, of course, that light pollution or other atmospheric conditions may obscure your view. Go-to technology is wonderful, but it can't work miracles. It will point you in the right direction, but it can't guarantee that you'll always see what you're looking for.

If your instrument does not have a go-to controller, glance back at Chapter 1, which introduces the idea of celestial coordinates and altazimuth coordinates as well as the utility of constellations as celestial landmarks. Later chapters have more to say about finding specific objects. What you should familiarize yourself with now, however, is the finderscope affixed to the side of your telescope. Unless you have a rich-field telescope, commanding about a three- or four-degree slice of the sky, you will find it almost impossible to locate with the main telescope anything you happen to see with your naked eye. ("There's Venus! But why can't I find it with this #^$%@% telescope!?")

Take the time and effort to follow what your instruction manual says about adjusting the finderscope so that it can be used to locate objects quickly. This adjustment should take just a few minutes and can be done in daylight; once it's done, it's done (at least until you or someone else bumps the finder out of alignment). In any case, the alignment process is far less tedious and frustrating than trying to sight with your naked eye along the telescope tube and then just hoping you can finally find what you're looking for.

Another option is called a Telrad Reflex Sight. Many amateurs use one of these—an inexpensive "bullseye" on the sky. In many ways, this product is even more helpful than a finderscope.

Close Encounter

Carefully read the owner's manual that comes with your telescope. Make yourself thoroughly familiar with all the controls, especially those relating to equatorial mount adjustment. This process is important because it brings your telescope's axis in alignment with the axis of the rotating Earth. Any good telescope will come with a manual that describes this process in detail. The basic idea is that you want the polar axis of your telescope to point directly at the North Star. That is, when your telescope is pointing at a declination of 90 degrees, Polaris had better be in your field of view. If not, your telescope will not be polar aligned, and you will lack a reliable reference point from which to find *anything*.

Learning to See

Understandably, you will be eager to try out your new telescope. Here are a few words of advice: Expect to be thrilled—immediately—by the spectacle of the moon, with its sharply delineated craters and mountains. Point your telescope elsewhere, however, and you may be disappointed—at least until you learn more about what to look for. We have become spoiled by dazzling images from the *Hubble Space Telescope,* orbiting above our atmosphere and toting the most sophisticated instruments available. No, your telescope won't duplicate the performance of *Hubble.* But the point is that it is *your* telescope, and the photons of light that left the Orion Nebula are striking *your* retina. The experience is *yours.*

Your first impulse may be to blame any disappointment you feel on your telescope.

Resist the impulse. As you learn what to look for—and as you come to appreciate the significance of what you see—you will derive great satisfaction from your instrument.

Low-Light Adjustment

You have some learning to get under your belt, but right now, neophyte that you are, you can do something to enhance your experience. Unless you are looking at the bright moon, don't rush to the eyepiece until you have allowed your vision to become "dark adapted." This natural adjustment will greatly enhance your ability to see faint objects—and it will make brighter objects that much more exciting.

Adapting your eyes to the dark requires about 15 minutes away from sources of light. If somebody shines an uncovered flashlight in your eyes, you'll have to become dark adapted all over again. Red light, however, will not reverse dark adaptation. For those on liberal budgets, there are specially made, compact flashlights with red bulbs. For the rest of us, either equip your flashlight with a dark red filter (you can use red acetate purchased from a hobby shop) or (less effectively) simply put a red sock over the flashlight. This way, you'll be able to see what you are doing and even consult star maps without spoiling your dark adaptation.

Don't Look Too Hard

Next, relax. Don't look too hard. We mean this as sincere and literal advice. Your eye's sharpest color vision is in the center of your field of view. This is where color-receptor neurons known as cones are most densely concentrated. However, so-called rods, the visual receptors sensitive to black, white, and shades of gray, while insensitive to color, are *more* sensitive than cones to low levels of light. This means you can actually better see fainter objects with your *peripheral* vision than with your *center-field* vision. Learn to look askance at the stars. This practice is sometimes called "averted vision." Using it, you will typically see fainter stars.

Peering through a telescope for extended periods is fun, but it can also be fatiguing. Don't squint. Don't peer. Step away from your telescope periodically to walk around. Relax and enjoy.

You'll enjoy your astronomy sessions more, as well as reduce fatigue, if you practice keeping both eyes open when you look through the eyepiece. If you can't resist the urge to close one eye, buy a pirate's eye patch from the local toy store or costume shop. Then you can keep both eyes open without distraction and even feel like a real celestial navigator. A parrot on the shoulder is optional.

Astronomer's Notebook

When observing in cold, humid weather, beware of condensation, which can fog your optics, and of distortion caused by the contraction of various telescope parts and adjustments. Let your telescope cool down before you do any serious observing. Usually, Schmidt-Cassegrain and Maksutov-Cassegrain telescopes lack a dew shield, the cuff-like shade at the front of the telescope. You might want to improvise one with a large cylindrical oatmeal box or ice cream carton. This will help keep condensation from forming on the objective lens.

The Least You Need to Know

➤ A good intermediate step between naked-eye viewing and investing in a telescope is purchasing a decent pair of binoculars. Just remember that good binoculars can be as expensive as a decent starter telescope.

➤ In general, expect to spend at least $300 to $400 for a basic new telescope. Except for rich-field reflectors (some good ones can be had for $300), instruments priced below this range are likely to produce disappointing results. Also consider buying a good used telescope from a member of your local astronomy club.

➤ Although the refracting telescope is a good choice for those especially interested in viewing bright objects (such as the Moon or Venus), a Newtonian reflector, Schmidt-Cassegrain, or Maksutov-Cassegrain is probably best for all-around viewing.

➤ Invest in the biggest aperture (primary lens or mirror diameter) and the best tripod mount you can afford. Don't worry about high magnification, which bargain manufacturers often tout.

➤ Consider buying a telescope with a go-to controller, which makes finding your way around the sky easier.

➤ Take the necessary time to understand your telescope and learn about the sky. Being informed and realistic in your expectations will help you avoid disappointment with your first telescopic explorations.

Over the Rainbow

The light we receive from distant sources is generated on the tiniest of scales. To explore the largest objects, such as galaxies, we have to first understand the smallest of objects, atoms and the particles making up atoms. The photons that we detect with our eyes and catch with our telescopes were generated in many different ways: sometimes by electrons hopping between different orbital levels in an atom, or other times by the energetic collisions of atomic nuclei. We now explore the ways in which photons of light arise, how they get from there to here, and what they can tell us about the objects that we observe.

We have concentrated thus far on optical photons (the ones that we can see with our eyes). As it turns out, our eyes respond to "visible" wavelengths because that is where the peak of the emission from the sun is located in the electromagnetic spectrum. If our eyes were most sensitive to infrared radiation, for example, we would see some things we can't now see (body heat), but would miss a lot of other useful stuff.

In this chapter, we're going to talk more about visible light and the electromagnetic spectrum, of which visible light is a tiny subset. Think of it this way: If the electromagnetic spectrum is represented by a piano keyboard, then the visible part of the spectrum is but a single key or note. In the cosmic symphony, there are many notes, and we want to be able to hear them all. If you're concerned that this sounds more like physics than astronomy, you're right. But don't be intimidated. Most of astronomy involves applications of physics principles, and we are convinced that understanding what you are seeing when you look at a star greatly enhances the experience of looking. Remember this astounding fact from Chapter 5, "The Art of Collecting Light (with a Telescope)": When you look at the light from our sun or a distant star, you are witnessing the product of nuclear fusion reactions that are, every second, releasing more energy than any atomic explosion Earth has ever witnessed. Yet it is not just brute energy, but also information from the sky. Let's take a closer look.

Making Waves

Electromagnetic radiation sounds like dangerous stuff—and, in fact, some of it is. But that the word *radiation* need not set off sirens in your head. It just describes any way energy is transmitted from one place to another without the need for a physical connection between the two places. We use it as a general term to describe any form of light. It is important that radiation can travel without any physical connection, because space is essentially a vacuum; that is, much of it is empty. If you went on a space walk clicking a pair of castanets, no one, including you, would hear your little concert. Sound is transmitted in waves, but not as radiation. Sound waves require some medium to travel in. So despite what most science fiction movies would lead you to believe, explosions in space are silent. Light (and other forms of electromagnetic radiation) requires no such medium to travel, although many physicists tried in vain to detect a medium, which they called the ether. We'll talk more about this fact in a moment.

The electromagnetic part of the phrase denotes the fact that the energy is conveyed in the form of fluctuating electric and magnetic fields. These fields require no medium to support or sustain them.

Anatomy of a Wave

We can understand how electromagnetic radiation is transmitted through space if we appreciate that it involves waves. What is a wave? The first image that probably jumps to mind is that of ocean waves. And ocean waves do have some aspects in common with the kind of waves that we use to describe electromagnetic radiation. One way to think of a wave is that it is a way for energy to be transmitted from one place to another without any physical matter being moved from place to place. Or you may think of a wave as a disturbance that carries energy and that occurs in a distinctive and repeating pattern. A row boat out in the ocean will move up and down in a regular way as waves pass it. The waves do transmit energy to the shore (think of beach erosion), but the row boat will stay put.

That regular up-and-down motion that the rowboat experiences is called harmonic motion. But there are two important differences with electromagnetic radiation: The sources of waves are things on atomic scales (electrons and the nuclei of atoms), and no medium is required for electromagnetic waves to travel through space. The "pond" of space consists only of electric and magnetic fields, and photons of light are ripples in that ghostly pond.

Waves come in various shapes, but they all have a common anatomy. They have *crests* and *troughs,* which are, respectively, the high points above and low points below the level of an undisturbed state (for example, calm water). The distance from crest to crest (or trough to trough) is called the *wavelength* of the wave. The height of the wave—that is, the distance from the level of the undisturbed state to the crest of the wave—is its *amplitude.* The amount of time it takes for a wave to repeat itself at any point in space is its *period.* In other words, the period is the time between the passage of wave crests as seen by an observer in the bobbing row boat. The number of wave crests that pass a given point during a given unit of time is called the *frequency* of the wave. If many crests pass a point in a short period of time, we have a high-frequency wave. If few pass that point in the same amount of time, we have a low-frequency wave. The frequency and wavelength of a wave are inversely proportional to one another, meaning that as one gets bigger, the other gets smaller. *High* frequency radiation has *short* wavelengths.

Astronomer's Notebook

Recall from Chapter 5 that wave frequency is expressed in a unit of wave cycles per second, called the hertz, abbreviated Hz. Wavelength and frequency are inversely related; that is, if you double the wavelength, you automatically halve the frequency, and if you double the frequency, you automatically halve the wavelength. Multiply wavelength by frequency, and you get the wave's velocity. For electromagnetic radiation, wavelength multiplied by frequency is always *c*, the speed of light.

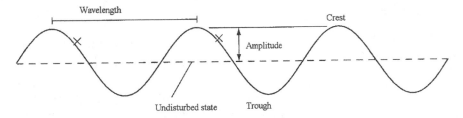

The parts of a wave.

(Image from the authors' collection)

New Wave

If you don't happen to like math, don't panic. Just visualize stone-generated waves rippling across a pond, and you'll understand the basic concept of waves.

But wait a minute. There is something wrong with our ripples in a pond as a model of electromagnetic radiation.

Water is a medium, a substance, something through which waves are transmitted. Space, we have said, is very nearly a vacuum, nothing. How, then, do waves move through it?

This is a question that vexed physicists for centuries. They understood the concept of waves. But they also understood that sound, a wave, could not travel through a vacuum, whereas light, also a wave, could.

Why?

At first, most scientists believed that the very fact that light is transmitted through space means that space must not be empty. They knew it didn't have air, as on Earth, but they suggested it was filled with another substance, which they called the ether.

But this fictitious substance did not long vex physicists. A series of experiments in the late nineteenth century made it clear that ether didn't exist and that although light could be studied as a wave, it was a different kind of wave than, say, sound.

Big News from Little Places

The Greek philosopher Democritus (ca. 460–ca. 370 B.C.E.) was partially right: matter does consist of atoms. But he would have been fascinated to know that the story doesn't end there. Atoms can be further broken down into electrons, protons, and neutrons, and the latter two are made of even smaller things called quarks. Electrons carry a negative electric charge, and protons a positive charge. Neutrons have a mass almost equal to a proton, but as their name implies, neutrons are neutral, with no positive or negative charge. Charged particles (like protons and electrons) that are not moving are surrounded by what we call an electric field; those in motion produce electromagnetic radiation.

James Clerk Maxwell (1831–1879) first explored what would happen if such a charged particle were to oscillate, or move quickly back and forth. He showed that a moving charged particle created a disturbance that traveled through space—*without the need for any medium*. Particles in space are getting banged around all the time. Atoms collide, electrons are accelerated

Astro Byte

Many people find it difficult to accept the idea of forces acting without any physical connection between them. Some scientists accused Newton of absurdity and even madness for having set forth a theory of gravity that proposed such action.

by magnetic fields, and each time they move, they pull their fields along with them, sending "electromagnetic" ripples out into space.

In short, information about the particle's motion is transmitted through space by a changing electric and magnetic field. But a field is not a substance. It is a way in which forces can be transmitted over great distances without any physical connection between the two places. The force of gravity, which we have discussed, can also be thought of as a field.

Let's turn to a specific example: A star is made up of innumerable atoms, most of which at unimaginably hot stellar temperatures are broken into innumerable charged particles. A star produces a great deal of energy (by nuclear fusion, which we'll discuss in Chapter 16, "Our Star"). This energy causes particles to be in constant motion. In motion, the charged particles are the center points of electromagnetic waves (disturbances in the electromagnetic field) that move off in all directions. A small fraction of these waves reaches the surface of the earth, where they encounter other charged particles. Protons and electrons in our eyes, for instance, oscillate in response to the fluctuations in the electric field. As a result, we perceive light: an image of the star.

If we happened to have, say, the right kind of infrared-detecting equipment with us, electrons in that equipment would respond to a different wavelength of vibrations originating from the same star. Similarly, if we were equipped with sufficiently sensitive radio equipment, we might pick up a response to yet another set of proton and electron vibrations.

Remember, it is not that the star's electrons and protons have traveled to Earth, but that the waves they generated so far away have excited other electrons and protons here. Call it an interstellar handshake.

Astronomer's Notebook

How fast do electromagnetic waves move? All of them—whether visible light, invisible radio waves, x-rays, or gamma radiation—move (in a vacuum) at the speed of light, approximately 186,000 miles per second (299,792,458 meters per second). Fast, but hardly an infinite, unlimited speed. Remember the Andromeda galaxy from Chapter 5? We can see it, but the photons of light we just received from the galaxy are 2 million years old. Now, that's a long commute.

Full Spectrum

Often, when people get excited, they run around, jump up and down, and shout without making a whole lot of sense. But when atomic particles get excited, they can produce energy that is radiated at a variety of wavelengths. In contrast to the babble of an excited human throng, this electromagnetic radiation can tell you a lot, if you have the instruments to interpret it.

Our eyes, one such instrument, can interpret electromagnetic radiation in the 400 to 700 *nanometer* (or 4000 to 7000 *Angstrom*) wavelength range. A nanometer (abbreviated nm) is one billionth of a meter, or 10^{-9} meter. An Angstrom (abbreviated A) is 10 times smaller, or 10^{-10} meter. But that is only a small part of the spectrum. What about the rest of the "keyboard"?

The Long and the Short of It

The senior author of this book is a radio astronomer. When he tells people this, he is often met with a blank stare. The second most common response is, "Oh, so you listen to the stars." This popular misconception is reinforced by images of Jodie Foster, star of the sci-fi blockbuster *Contact*, "listening" to the pulses of a pulsar. In fact, you *can* "hear" radio sources on your radio (we'll tell you how in Chapter 8, "Seeing in the Dark"), and when most people hear the word *radio*, they think of the box in their car that receives radio signals broadcast from local towers, amplifies them, then uses them to drive a speaker, producing sound waves heard with the ears. But radio waves themselves are as silent as optical light or x-rays or gamma rays.

Close Encounter

Radio waves, visible light, gamma rays—the only difference is wavelength. But it is quite a difference. Some of the radio waves received by the Very Large Array radio telescope in New Mexico have wavelengths as large as a yardstick, whereas gamma rays have wavelengths the size of an atomic nucleus.

The energy of a particular wavelength of electromagnetic radiation is directly proportional to its frequency. Thus, photons of light that have high frequencies and short wavelengths (such as x-rays and gamma rays) are the most energetic, and photons that have low frequencies (radio waves) are the least energetic. Ever wonder why the origin stories of comic-book superheroes often involve gamma-rays? Those gamma ray photons carry a lot of energy, and it takes a lot of energy to make a superhero, apparently.

Since we are mere mortals, however, we are highly fortunate that the earth's atmosphere absorbs most of the high-energy photons that strike it. Energetic photons tend to scramble genetic material, and the human race wouldn't last long without the protective blanket of the upper atmosphere. If a massive star were to explode somewhere near the earth in the future, the most harmful effect would be the high-energy photons that would cause a wave of mutations in the next generation.

Radio waves are simply a form of electromagnetic radiation that has very long wavelengths.

So the only difference between radio waves and light waves is the length of the wave (or the frequency, which is always inversely related to wavelength). Indeed, all of the forms of electromagnetic radiation represented across what we call the electromagnetic spectrum—radio waves, infrared, visible light, ultraviolet, x-rays, and gamma rays—are transmitted at the speed of light as waves differentiated only by their wavelength or frequency. Radio waves are at the low end of the spectrum, which means their waves are big (on the scale of millimeters to meters in size) and their frequency, therefore, low—in the mega- (million) to giga- (billion) Hertz range. Look on a radio tuner. You should notice that the FM band is rated in MHz, and the AM band in kHz, or 1000 (kilo-) hertz. Gamma rays, in contrast, are at the high end of the spectrum, with very short wavelengths and very high frequencies.

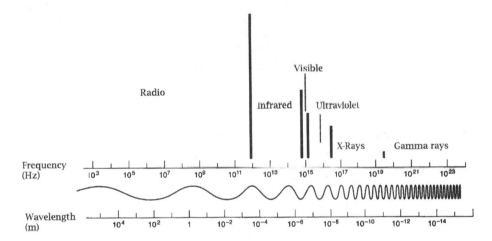

The electromagnetic spectrum.

(Image from the authors' collection)

What Makes Color?

Visible light, we have said, is defined as light with wavelengths from 400 to 700 nm in length, about the dimensions of an average-sized bacterium.

Within such a tiny range of wavelengths all colors are contained. Just as wavelength (or frequency) determines whether electromagnetic radiation is visible light or x-rays or something else, so it determines what color we see within this tiny range. Our eyes respond differently to electromagnetic waves of different wavelengths. Red light, at the low-frequency end of the visible spectrum, has a wavelength of about 7.0×10^{-7} meters (and a frequency of 4.3×10^{14} Hz). Violet light, at the high-frequency end of the visible spectrum, has a wavelength of 4.0×10^{-7} meters (and a frequency of 7.5×10^{14} Hz). All of the other colors fall between these extremes, in the familiar order of the rainbow: orange just above red; yellow above orange; then green, blue, and violet.

So-called white light is a combination of all the colors of the visible spectrum. Light is different from paint. Dump together a lot of different colors of paint, and you'll end up with a brownish gray. However, dump together all of the colors of the rainbow, and you'll get "white light." When sunlight is refracted (or bent) as it passes through water droplets in the air, different colors of light are bent by different amounts (the same effect that gives rise to chromatic aberration in refracting telescopes, as we saw in Chapter 5). The "divided light" then reflects on the back side of the raindrop, and you see the result of this process occurring in a myriad of water droplets as a spectacular rainbow.

Astro Byte

For 25 years, cosmic gamma ray bursts (GRBs) have been one of the great mysteries of modern astronomy. GRBs have given us several clues as to what they might be. They were seen to occur frequently and appeared to be spread evenly over the sky. Their distribution on the sky indicated that they were the result of events happening either very close (perhaps in the Oort cloud), or very far (in other galaxies). The events were not, for example, seen to be concentrated in the plane of our Galaxy.

Concentrated study and follow-up observations have shown that GRBs appear to arise in distant galaxies. If they are distant and very bright, then the source of the GRB must be a very energetic event. One possible explanation of GRBs is that they are the result of the merger of two neutron stars, the dense remnant cores of exhausted massive stars. This explanation has accounted for many of the known characteristics of GRBs.

Heavenly Scoop

As we said earlier in this chapter and in Chapter 5, we see celestial objects because they are producing energy, and that energy is transmitted to us in the form of electromagnetic radiation. As we will see in later chapters, different physical processes produce different wavelengths (energies) of light. Thus the portion of the spectrum from which we receive light itself is an important piece of information.

Atmospheric Ceilings and Skylights

The information—the news—we get from space is censored by the several layers of Earth's atmosphere. In effect, our Earth is surrounded by a ceiling pierced by two skylights. A rather broad range of radio waves readily penetrates our atmosphere, as does a portion of infrared and most visible light, in addition to a small portion of ultraviolet. Astronomers speak of the atmosphere's *radio window* and *optical window*, which allow passage of electromagnetic radiation of these types. To the rest of the spectrum—lower-frequency radio waves, some lower-frequency infrared, and, fortunately for us, most of the energetic ultraviolet rays, x-rays, and gamma rays—the atmosphere is opaque, an impenetrable ceiling.

In many ways, the partial opacity of our atmosphere is a very good thing, since it protects us from x-ray and gamma radiation. An atmosphere opaque to these wavelengths, but transparent to visible light and some infrared, is a big reason why life can survive at all on Earth.

For astronomers, however, there is a downside to the selective opacity of the earth's atmosphere. Observations of ultraviolet, x-ray, and gamma ray radiation cannot be made from the surface of the earth, but must be made by means of satellites, which are placed in orbit well above the atmosphere. No wonder that the advent of the space age has led to such an explosion in the amount of information that we have about the universe.

Close Encounter

Has this ever happened to you? You're on a cross-country car trip, pass through a city, and find an FM radio station you really like, only to lose it within a few miles. On the other hand, many AM stations (with programming that can range from dull to downright frightening) seem to go on forever. The reason? FM waves are at a higher frequency and correspondingly shorter wavelength than AM. The earth's atmosphere is usually transparent to high frequency FM radio waves, but opaque to lower frequency AM. As a result, FM signals do not follow the curvature of the earth, but go off on a tangent and into outer space, whereas AM signals can be reflected downward by the atmosphere and tend to follow the earth's curve for a fair distance. If alien life forms are picking up our radio signals, they're getting mostly FM and television (which broadcasts in the upper end of the FM band), but not AM. They are, therefore, mostly denied the "pleasure" of Rush Limbaugh and Howard Stern.

The Black-Body Spectrum

As Maxwell first described in the nineteenth century, all objects emit radiation at all times because the charged atomic particles of which they are made are constantly in random motion. As these particles move, they generate electromagnetic waves. Heat an object, and its atomic particles will move more rapidly, thereby emitting more radiation. Cool an object, and the particles will slow down, emitting proportionately less electromagnetic radiation. If we can study the spectrum (that is, the intensity of light from a variety of wavelengths) of the electromagnetic radiation emitted by an object, we can understand a lot about the source. One of the most important quantities we can determine is its temperature. Fortunately, we don't need to stick a thermometer in a star to see how hot it is. All we have to do is look at its light carefully.

But how?

All objects emit radiation, but no natural object emits all of its radiation at a single frequency. Typically, the radiation is spread out over a range of frequencies. If we can determine how the *intensity* (amount or strength) of the radiation emitted by an object is distributed across the spectrum, we can learn a great deal about the object's properties, including its temperature.

Star Words

A **black body** is an idealized (imaginary) object that absorbs all radiation that falls on it and perfectly re-emits all radiation that it absorbs. The spectrum (or intensity of light as a function of wavelength) that such an object emits is an idealized mathematical construct called a **black-body curve,** which can serve as an index to measure the peak intensity of radiation emitted by a real object. Some astronomical sources (like stars) can be approximated as black bodies.

Physicists often refer to a *black body,* an imaginary object that absorbs all radiation falling upon it and re-emits all the radiation that it absorbs. The way in which this re-emitted energy is distributed across the range of the spectrum is drawn as a *black-body curve.*

Now, no object in the physical world absorbs and radiates in this ideal fashion, but the black-body curve can be used as a reference index against which the peak intensity of radiation from real objects can be measured. The reason is that the peak of the black-body curve shifts toward higher frequencies (and shorter wavelengths) as an object's temperature increases. Thus, an object or region that is emitting very short wavelength gamma ray photons must be much hotter than one producing longer wavelength radio waves. If we can determine the wavelengths of the peak of an object's electromagnetic radiation emissions, we can determine its temperature.

Astronomers measure peak intensity with sophisticated scientific instruments, but we all do this intuitively almost every day. You have an electric kitchen range, let's say. The knob for one of the heating elements is turned to *off.* The heating element is black in color. This tells you that it may be safe to touch it.

But if you were to turn on the element, and hold your hand above it, you would feel heat rising, and would know that it was starting to get hot. If you had infrared vision, you would see the element "glowing" in the infrared. As the element grows hotter, it will eventually glow red, and you would know that it was absolutely a bad idea to touch it (regardless of where the control knob happened to be pointing).

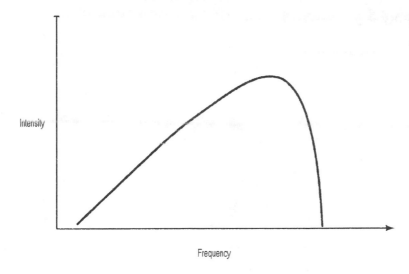

A black-body curve. The peak frequency gives the temperature of the object.

(Image from the authors' collection)

At room temperature, the metal of the heating element is black, but as it heats up, it changes color: from dull red to bright red. If you had a very high-voltage electric range and a sufficiently durable heating element, you could crank up the temperature so that it became even hotter. It would emit most of its electromagnetic radiation at progressively higher frequencies.

Now, an object that omits *most* of its radiation at optical frequencies would be *very* hot. And a range will never (we hope) reach temperatures of 6000 K, like the sun. The red color you see from the range is in the "tail" of its black-body spectrum. Even when hot, it is still emitting most of its radiation in the infrared part of the spectrum.

Watch Your Head, Here Comes an Equation

Physicists have boiled this black-body curve business down to a couple of laws. Wien's law states that the wavelength of peak emission is proportional (the symbol α means "is proportional to") to 1 over absolute temperature as expressed in Kelvins:

$$\text{Wavelength of peak emission } \alpha \ \frac{1}{\text{temperature}}$$

This law puts into an equation what we have been saying in words. As something gets hotter, its peak wavelength gets smaller. The two are inversely proportional.

Don't worry, we'll discuss absolute temperature and *Kelvins* in Chapter 16. For now, just know that the hotter an object, the bluer its radiation, and the cooler an object, the redder its radiation.

It is also true that the total energy an object radiates is proportional to its temperature—in fact, it is proportional to the fourth power of the object's temperature:

Total energy radiated α temperature^4

In other words, if you double the temperature of an object, say from 1000 K to 2000 K, it will radiate not twice the energy, but *16* times the energy (2^4, or $2 \times 2 \times 2 \times 2 = 16$). This relationship between energy radiated and temperature is called the Stefan-Boltzmann law.

Star Words

The **Kelvin** (K) temperature scale is tied to the Celsius (C) temperature scale, and is useful because there are no negative Kelvin temperatures. 0 C is the temperature at which water at atmospheric pressure freezes. 100 C is the temperature at which water boils. Absolute zero (0 K) is the coldest temperature that matter can attain. At this temperature, the atoms in matter would stop jiggling around all together. 0 K corresponds to approximately –273 C, and under laboratory conditions, temperatures in the milliKelvin (thousandths of a Kelvin) range can be attained. For a sense of scale, stars like the Sun have surface temperatures of about 6000 K. A warm day on Earth is about 300 K.

Read Any Good Spectral Lines Lately?

Using the spectrum and armed with the proper instrumentation, then, astronomers can accurately read the temperature of even very distant objects in space. And even without sophisticated equipment, you can startle your friends by letting them know that Betelgeuse (a *reddish* star) must have a lower surface temperature than the *yellow* sun.

Astronomers also use the spectrum to learn even more about distant sources. A *spectroscope* passes incoming light through a narrow slit and prism, splitting the light into its component colors. Certain processes in atoms and molecules give rise to emission at very particular wavelengths. Using such a device, astronomers can view these individual spectral lines and glean even more information about conditions at the source of the light.

While ordinary white light simply breaks down into a continuous spectrum—the entire rainbow of hues, from red to violet, shading into one another—light emitted by certain substances produces an emission spectrum with discrete *emission lines,* which are, in effect, the fingerprint of the substance.

Hydrogen, for example, has four clearly visible spectral lines in the visible part of the spectrum (red, blue-green, violet, and deep violet). The color from these four lines (added together as light) is pinkish. These four spectral lines result from the electron that is bound to the proton in a hydrogen atom jumping between particular energy levels. There are many other spectral lines being emitted; it just so happens that only four of them are in the visible part of the spectrum. Hot hydrogen gas is the source of the pinkish emission from regions around young stars like the Orion nebula (Chapter 18, "Stellar Careers").

In our hydrogen atom example, a negative electron is bound to a positive proton. The electron, while bound to the proton, can only exist in certain specific states or *energy levels.* Think of these energy levels as rungs on a ladder. The electron is either on the first rung or the second rung. It can't be in between. When the electron moves from one energy level to another (say, from a higher one to a lower one), it gives off energy in the form of a photon. Since the levels that the electron can inhabit are limited, only photons of a few specific frequencies are given off. These particular photons are apparent as bright regions in the spectrum of hydrogen: the element's spectral emission lines.

Star Words

A **spectroscope** is an instrument that passes incoming light through a slit or prism, splitting it into its component colors. All substances produce characteristic **spectral lines,** which act as the "fingerprint" of the substance, enabling identification of it.

Star Words

Emission lines are narrow regions of the spectrum where a particular substance is observed to emit its energy. These lines result from basic processes occurring on the smallest scales in an atom, electrons bound to a nucleus moving between different energy states.

Star Words

Spectral lines have a certain rest frequency; that is, there is a certain frequency at which we expect to see the line. If the line is somewhere else, or if a series of lines are all systematically moved, then the source of the lines might be in motion relative to us. If the source is coming toward us, the lines will occur at a higher frequency (known as a blue shift), and if the source is moving away from us, the lines will occur at a lower frequency (known as a red shift). The **Doppler shift** is most familiar in sound waves. We have all heard the shift in tone (frequency) that occurs when a car drives past us, especially if the driver leans on the horn as he passes. Doppler shifts are a property of all waves, light or sound.

Astronomer's Notebook

Another way spectral lines arise is when molecules rotate or vibrate. Think of it this way: Your fan has three settings, which are low, medium or high. Molecules appear to have settings as well. They can only rotate (like the fan) at particular speeds, and when the switch is turned from high to medium, they give off a photon of light. Spectral lines from molecules are very important ways to study our own Galaxy, as we'll see in Chapter 25, "What About the Big Bang?"

Depending on how we view an astronomical source, we will see different types of spectra. A black-body source that is viewed directly will produce a continuous spectrum. But if the photons from the source pass through a foreground cloud of material, certain energies will be absorbed by the cloud (depending on its composition), and we will see a black-body spectrum with certain portions of the spectrum missing or dark. This is called an absorption spectrum. If a cloud of material absorbs energy and then re-emits it in a different direction, we see the result as emission lines, or bright regions in the spectrum. The clouds of hot gas around young stars produce such emission lines.

The light that reaches us from stars carries a lot of information. The color of the object can tell us its temperature, the wavelength of the light reaching us can tell us about the energies involved, and the presence (or absence) of certain wavelengths in a black-body spectrum can tell us what elements are present in a given source.

And all of this information reaches us in the comfort of our home planet. Who knew we could learn so much without actually going anywhere?

The Least You Need to Know

➤ Visible light is a rather small subset of the electromagnetic spectrum.

➤ The difference between visible light and other electromagnetic waves, say, radio waves, is a matter of wavelength or frequency.

➤ Unlike sound waves, light waves can travel through a vacuum—empty space— because these waves are disturbances in the electromagnetic field and require no medium (substance) for transmission.

➤ The peak wavelength in an object's spectrum tells us its temperature.

➤ One way spectral lines arise is by the specific energies given off when electrons jump between energy levels in an atom, or when molecules spin at different rates.

➤ Astronomers use spectroscopes to read the spectral "fingerprint" produced by the light received from distant objects and thereby determine the chemical makeup of an object.

Chapter 8

Seeing in the Dark

In This Chapter

➤ The accidental invention of radio astronomy

➤ How radio telescopes work

➤ Interferometers: what they are and what they do

➤ The purpose of radio astronomy

➤ Amateur radio astronomy

➤ How you can "listen" to meteor showers

➤ Monitoring solar flares and Jupiter's radio noise

➤ Other nonvisible astronomy techniques

"What's an astronomer, Daddy?"

Spending much time around a little boy or girl can be pretty exhausting. All those questions! At least this one has a quick answer: "An astronomer is a person who looks at the sky through a telescope."

"But, Daddy, the visible spectrum is squeezed between 400 and 700 nm. What about the rest of the electromagnetic spectrum?"

Smart kid.

Until well into the twentieth century, astronomers had no way to "see" most of the nonvisible electromagnetic radiation that reached Earth from the universe. Then along came radio astronomy, which got its accidental start in 1931–1932 and was cranking into high gear by the end of the 1950s. Over the past 40 years or so, much of our current knowledge of the universe has come about through radio observations.

Radio astronomy is simply the study of the universe at radio wavelengths. Astronomers used to categorize themselves by the wavelength of the observations that they made: radio astronomer versus optical astronomer. Increasingly, though, astronomers define their work more by what they study (pulsars, star formation, galactic evolution) than by what wavelength they use. The reason for this change is that, in recent years, new instruments have opened the electromagnetic spectrum to an unprecedented degree. Astronomers now have the ability to ask questions that can be answered with observations at many different wavelengths.

This chapter explains what professional radio astronomers do and how amateurs can make their own observations in this low frequency part of the spectrum.

Dark Doesn't Mean You Can't See

On a clear night far from urban light pollution, the sky is indeed dazzling. Just remember that the electromagnetic information your eyes are taking in, wondrous as it is, comes from a very thin slice of the entire spectrum. As we mentioned in the last chapter, the earth's atmosphere screens out much of the electromagnetic radiation that comes from space. It allows only visible light and a bit of infrared and ultraviolet radiation to pass through a so-called optical window and a broad portion of the radio spectrum to pass through a radio window.

Two windows.

If your house had two windows, would you look through only one?

A Telephone Man Tunes In

The first true radio astronomer was not trained as an astronomer at all. Even to this day, many astronomers who work in the radio regime were trained as physicists and electrical engineers, and later learned to apply their knowledge to astronomy. Karl Jansky, the son of a Czech immigrant who settled in Oklahoma (where Karl was born in 1905), took a degree in physics at the University of Wisconsin. After graduating, Karl went to work in 1928 not as an astronomer, but as a telephone engineer with Bell Labs. The phone company was looking for ways to make telephone communications possible with shortwave radio, but the transmissions were bedeviled by all sorts of interference.

Now most people hadn't given much thought to radio static. After all, static was something to be avoided if possible—meaningless noise that only interfered with communications. Jansky was given the assignment of studying sources of static at a wavelength of 14.6 m in an effort to track down the precise sources of radio interference and eliminate them.

On a farm in Holmdel, New Jersey, not far from Bell Labs, Jansky set up a very un-gainly looking device, which he called a merry-go-round. It was a large directional antenna, which looked rather like the biplane wing of the Wright brothers' first air-plane. It was mounted on some discarded Model T Ford wheels and could be rotated through 360 degrees by means of a motor. Using this contraption, Jansky was soon able to identify all the known sources of radio interference except one.

Jansky tracked the stubborn and mysterious interference. When amplified and sent to a speaker, the interference sounded like a faint hiss. The source *seemed* to be in the sky, since Jansky could track it rising and setting with the stars.

But it wasn't coming from just anywhere in the sky. By the spring of 1932, Jansky traced the primary source of radio noise to the direction of the constellation Sagit-tarius, which astronomers Harlow Shapley and Jan H. Oort had identified (from the distribution of globular clusters in the Galaxy) as the direction of the center of the Milky Way Galaxy. Using his merry-go-round antenna, Jansky had "discovered" the center of the much bigger merry-go-round that is our galaxy. There were other sources of radio noise in the sky as well, but Jansky noted that the sun itself was not an impressive source of radio noise. This observation was a bit surprising, since the sun is so close to us. He concluded that whatever the source of radio noise, it proba-bly wasn't distant stars.

Karl Jansky with the first radio telescope, Holmdel, New Jersey.

(Image from Lucent Technologies)

Jansky published his "discovery" late in 1932, and the detection of radio signals from space appeared in national newspapers by the following year. Strangely enough, Jansky himself didn't pursue the science he had accidentally created. As for most pro-fessional astronomers, they continued to look through only one of their two win-dows, the portion of the spectrum available to optical telescopes.

It took another nonastronomer, Grote Reber, to appreciate the possibilities of what Jansky had discovered. In today's image-conscious world, we might call Reber a nerd. But as the example of Bill Gates has shown us, some nerds go on to change the

world. Born in Wheaton, Illinois in 1911, he grew up tinkering with radio transmitters, building one powerful enough to communicate with other ham radio operators all over the world. Like many early radio astronomers, he became an electrical engineer, but never lost his interest in amateur radio, and when he read about Jansky's discovery, he tried, without success at first, to adapt his own shortwave receiver to pick up interstellar radio waves with wavelengths of 10 cm.

He tinkered with the electronics (trying longer wavelengths), and, in 1937, built a paraboloidal antenna 30 feet in diameter. With this, Reber not only confirmed Jansky's discovery of radio waves from the direction of Sagittarius, but found other sources in the direction of the constellations Cygnus, Cassiopeia, and elsewhere.

Astro Byte

Although Jansky didn't follow up on his discovery, he is considered the father of radio astronomy. Radio astronomers have honored him by naming the basic unit of radio brightness the Jansky (Jy).

Close Encounter

Until after World War II, Grote Reber's 30-foot dish was the only radio telescope in the world, and by 1942 he had completed the world's first preliminary radio maps of the sky. The end of WWII started a period of intense research in radio astronomy all over the world. A large number of former radar engineers (particularly in England and Australia) turned their talents to the pursuit of more distant sources using radio telescopes. After the war, Reber moved his radio telescope to Sterling, Virginia, and was given a government job in Washington, D.C. as chief of the Experimental Microwave Research Section. In 1951, he built a new radio telescope on an extinct Hawaiian volcano and mapped out low-frequency (long-wave) celestial signals (previously he had concentrated on shortwave signals). In pursuit of a bigger radio window, he moved in 1954 to Tasmania, Australia, a place where the earth's atmosphere is occasionally transparent to electromagnetic radiation more than 30 meters in wavelength. In 1957 he returned for a time to the United States to work at the newly created National Radio Astronomy Observatory in Green Bank, West Virginia, but went back to Tasmania in 1961 to continue mapping long-wave radio sources.

Reber confirmed that the radio signals did not coincide with the positions of visible stars. Directing his dish toward such bright stars as Sirius, Vega, or Rigel, he detected nothing. But looking toward a starless area in Cassiopeia, he picked up strong radio waves. He had unknowingly detected a supernova remnant known as Cassiopeia A.

Anatomy of a Radio Telescope

The basic anatomy of a radio telescope hasn't changed all that much from Reber's dish—though the instruments have become much larger and the electronics more sophisticated. A radio telescope works just like an optical telescope. It is just a "bucket" that collects radio frequency waves, and focuses them on a detector. A large metal dish—like a giant TV satellite dish—is supported on a move-able mount (either equatorial or altazimuth). A detector, called a receiver horn, is mounted on legs above the dish (prime focus) or below the surface of the dish (Cassegrain focus). The telescope is pointed toward the radio source, and its huge dish collects the radio waves and focuses them on the receiver, which amplifies the signal and sends it to a computer. Since the radio spectrum is so broad, astronomers have to decide which portion of the radio spectrum they will observe. Different receivers are used for observations at different frequencies. Receivers are either swapped in and out, or (more typically) the radio signal is directed to the correct receiver by moving a secondary reflecting surface (like the secondary mirror in an optical telescope).

Astronomer's Notebook

Why do radio telescopes have to be so big? A 30-foot dish seems excessive. The resolution is actu-ally determined by the ratio of the wavelength being observed to the diameter of the tele-scope. So optical telescopes (which detect short-wavelength optical photons) can be much smaller than radio telescopes, which are trying to detect long-wavelength radio waves, and have the same resolution.

Bigger Is Better: The Green Bank Telescope

In the case of radio telescopes, size really does matter. The resolution of a telescope depends not only on its diameter, but the wavelength of the detected radiation (the ratio of wavelength to telescope diameter determines the resolution). Radio waves are big (on the order of centimeters or meters), and the telescopes that detect them are correspondingly huge. Also, the radio signals that these instruments detect are very faint, and just as bigger optical telescope mirrors collect more light than smaller ones, bigger radio telescopes collect more radio waves and image fainter radio signals than smaller ones.

Collecting radio signals is just part of the task, however. You may recall from Chap-ter 5, "The Art of Collecting Light (with a Telescope)," that, for practical purposes, very good optical telescopes located on the earth's surface can resolve celestial objects to 1" (1 arcsecond—$\frac{1}{60}$ of 1 arcminute, which, in turn, is $\frac{1}{60}$ of 1 degree). The best angular resolution that a very large single-dish radio telescope can achieve is about 10 times coarser than this, about 10", and this, coarse as it is, is possible only with the very largest single dish radio telescopes in the world. The National Radio Astronomy Observatory has just commissioned the world's largest fully steerable radio telescope. The 100 m dish will have a best resolution of 14".

Star Words

A **radio interferometer** is a combination of radio telescopes linked together electronically to create the equivalent (in terms of resolution) of a giant radio telescope. The use of an interferometer greatly increases resolving power at radio frequencies. Earth-based radio interferometers have much better resolution than any conventional Earth-based optical telescope. Of course, optical interferometers (which are being tested) will exceed radio interferometers in resolution when they come on line because of the higher frequency of optical radiation.

The world's largest nonsteerable single-dish radio telescope was built in 1963 at Arecibo, Puerto Rico, and has a dish, 300 meters (984 feet) in diameter sunk into a natural valley. While its great size makes this the most sensitive radio telescope, the primary surface is nonsteerable—totally immobile—and, therefore, is limited to observing objects that happen to pass roughly overhead (within 20 degrees of zenith) as the earth rotates.

The Arecibo Radio Telescope, Arecibo, Puerto Rico.

(Image from National Astronomy and Ionosphere Center)

Interference Can Be a Good Thing

There is a way to overcome the low angular resolution due to the size of radio waves: link together a lot of smaller telescopes so they act like one giant telescope. A *radio interferometer* is a combination of two or more radio telescopes linked together electronically to form a kind of virtual dish, an array of antennas that acts like one gigantic antenna. It is as if we had small pieces of a very large optical telescope (imagine a giant mirror with a lot of its surface area punched out), so that while an interferometer has the *resolution* of a very large telescope, it does not have the surface area or sensitivity to faint sources of a truly gigantic telescope.

The National Radio Astronomy Observatory (NRAO) maintains and operates the Very Large Array (VLA) interferometer on a vast plain near Socorro, New Mexico, consisting of 27 large dishes arrayed on railroad tracks in a Y-shaped pattern. Each arm is 12.4 miles (20 km) long, and the largest distance between 2 of the antennas is 21.7 miles (35 km). As a result, the VLA has the resolving power of a radio telescope 21.7 miles across. The VLA recently celebrated its twentieth anniversary.

The Green Bank Telescope, dedicated August 2000.

(Image from NRAO/AUI)

123

For radio astronomers who want something even larger than "very large," there is *Very Long Baseline Interferometry* (*VLBI*), which can link radio telescopes in different parts of the world to achieve incredible angular resolutions better than a thousandth of an arcsecond (.001″). From its offices in Socorro, New Mexico, the NRAO also operates the VLBA (Very Long Baseline Array), which consists of 10 radio dishes scattered over the United States, from Mauna Kea, Hawaii, to St. Croix, U.S. Virgin Islands. In 1996, Japanese astronomers launched into Earth's orbit a radio telescope to be used in conjunction with the ground based telescopes in order to achieve the resolution of a telescope larger than the earth itself. For more information, see www.nrao.edu.

What Radio Astronomers "See"

Insomnia is a valuable affliction for optical astronomers, who need to make good use of the hours of darkness when the sun is on the other side of the earth. But as Karl Jansky discovered so many years ago, the sun is not a particularly bright radio source. In consequence, radio astronomers (and radio telescopes) can work night and day. The VLA, for example, gathers data (or runs tests) 24 hours a day, 363 days a year. Not only is darkness not required, but you can even make radio observations through a cloud-filled sky. The senior author of this book even observed a distant star-forming region in the midst of a storm during which lightning struck near the VLA and disabled it for a few minutes.

As the Dutch astronomer Jan Oort realized after reading Reber's work in the 1940s, radio waves opened new vistas into the Milky Way and beyond. Radio astronomers can observe objects whose visible light doesn't reach the earth because of obscuration by interstellar dust or simply because they emit little or no visible light. The fantastic objects known as quasars, pulsars, and the regions around black holes—all of which we will encounter later in this book—are often faint or invisible optically, but do emit radio waves.

The spiral form of our own Galaxy was first mapped using the 21 cm radio spectral line from neutral (cold) hydrogen atoms, and the discovery of complex molecules between the stars was made at radio frequencies. The very center of our own Milky Way Galaxy is hidden from optical probing, so that most of what we know of our galactic center has come from infrared and radio observation. Since radio interferometers are detecting an interference pattern, radio data has to be processed in ways different from optical data. But the end result is either a radio image, showing the brightness of the source on the sky, or a radio spectrum, showing a spectral line or lines. In Chapter 18, "Stellar Careers," we'll describe how radio spectral lines arise.

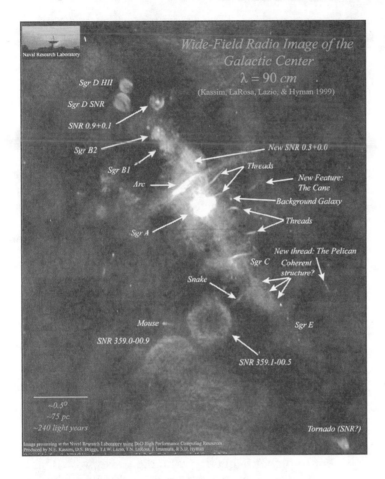

Image of our Galaxy, the Milky Way, as seen with the Very Large Array at 90 cm.

(Image from N. Kassim et al., NRL)

You Can Do This, Too!

Building a huge radio telescope like Arecibo or the Green Bank Telescope (GBT) takes a great deal of money, and so does operating one. Even if you had the cash, your neighbors (not to mention the local zoning board), might frown on your building even a modest 30-foot-diameter dish antenna in your backyard. However, remember that radio astronomy originated with nonastronomers, and there is still plenty of room in the field for amateurs, including amateurs of modest means.

A small but committed group of enthusiasts have formed the Society of Amateur Radio Astronomers (SARA). Most books for budding astronomers don't discuss amateur radio astronomy, though it is a fascinating and rewarding subject.

Astronomer's Notebook

You can contact the Society of Amateur Radio Astronomers (SARA) at its Web site: www. bambi.net/sara.html. The Web site of Radio-Sky Publishing also offers valuable information on amateur radio astronomy: www.radiosky.com.

Astronomer's Notebook

You can detect sporadic meteor events any day of the year, and you don't have to wait for clear skies or night to fall. However, meteor showers and even out-and-out meteor storms occur at certain times of the year, when the earth passes through the debris that litters the "orbital paths" of known comets. Two of the most intense are the Perseid shower in August and the Geminid shower in December. See Chapter 11, "Solar System Home Movie," for more information on when to expect peak meteor activity.

Amateur Radio Astronomy: No-Cost and Low-Cost Approaches

As you may recall from Chapter 6, "You and Your Telescope," a decent optical telescope costs at least $300 to $400. For free optical astronomy, all you need are your eyes. You *can* also do some radio astronomy for free—if you own an FM radio or a television set. We thank Tom Crowley of Atlanta Astronomy Club for many of the following ideas.

Have you ever witnessed a meteor shower? The streaks of light in the night sky can be quite spectacular. Meteors are the bright trails of ionized atmosphere behind tiny bits of cometary debris that enter the earth's atmosphere. Most meteorites are no larger than a pea. What if we told you that there was another way to watch a meteor shower—using a radio telescope otherwise known as an FM radio?

Meteor counts by radio are about ten times more accurate than visual observation—and, as with any radio observations, you can observe during the day or through clouds. It doesn't have to be dark or clear outside. You may want to supplement your optical meteor gazing with your radio on cloudy nights.

Recall from Chapter 7, "Over the Rainbow," that the earth's atmosphere is transparent to some forms of electromagnetic radiation and opaque to others. The upper atmosphere normally reflects low-frequency AM radio signals. In contrast, the atmosphere is transparent to higher frequency FM radio waves, which, as a consequence, have a shorter range. They usually penetrate and are not reflected by the atmosphere.

But something happens when a meteorite enters the atmosphere. Each piece of debris that tears into the atmosphere (at up to 40 miles per second), heats up the air around it and creates a tiny ionized (electrically charged) vapor trail in the upper atmosphere. These columns of charged particles can reflect even higher frequency FM radio waves. This temporary condition means that previously out-of-range FM broadcasts can (for a moment) be heard. During periods of known meteoric activity, stay at the low end of the FM dial and try to find FM radio stations that are from 400 to 1,300 miles away. You might call a distant friend to get the broadcast frequency of a few stations. When a distant station fades in for a second or two, you are indirectly observing a meteor. The trail behind it has momentarily reflected a distant radio signal into your receiver.

It helps if you can hook up your radio to an outdoor antenna, but if the meteor shower is fairly intense, you should detect many events even without such an antenna. You can also try tuning your TV set to the lowest unused VHF channel. Again, when a distant station, normally out of range, fades in and the signal becomes strong for a second or two, you know a meteor has entered the atmosphere. (Note that this works only with a television receiving signals from an outdoor antenna—not via cable or satellite!)

Just by tuning in your radio, you can do some meaningful radio astronomy. You can make it more interesting by recording the events on tape, or keeping a written record of the number of events you detect per hour.

But amateur radio astronomy need not be limited to listening for distant FM radio or TV stations. If you are an amateur radio operator—a ham—you already have much of the equipment required for more serious radio astronomy. If you aren't into amateur radio, you can get started for a highly variable but modest cost. The first step to take is to log onto SARA's World Wide Web site (www.bambi.net/sara.html) for overview information.

Essentially, amateurs can use either nonimaging or imaging radio astronomy techniques. Nonimaging techniques (which monitor radio emissions without pinpointing locations) require a simple shortwave receiver, usually modified to receive a narrow band of frequencies, and a simple antenna system. With such equipment, you can track radio emission from Jupiter, solar flares, and meteor events.

Imaging techniques (which provide more detailed information on the location and nature of the signal) require a more serious commitment of resources, including a much larger dish-type antenna, more sophisticated receiving equipment, reasonably elaborate recording equipment, and (probably) a rural location removed from most sources of radio interference. For purposes of this book, we'll restrict ourselves to the more approachable nonimaging techniques, which are more appropriate for beginners.

Solar Flares and Meteor Events

Solar flares, which we'll discuss in Chapter 16, "Our Star," are explosive events that occur in or near an active region on the sun's surface. Flares can be detected with a very low frequency (VLF) receiver operating in the 20- to 100-KHz (kilohertz, or thousand hertz) band. (This is below the region of the spectrum where AM radio stations broadcast.) Such a receiver can be homemade, using plans

Star Words

Solar flares are explosive events that occur in or near an active region on the sun.

supplied by such organizations as SARA. Solar flares can also be monitored in regular shortwave radio bands with a standard shortwave receiver.

Why do solar flares create radio signals? They generate x-rays that strike a part of our atmosphere called the ionosphere, greatly enhancing the electron count in this atmospheric region. These electrons generate the noise picked up by the radio.

ET Phone Home

If you've seen such sci-fi movies as *The Arrival, Independence Day,* or *Contact,* you already know about an organization called SETI (Search for Extra-Terrestrial Intelligence). It is an international group of scientists and others who, for the most part, use radio telescopes to monitor the heavens in search of radio signals generated not by natural phenomena, but broadcast artificially by intelligent beings from other worlds. So far, no clearly artificial extraterrestrial radio signals have been confirmed, but SETI personnel keep searching. The SETI project got a large boost recently when Paul Allen, one of the co-founders of Microsoft, committed $12.5 million to the project. The new instrument to be built exclusively for SETI will be called the Allen Telescope Array.

We'll take a closer look at SETI and other efforts to make contact in Chapter 23, "Moving Out of Town," but if you are interested in the search for extraterrestrial intelligence, you don't have to just read about it, you can actually participate in it.

A few highly committed amateur radio astronomers have built SETI-capable radio telescopes and spend time searching for artificial signals of extraterrestrial origin. If you're not up to making such a commitment, the SETI Institute is developing an alternative. In a project called SETI@home, a special kind of screensaver program (a program that, typically, puts up a pretty, animated picture on your PC monitor when the computer is idle) has been installed on over 1.5 million computers in 224 countries. When the computer is idle, this program will use the time to go to work analyzing data from four million different combinations of frequency bandwidth and drift rate recorded by the world's largest radio telescope at Arecibo, Puerto Rico. With thousands of computers crunching this data, the SETI Institute believes that it can analyze data more quickly, thereby increasing the chances of ferreting out a radio signal from an intelligent source. Information on SETI@home can be found on the Net at www.seti-inst.edu.

The Rest of the Spectrum

Optical astronomy with the naked eye is at least 5,000 years old and probably much older. Optical telescope astronomy is about 400 years old. Radio astronomy is a youthful discipline at about 70 years, if we date its birth from Jansky's work in the early 1930s. But it has been only since the 1970s that other parts of the electromagnetic spectrum have been regularly explored for the astronomical information they may yield. As might have been expected, each new window thrown open on the cosmos has brought in a fresh breeze and enriched our understanding of the universe.

New Infrared and Ultraviolet Observations

Telescopes need to be specially equipped to detect infrared radiation—the portion of the spectrum just below the red end of visible light. Infrared observatories have applications in almost all areas of astronomy, from the study of star formation, cool stars, and the center of the Milky Way, to active galaxies, and the large-scale structure of the universe. IRAS (the Infrared Astronomy Satellite) was launched in 1983 and sent images back to Earth for many years. Like all infrared detectors, though, the ones on IRAS had to be cooled to low temperatures so that their own heat did not overwhelm the weak signals that they were trying to detect. Although the satellite is still in orbit, it has long since run out of coolant, and can no longer make images. The infrared capability of the *Hubble Space Telescope* provided by NICMOS (Near-Infrared Camera and Multi-Object Spectograph) yielded spectacular results while in operation. The Next-Generation Space Telescope (NGST) will be optimized to operate at infrared wavelengths, and will be cooled passively (by a large solar shield).

Ultraviolet radiation, which begins in the spectrum at frequencies higher than those of visible light, is also being studied with new telescopes. Since our atmosphere blocks all but a small amount of ultraviolet radiation, ultraviolet studies must be made by high-altitude balloons, rockets, or orbital satellites. The *Hubble Space Telescope*, for instance, has the capability to detect ultraviolet (UV) photons as well as those with frequencies in the visible and infrared. Ultraviolet observations provide our best views of stars, and stars with surface temperatures higher than the sun's.

Chandrasekhar and the X-Ray Revolution

Electromagnetic radiation at the highest end of the spectrum can now be studied. Since x-rays and gamma rays cannot penetrate our atmosphere, all of this work must be done by satellite. Work began in earnest in 1978 when an x-ray telescope was launched, called the High-Energy Astronomy Observatory (later, the Einstein Observatory). The Röentgen Satellite (ROSAT) was next launched by Germany in 1990. The Chandra X-ray Observatory (named for astronomer Subrahmanyah Chandrasekhar) was launched into orbit in July 1999 and has produced unparalleled high-resolution images of the x-ray universe. The Chandra image of the Crab Nebula, home to a known pulsar, showed never before seen details of the environment of an exploded star. For recent images, go to www.chandra.harvard.edu. X-rays are detected from very high energy sources, such as the remnants of exploded stars (supernova remnants) and jets of material streaming from the centers of galaxies. We'll tell you more about some of these phenomena in Part 4, "To the Stars." Chandra is the premier x-ray instrument, doing in this region of the spectrum what the *Hubble Space Telescope* has done for optical observations.

In 1991, the Gamma Ray Observatory (GRO) was launched by the space shuttle. It is revealing unique views of the cosmos, especially in regions where the energies involved are very high: near black holes, at the centers of active galaxies, and near neutron stars. We'll have more to say about many of these phenomena in Part 4.

*Chandra image of
the Crab Nebula.*

*(Image from
NASA/JHU/AUI)*

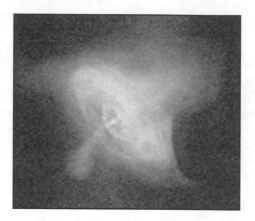

Capturing the Full Spectrum

Once satellites are launched, any astronomer can make a proposal to obtain observing time. There are always many more proposals than time available, but this practice makes some of the world's most advanced technologies available to astronomers at institutions worldwide. Today's astronomers have the unique ability to turn to almost any region of the electromagnetic spectrum for answers to their questions. They can play with a full keyboard, instead of plunking away on a single key.

The Least You Need to Know

➤ Electrical engineer Karl Jansky invented radio astronomy in the 1930s, when he was tracing sources of shortwave radio interference for the telephone company.

➤ A radio telescope typically consists of a large parabolic dish that collects and focuses very weak incoming radio signals on a receiver. The signal is then amplified and processed using sophisticated electronics and a computer.

➤ Radio telescopes can detect objects that are blocked by interstellar material or that emit little or no visible radiation.

➤ Although professional radio astronomy typically involves elaborate equipment and massive facilities, amateur radio astronomers can use equipment as modest as an ordinary FM radio.

➤ In recent years, astronomers have launched instruments into orbit that can detect all segments of the electromagnetic spectrum, from infrared, through visible, and on to ultraviolet, x-rays, and gamma rays. The highest frequency radiation (x-rays and gamma rays) comes from some of the most energetic and exotic objects in the universe.

Chapter 9

Space Race: From *Sputnik* to the International Space Station

---

In This Chapter

➤ The early development of rockets

➤ *Sputnik:* The space race begins

➤ The first men in space

➤ Lunar missions

➤ Planetary probes

➤ The Space Shuttle and space stations

While countless human beings have gazed up at the sky with wonder, a few were never content just to look. They didn't want to wait for the information to get here, they wanted to go *there*. In the second century C.E., the Greek satirist Lucian wrote the first account we have of a fictional trip from the earth to the moon. Doubtless, someone had thought about such a trip before Lucian, and certainly many contemplated space travel after him. It was not until the eighteenth century that people were first lofted into the air by hot-air balloons. And while the airplane made its debut in 1903, human spaceflight—in which a human ventured beyond the earth's protective atmospheric blanket—did not come about until the 1961 flight of a Soviet cosmonaut Yuri Alekseyevich Gagarin.

This chapter reviews manned as well as unmanned spaceflight, chiefly from the point of view of astronomy.

This Really *Is* Rocket Science

While spaceflight was the subject of many centuries of speculation, three men worked independently to lay its practical foundation. Konstantin Eduardovich Tsiolkovsky (1857–1935) was a lonely Russian boy, almost totally deaf, who grew up in retreat with his books. He became a provincial schoolteacher, but his consuming interest was flight, and he built a wind tunnel to test various aircraft designs. Soon he became even more fascinated by the thought of space travel, producing the first serious theoretical books on the subject during the late nineteenth and early twentieth centuries.

Another quiet, introspective boy, this one a New Englander, Robert Hutchings Goddard (1882–1945), was captivated by H. G. Wells's science-fiction novel *War of the Worlds,* which he read in an 1898 serialization in the *Boston Post.* On October 19, 1899 (as he remembered it for the rest of his life), young Goddard climbed a cherry tree in his backyard and "imagined how wonderful it would be to make some device which had even the possibility of ascending to Mars."

From that day, the path of his life became clear to him. Goddard earned his Ph.D. in physics in 1908 from Clark University in his hometown of Worcester, Massachusetts, and, working in a very modest laboratory, he showed experimentally that thrust and propulsion can take place in a vacuum (this follows from Newton's Laws of motion—the expelled gases pushing forward on the rocket). He also began to work out the complex mathematics of energy production versus the weight of various fuels, including liquid oxygen and liquid hydrogen. These are the fuels that would ultimately power the great rockets that lofted human beings into orbit and to the moon—and still power the launch of many rockets today. Goddard was the first scientist to develop liquid-fuel rocket motors, launching the inaugural vehicle in 1926, not from some governmental, multimillion-dollar test site, but from his Aunt Effie's farm in Auburn, Massachusetts. Through the 1930s and 1940s, he tested increasingly larger and more powerful rockets, patenting a steering apparatus and the idea of what he termed "step rockets"—what would later be called multistage rockets—to gain greater altitude.

Goddard's achievements were little recognized in his own time, but, in fact, he had single-handedly mapped out the basics of space-vehicle technology, including fuel pumps, self-cooling rocket motors, and other devices required for an engine designed to carry human beings, telecommunications satellites, and telescopes into orbit.

Hermann Oberth (1894–1989), born in Austria, was destined for a medical career, like his father, but his medical studies were interrupted by World War I. Wounded, he studied physics and aeronautics while recovering. While he was still in the Austrian army, he performed experiments to simulate weightlessness, and designed a long-range, liquid-propellant rocket. The design greatly impressed Oberth's commanding officer, who sent it on to the War Ministry, which summarily rejected it. After the war, University of Heidelberg faculty members likewise rejected Oberth's dissertation concerning rocket design. Undaunted, Oberth published it himself—to great acclaim—as *The Rocket into Interplanetary Space* (1923). In 1929, he wrote *Ways to Spaceflight,* winning a prize that helped him finance the creation of his first liquid-propellant rocket, which he launched in 1931.

During World War II, Oberth became a German citizen and worked with Wernher von Braun to develop rocket weapons.

From Scientific Tool to Weapon and Back Again

From the early 1900s through the 1930s, peacetime governments and the scientific community showed relatively little interest in supporting such pioneers as Tsiolkovsky, Goddard, and Oberth. Unfortunately, it took war in Europe, and a desire to launch bombs onto other nations, to spur serious, practical development of rockets. The research and development took place almost exclusively in Germany.

During the late 1930s, under the militaristic regime of Adolf Hitler, two rocket weapons were created. The first, known as the V-1, was more a pilotless jet aircraft than a rocket. About 25 feet long, it carried a 2,000-pound bomb at 360 miles per hour for a distance of about 150 miles. It was a fairly crude device: When it ran out of fuel, it crashed and exploded. Out of about 8,000 launched, some 2,400 rained down on London from June 13, 1944, to March 29, 1945, with deadly effect.

In contrast to the V-1, the V-2 was a genuine rocket, powered not by an air-breathing jet engine, but by a rocket engine burning a mixture of alcohol and liquid oxygen. The V-2 had a range of about 220 miles and also delivered 2,000 pounds of high explosives to its target. From September 8, 1944, to March 27, 1945, about 1,300 V-2s were launched against Britain. Scientists of every stripe spent the years from 1939 to 1945 directing their energies toward the defeat of the enemy. Many of the techniques developed during the war (radar technology and rocket engines, to name two) would become crucial to astronomy in the decades after WWII.

Astronomer's Notebook

The chief difference between a jet engine and a rocket engine is that a jet engine requires the intake of air to produce combustion and thrust, whereas a rocket engine uses an on-board fuel system that requires no external source of oxygen for combustion. Such a self-contained system is an absolute requirement once a craft leaves the earth's atmosphere.

Astro Byte

The V-2 rocket was 47 feet long and weighed about 29,000 pounds. It developed 60,000 pounds of thrust, which lofted it about 60 miles into the air.

During the last days of the war in Europe, as U.S. forces invaded Germany from the west and Soviet forces invaded from the east, both sides scrambled to capture V-2s and, with them, German rocket scientists, such as Wernher von Braun. Both sides saw the potential in being able to deliver bombs over long distances. These rockets and the scientists who made them were at the center of the Cold War and the Space Race—a period of competition in politics and high technology between the two superpowers that dominated the postwar world.

Playing with Balloons

While the V-2 had achieved great altitude by the 1940s, scientists were still a long way from attempting a human ascent. These early rockets were intended to explode at the end of the journey. If an instrument or a human were on board, explosions were to be avoided at all costs. In fact, another technology, the balloon, would be the first to take human beings into the upper stratosphere, the frontier of space.

Auguste Piccard (1884–1962), a Swiss-born Belgian physicist, built a balloon in 1930 to study cosmic rays, which the earth's atmosphere filters out. Piccard developed revolutionary pressurized cabin designs, which supported life at high altitudes, and, in 1932, reached an altitude of 55,563 feet. The following year, balloonists in the Soviet Union used Piccard's design to reach 60,700 feet, and an American balloonist topped that later in the year at 61,221 feet.

The Battle Cry of Sputnik

Impressive as the achievements of Piccard and others were, balloons could never move beyond the frontier of space. They needed the earth's atmosphere to loft them. After the war, scientists in America and the Soviet Union began experimenting with so-called *sounding rockets* developed from the V-2s, in part to probe (sound) the upper atmosphere. While a sounding rocket was accelerated to speeds of up to 5,000 miles per hour, it would run out of fuel by about 20 miles up. This acceleration gave the rockets sufficient velocity to continue their ascent to about a hundred miles, after which the rocket fell to Earth. Any instrumentation it carried had to be ejected, parachuted to safety and recovered, or the information had to be transmitted to a ground station by radio before the rocket crashed.

The goal of rocket science at this point was not only to reach higher altitudes, but to achieve a velocity that could launch an artificial satellite into orbit around the earth. Imagine a rock thrown into the air. The force of gravity causes it to travel in a parabola and return to the earth. If the ball were thrown at a greater and greater velocity, it would travel farther and

Star Words

Sounding rockets were early sub-orbital rocket probes launched to study the earth's upper atmosphere.

farther until it returned to the earth. At some velocity, however, the rock would never return to the earth, but continually fall toward it (this is what the moon is doing: *orbiting* the earth). It was no mean trick to get a satellite going fast enough to make it orbit the earth.

A single-stage rocket, like the V-2, exhausted its fuel supply before it reached sufficient altitude and velocity to achieve orbit. It lacked the necessary thrust. To build a more powerful rocket required a return to Goddard's idea of a "stepped" or staged rocket. A staged rocket jettisoned large parts of itself as fuel in each lower part—or stage—ran out. Thus the rocket became progressively less massive as it ascended, both by burning fuel and by discarding the empty fuel tanks.

During the early and mid-1950s, there was much talk of putting a satellite into orbit, and both the United States and the Soviets declared their intention to do so.

In the Cold War atmosphere of the time, it came as a great shock to Americans when the USSR was the first to succeed, launching *Sputnik I* (Russian for "satellite") into orbit on October 4, 1957. The 185-pound (83.25 kg) satellite had been lofted to an altitude of about 125 miles (201 km) and had achieved the required Earth orbital velocity of some 18,000 miles (28,980 km) an hour.

Astro Byte

The orbit of *Sputnik* I reached an apogee (farthest point from Earth) of 584 miles (940 km) and a perigee (nearest point) of 143 miles (230 km). It circled Earth every 96 minutes and remained in orbit until early 1958, when it fell back into the earth's atmosphere and burned up.

The first *Sputnik* was a primitive device by today's standards. It did nothing more than emit a radio beep to tell the world it was there.

But it didn't have to do more than that. The point was made, the Space Age was born, and the space race had begun.

Early Human Missions

While both the Soviet Union and the United States launched a series of artificial satellites, the major goal quickly became "putting a man in space." This objective was less scientific than psychological and political. The Soviet communists were determined to demonstrate the superiority of their technology generally and, in particular, the might of their ballistic missiles. At the time, their rockets were more powerful than what the United States had. The Soviets were eager to demonstrate that they were capable of lofting a person (and all of the machinery necessary to support a person) into space—or a warhead onto an American city.

Just as the Soviets had been first into orbit with *Sputnik,* so, on April 12, 1961, they were first to put a person, Yuri Alekseyevich Gagarin, into space—and into Earth orbit, no less. The first woman, Valentina Tereshkova of the USSR followed in 1963. It took America 20 more years to achieve this landmark. Through the rest of the 1960s, the Soviets and the Americans sent cosmonauts and astronauts into orbit and even had them practice working outside of their spacecraft in what were termed "extravehicular activities" or, more familiarly, space walks.

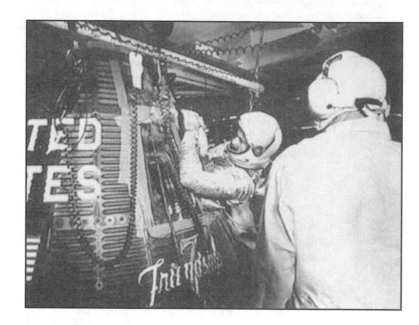

Col. John H. Glenn Jr., climbs into his Friendship 7 *capsule before becoming the first American to orbit the earth on February 20, 1962.*

(Image from NASA)

Satellites and Probes

Astronomers and other scientists were not always enthusiastically supportive of the manned space program, many of them feeling that it stole both public attention and government funding away from more useful data-gathering missions that could be carried out much more efficiently and inexpensively by unmanned satellites and probes. There is much truth to this sentiment. However, at least in the 1960s, unmanned exploratory missions continued to have high priority, and did not really suffer from the parallel development of the manned space program.

The Explorers

The first space satellite the United States sent into orbit was *Explorer 1,* launched on January 31, 1958. While the satellite didn't beat *Sputnik 1* into space, it accomplished considerably more than the Soviet probe. *Explorer 1* carried equipment that discovered the innermost of the Van Allen radiation belts, two zones of charged particles that surround the earth. By 1975, when the *Explorer* series of missions ended, 55 satellites had been launched, including *Explorer 38* (July 4, 1968), which detected galactic radio sources, and *Explorer 53* (May 7, 1975), which investigated x-ray emission inside and beyond the Milky Way.

Observatories in Space

In 1962, the United States launched its first extraterrestrial observatory, the *Orbiting Solar Observatory* (*OSO*). It was the first of a series of solar observatories, designed to gather and transmit such data as the frequency and energy of solar electromagnetic radiation in ultraviolet, x-ray, and gamma ray regions of the spectrum—all regions to which our atmosphere is partially or totally opaque.

JFK's Challenge

On May 5, 1961, about three weeks after the Russians put a man into a single orbit, U.S. Navy commander Alan B. Shephard was launched on a 15-minute suborbital flight into space. Americans were proud of this achievement, to be sure, but the Soviets had clearly upstaged it. Just 20 days later, on May 25, President John F. Kennedy spoke to Congress: "I believe this nation should commit itself to achieving the goal, before the decade is out, of landing a man on the moon and returning him safely to Earth. No single space project in this period will be more impressive to mankind, or more important for the long-range exploration of space, and none will be so difficult or expensive to accomplish."

Lunar Probes

There were voices raised in protest, both in the political and scientific communities. Why try to put men on the moon, when unmanned probes could tell as much or more—and accomplish the mission with far less expense and danger?

The Russians had successfully launched the first lunar probe, *Luna 2,* on September 12, 1959, targeting and hitting the moon with it. *Luna 3,* launched the following month, on October 4, 1959, made the first circumnavigation of the moon and transmitted back to Earth civilization's first photographs of the Moon's mysterious far side. Another Soviet lunar first would come on January 31, 1966, when *Luna 9* made a successful lunar soft landing—as opposed to a destructive impact.

In 1961, the United States launched the first of the *Ranger* series of nine unmanned lunar probes, hitting the moon with *Ranger 4* in 1962 and orbiting it, with *Rangers 7, 8,* and *9,* during 1964–1965. These last three missions generated some 17,000 high-resolution photographs of the lunar surface, not only valuable as astronomy, but indispensable as a prelanding survey.

Astro Byte

Lunar Orbiter photographs enabled astronomers to create moon maps at least 100 times more detailed than the best available from Earth-based observations. This improvement was achieved by simply getting the camera a lot closer to the subject.

From 1966 to 1968, seven *Surveyor* probes made lunar landings (not all successful), took photographs, sampled the lunar soil, and performed environmental analysis. *Surveyor 6* (launched on November 7, 1967) landed on the lunar surface, took photographs, then lifted off, moved eight feet, landed again, and took more photographs. It was the first successful lift-off from an extraterrestrial body.

The *Lunar Orbiter* series, five orbital missions launched during 1966–1967, mapped much of the lunar surface in 1,950 wide-angle and high-resolution photographs. These images were used to select the five primary landing sites for the manned *Apollo* missions.

The Apollo Missions

The data from the unmanned probes and orbiters was overwhelming in its volume and detail. Some critics continued to argue: Why send human beings?

The manned missions clearly captured public attention, beginning with the Soviet *Vostok* series (1961–1963, including *Vostok 6,* which carried the first woman into space, Valentina V. Tereshkova) and the U.S. *Mercury* series (1961–1963). The *Mercury* series included two suborbital flights and the first U.S. manned flight in orbit, *Friendship 7,* commanded by John H. Glenn Jr., and launched on February 20, 1962. (On October 29, 1998, 77-year-old Senator John Glenn boarded the *Space Shuttle Discovery* and became the oldest man in space. He returned from the mission on November 7.)

The U.S. *Gemini* program came next, twelve two-man spaceflights launched between 1964 and 1967. The *Gemini* flights were intended very specifically to prepare astronauts for the manned lunar missions by testing their ability to maneuver spacecraft, to develop techniques for orbital rendezvous and docking with another vehicle—essential procedures for the subsequent *Apollo* Moon-landing program—and to endure long spaceflights. The eight-day *Gemini 5* mission, launched August 21, 1965, was the longest spaceflight to that time. The Soviets also developed larger launch vehicles and orbiters. *Voskhod 1,* launched on October 12, 1964, carried three "cosmonauts" (as the Russians called their astronauts) into Earth orbit.

The U.S. *Apollo* lunar missions not only made up the most complex space exploration program ever conceived, but were perhaps the most elaborate scientific and technological venture in the history of humankind. Today, even if we had the desire, we no longer have the launch vehicles required to bring astronauts to the moon.

According to the mission plan, a *Saturn V* multistage booster (rocket) would lift the 3-man Apollo spacecraft on its 2¼-day voyage to the moon, leaving behind the launch stages in pieces as it left the earth. After its journey, the small remaining piece of the initially launched craft would become a satellite of the moon, and the Lunar Module, with two men aboard, would separate from the orbiting Command Module and land on the moon. After a period of exploration on the lunar surface, the astronauts would climb back into the Lunar Module, lift off, and dock with the orbiting

Command Module, which would fire its rockets to leave its lunar orbit and carry the three astronauts back to Earth.

After several preliminary missions, including Earth- and Moon-orbital flights, *Apollo 11* was launched on July 16, 1969. On board were Neil A. Armstrong, Edwin E. "Buzz" Aldrin Jr., and Michael Collins. While in lunar orbit, Armstrong and Aldrin entered the Lunar Module, separated from the Command Module, and landed on the Moon, July 20, 8:17 P.M. Greenwich Mean Time.

Apollo 11 *Astronaut Edwin "Buzz" Aldrin on the Moon—a momentous snapshot by Neil Armstrong.*

(Image from NASA)

"That's one small step for [a] man," Armstrong declared, "one giant leap for mankind." And perhaps that sentence expressed the rationale for the effort, which went beyond strictly scientific objectives and spoke of and to the human spirit.

Not that science was neglected. During their stay of 21 hours and 36 minutes, the astronauts collected lunar soil and moon rocks and set up solar-wind (see Chapter 11, "Solar System Home Movie") and seismic experiments.

Apollo 12 landed on the Moon on November 19, but *Apollo 13,* launched April 11, 1970, had to be aborted because of an explosion, and the astronauts, as recounted in a recent film through their great skill, resourcefulness, and courage, barely escaped death. *Apollo 14* (launched January 31, 1971), *Apollo 15* (July 26, 1971), *Apollo 16* (April 16, 1972), and *Apollo 17* (December 7, 1972) all made successful lunar landings. Budgetary constraints, declining public interest, and the improving capabilities of unmanned missions eventually brought an end to the *Apollo* missions.

Close Encounter

The *Apollo* program was delayed and nearly aborted by an angry Congress after an electrical system failure ignited a catastrophic fire in the pure oxygen environment inside the *Apollo 1* spacecraft cabin during what should have been a relatively risk-free ground-based launchpad test on January 27, 1967. Astronauts Virgil I. (Gus) Grissom, Edward H. White, and Roger B. Chaffee were killed.

If any good can be said to have come from this tragedy, it is that NASA policymakers, engineers, and scientists refocused their attention on safety—something that had been at least partially compromised in the effort to beat the Russians.

Astro Byte

The *Viking* missions performed robotic experiments on the surface of Mars, including tests designed to detect the presence of simple life forms such as microbes. Great excitement was created when these tests yielded positive results. Subsequent analysis revealed that the results were false positives, resulting from chemical reactions in the Martian soil.

Planetary Probes

The lunar missions did not spell the end of unmanned exploration. Concurrent with the manned efforts, the United States launched an important series of planetary probes, beginning in the late 1950s and continuing today.

Mariners *and* Vikings

The U.S. *Mariner* program launched probes designed to make close approaches to Mars, Venus, and Mercury. *Mariner 2* (1962) and *Mariner 5* (1967) analyzed the atmosphere of Venus. *Mariner 4* (1964) and 6 and 7 (both 1969) photographed the Martian surface, as well as analyzed the planet's atmosphere. *Mariners 6* and 7 also used infrared instruments to create thermal maps of the Martian surface, and, in 1971, *Mariner 9,* in orbit around Mars, transmitted television pictures of the planetary surface. *Mariner 10,* launched in 1973, was the first spacecraft to make a close approach to Mercury and photograph its surface.

But even more exciting were the two *Viking* missions, launched in 1975. The following year, both made successful soft landings on Mars and conducted extensive analysis of the Martian surface.

The Viking 1 *spacecraft consisted of an orbiter and a lander. The orbiter studied the planet and its atmosphere from orbit. The lander actually touched down on the Martian surface, where it studied atmosphere, wind, and soil composition.*

(Image from NASA)

Pioneers *and* Voyagers

In the fall of 1958, *Pioneer 1* was launched into lunar orbit as a dress rehearsal for the planetary probes that followed. The rest of the *Pioneer* craft probed the inner solar system for planetary information, and *Pioneers 10* (1972) and 11 (1973) explored Jupiter and Saturn, the giants at the far end of our solar system. Later, in 1978, *Pioneer Venus 1* and *Pioneer Venus 2* orbited Venus to make surveys of that planet's lower atmosphere and, using radar imaging, penetrated thick gaseous clouds in order to reveal the spectacular and forbidding landscape below.

Magellan, Galileo, *and* Ulysses

More recent U.S. planetary probes have been increasingly ambitious. *Magellan* was launched in May 1989 and ultimately placed into orbit around Venus. Using high-resolution radar imaging, *Magellan* produced images of more than 90 percent of the planet, yielding more information about Venus than all other planetary missions combined. The spacecraft made a dramatic conclusion to its four-year mission when it was commanded to plunge into the planet's dense atmosphere on October 11, 1994, in order to gain data on the planet's atmosphere and on the performance of the spacecraft as it descended.

Astronomer's Notebook

Do you want to find out what *Galileo* is doing this week? Log onto the *Galileo* Web site at jpl.nasa.gov/galileo. The voyage of *Ulysses* can be followed on the Web at ulysses.jpl.nasa.gov.

On October 18, 1989, *Galileo* was launched on a journey to Jupiter and transmitted data on Venus, the earth's moon, and asteroids before reaching Jupiter on July 13, 1995, and dropping an atmospheric probe, which gathered data on Jupiter's atmosphere. After an extended analysis of the giant planet, *Galileo* began a mission to study Jupiter's moons, beginning with Europa. The so-called Galilean moons were discovered by the mission's namesake, Galileo Galilei in 1610.

The *Ulysses* probe was delivered into orbit by the shuttle *Discovery* on October 6, 1990. A joint project of NASA and the European Space Agency (ESA), *Ulysses* gathers solar data and studies interstellar space as well as the outer regions of our own solar system. Much of the spacecraft's instrumentation is designed to study x-rays and gamma rays of solar and cosmic origin.

Astronomer's Notebook

Check in on *Mars Pathfinder* pictures at the project's Web site, mars.jpl.nasa.gov.

Galileo *being deployed by the Space Shuttle, which lifted it into orbit. The* Galileo *mission sent the spacecraft to Jupiter and the planet's four largest moons.*

(Image from NASA)

Mars Observer, Surveyor, *and* Pathfinder

Mars Observer, launched on September 25, 1992, was to conduct extensive imaging work while orbiting Mars, but contact was lost with the spacecraft on August 22, 1993, as the satellite was establishing an orbit around the red planet. It is possible that a fuel tank exploded, destroying the spacecraft. *Mars Global Surveyor* was launched on November 7, 1996, and is continuing a long project of (among other things) detailed low-altitude mapping of the Martian surface. Unexpected oscillations in its solar panels while coming into a circular orbit around the planet caused the start of the major surface mapping program to be delayed by almost a year.

Although the *Global Surveyor* project is extraordinarily ambitious, the public may have been more excited by the mission of the *Mars Pathfinder*. The craft was launched on December 4, 1996, and landed on Mars the following summer, using a combination parachute and rocket-braking system, as well as an air bag system to ensure a soft, up-right landing. A "micro-rover" vehicle was deployed, which began transmitting extraordinary panoramic and close-up pictures of the Martian landscape.

It is little wonder that *Pathfinder* has caused such a stir. We've always been fascinated by Mars, which, of all the planets, seems most like Earth and has often been thought of as possibly harboring life—even civilization. Chapter 13, "So Close and Yet So Far: The Inner Planets," includes a discussion of some of these ideas, as well as a *Pathfinder* image from the Martian surface.

Artist's conception of Mars Pathfinder *landed and set up on Mars.*

(Image from NASA)

A More Distant Voyager

The *Cassini-Huygens* mission, a joint undertaking of NASA, the European Space Agency (ESA), the Italian Space Agency (ASI), and several other organizations, was sent on its way October 15, 1997, to investigate Saturn as well as Titan (one of Saturn's moons). Some scientists believe that Titan might support life or, at least, offer conditions in which life could develop. The mission was named *Cassini,* in honor of the seventeenth-century French-Italian astronomer Jean Dominique Cassini, who discovered the prominent gap in Saturn's main rings; and *Huygens,* after the Dutch scientist Christiaan Huygens, who discovered the Saturn moon Titan in 1655, as well as the rings of Saturn. It recently transmitted dramatic images of Jupiter as it sped past on its way to Saturn.

Space Shuttles and Space Stations

The flight of *Apollo 17* in 1972 was the last manned lunar mission, but not the end of the U.S. manned space program. On April 12, 1981, the first Space Shuttle, a reusable spacecraft (the previous space capsules had been one-shot vehicles) was launched. The Shuttle was intended to transport personnel and cargo back and forth from a manned space station, planned for Earth orbit. So far, Shuttle missions have carried out a variety of experiments, have delivered satellites into orbit, and have even repaired and upgraded the *Hubble Space Telescope* (see Chapter 5, "The Art of Collecting Light (with a Telescope)"). In 1999, it started its most ambitious mission: the construction of an international space station, to be built in conjunction with Russia, Japan, and the European Space Agency (ESA). The realities of politics and economics mean that, in the twenty-first century, countries will be much more likely to cooperate in the race to space.

Close Encounter

The Shuttle program suffered a tragic setback on January 28, 1986, when Shuttle *Challenger* exploded 73 seconds after lift-off, killing its entire 7-member crew, including a high-school teacher, the first ordinary citizen to fly aboard the craft. The accident, caused by flawed booster rocket seams (O-rings), resulted in the suspension of flights until the problem could be corrected. The celebrated physicist Richard Feynman was on the panel that investigated the *Challenger* disaster, and demonstrated with graphic simplicity how cold O-rings—he placed a small O-ring in ice water—would lose their required flexibility. Shuttle flights were resumed in 1988 with the three remaining Shuttle craft: *Discovery, Columbia,* and *Atlantis.*

Skylab

On May 14, 1973, the United States launched its first orbiting space station, *Skylab,* designed to accommodate teams of astronauts to conduct a variety of experiments in geography, engineering, Earth resources, and biomedicine. Such work was carried out during 1973 and the beginning of 1974. In 1974, the craft's orbit was adjusted to an altitude believed sufficient to keep *Skylab* in orbit until 1983, when a visit from the Space Shuttle was contemplated. At that time, the orbit would again be adjusted. Unfortunately, *Skylab* wandered out of orbit prematurely, in June 1978, and ultimately disintegrated and fell into the Indian Ocean on July 11, 1979.

The Demise of Mir

After several years of mishaps and close calls, the decision was made to discontinue use of the *Mir Space Station* and to concentrate on the collaboration with the international community on the International Space Station. Early in 2001, the *Mir Space Station* was de-orbited and allowed to crash into the South Pacific Ocean. At 12:55 A.M. EST on March 22, 2001 (05:55 Greenwich Mean Time), the *Mir* station was 50 km (31 mi) above Earth's surface. At 12:58 A.M. EST (05:58 GMT, 8:55 A.M. Moscow time) fragments of the station hit the ocean.

Alix Bowles, Project Coordinator for MirReentry.com watched the space station break into pieces as it streaked through the sky from a beach in Fiji. "It was a stunning blue steak followed by a sonic boom," he said. "The pieces had a blue incandescence to them. There was something very peaceful about it," he added.

In its later years, the *Mir* station had become the butt of late-night television jokes, but, in fact, it was a productive scientific instrument and an important test bed for technology used on the International Space Station. *Mir* lasted years far longer than its designers had envisioned.

International Space Station: The Latest

After years of planning and design, the largest, most expensive international collaboration in space is now under construction. As of early 2001, the heart of the International Space Station (ISS) had been assembled, and the first crew exchange had taken place after four and a half months. The international effort now includes the United States, Canada, Japan, Russia, Belgium, Denmark, France, Germany, Italy, the Netherlands, Norway, Spain, Sweden, Switzerland, and the United Kingdom. As expected, the project has experienced growing pains. There were concerns early on when the power-generating solar panels failed to deploy properly, but with some human intervention, they were coaxed into place.

The projected completion date of the ISS is 2006, and budgetary constraints have recently reduced the size and cost of the space station. It is estimated that the cost of the space station is $8 per year per citizen—the cost of a movie (a matinee, if you live in New York City).

The International Space Station.

(Image from NASA)

The Least You Need to Know

➤ While rockets and space travel had been the subject of centuries of speculation, three men at the end of the nineteenth and beginning of the twentieth centuries—Konstantin Eduardovich Tsiolkovsky, Robert H. Goddard, and Hermann Oberth—were most responsible for laying the practical foundations of modern rocketry.

➤ After the Soviets launched *Sputnik* in 1957, the United States and the USSR became rivals in an often frenetic space race.

➤ Manned spaceflight became the goal of both the United States and USSR, despite protests from some in the scientific community, who believed that unmanned missions were safer, cheaper, and of more scientific value. These debates continue to this day.

➤ The Soviets orbited the first man in space, Yuri Gagarin, on April 12, 1961, but the United States was the first—and so far the only—nation to land men on the Moon, beginning with Neil Armstrong and Edwin "Buzz" Aldrin on July 20, 1969.

➤ The manned space program did not bring the unmanned program to an end. NASA has launched, and will continue to launch, a number of exciting planetary probes to Mars, Venus, Jupiter, and Saturn.

Part 3

A Walk Around the Block

This part is an extended tour of the solar system, beginning with our nearest neighbor, the Moon. After some lunar exploration, we venture to an overview of how the solar system was born and has developed (Chapter 11) and how it looks and behaves today (Chapter 12). We even find time to linger among the comets and the asteroids.

The remaining three chapters of this part are devoted to the planets and the moons that cruise our "neighborhood." First, we investigate the planets most like our own, the terrestrials Mercury, Venus, and Mars. Then we go on to the distant jovian worlds—the outer giants Uranus, Neptune, Jupiter, and Saturn. Finally, we take a closer look at the moons and rings of the jovians—and at the last planet discovered in our solar system, Pluto, which, as we'll see, seems more like a moon than a planet.

Chapter 10

The Moon: Our Closest Neighbor

> **In This Chapter**
>
> ➤ Galileo observes the moon
>
> ➤ How to observe the moon
>
> ➤ Peeking at the far side of the moon
>
> ➤ How the moon affects our tides
>
> ➤ Lunar geology
>
> ➤ Who needs the moon?
>
> ➤ Some theories on how the moon was formed

It has been more than 30 years since Neil Armstrong stepped from the *Apollo 11* Lunar Lander onto the surface of the moon. The moon is still the only celestial body other than the Earth where humans have stood. But why did we go there?

Columbus had sailed to a place promising great riches to exploit. The moon, in contrast, was and is a lifeless orb, devoid of water (mostly!), air, sound, weather, trees, or grass. While Columbus's voyages had their tight moments (he once had to "predict" a solar eclipse to impress the natives), on his return from the fourth and final voyage to the New World, Columbus announced that he had indeed found an Earthly paradise.

But the moon?

From the pictures we've all seen, the lunar landscape is one of rock, dust, and desolation. And although the astronauts were seen skipping across its surface, they were clearly happy to return to mother Earth. Why on earth did our nation expend such effort, treasure, and risk to send astronauts to the lunar surface? What have we learned from the information that we brought back? Why does the moon so fascinate us? We'll explore these issues and more in this chapter.

What If We Had No Moon?

It seems like a reasonable question to ask. What if we had no moon? Would it matter? What has the moon done for me lately?

It turns out that the presence of such a large moon as we have is unusual for a terrestrial planet. Mercury and Venus have no moons, and Mars has two tiny moons, Phobos and Deimos. To have a moon roughly ⅓ the size of the planet is unique in the inner solar system. Our Moon, for example, is as large as some of the moons of the giant gas planets in the outer solar system. If there were no moon, we would have no ocean tides, and the rotation rate of the earth would not have slowed to its current 24 hours. It is thought that early in the life of the Earth, it rotated once every 6 hours. The moon also appears to stabilize the rotational axis of the Earth. The Moon, in periodically blocking the light from the Sun's photosphere gives us a view of the outer layers of the Sun's atmosphere, and it also gave early astronomers clues to the distribution of objects in the solar system.

Close Encounter

Did you ever wonder why the sun and the moon are the same size in the sky? In actual physical size, the sun dwarfs the moon, but the sun is so much farther away, that the two appear the same size. Try this: Hold a dime and a quarter in front of your face. Now move them back and forth until they appear to be the same size. Which do you have to hold farther away for this to be true? The bigger one, of course! We happen to be on the earth at a time when the moon exactly blocks the light from the sun's photosphere during a solar eclipse. As the moon slowly drifts away from the earth, it will get smaller and smaller in the sky, and people will be right when they eventually say, "Solar eclipses just aren't what they used to be."

Lunar Looking

While the world greeted Jules Verne's 1865 book *De la Terre à la Lune* (translated in 1873 as *From the Earth to the Moon*) with acclaim and wonder, it was hardly the first fictional speculation about a voyage to our nearest cosmic neighbor. The Greek satirist Lucian had written about such a flight as early as the second century C.E. and the moon, our constant companion, has always been an object of intense fascination. Its reflected silvery glow bathes the Earth with romance and mystery. Its changing face, as it travels through its monthly cycle, has always commanded our attention, as have its peculiarly human qualities: Unlike the stars, it is pocked, mottled, imperfect. Almost all cultures at all times have seen some sort of face or figure in the features of the moon. Only rather recently have we realized just how important the moon has been in the evolution of our planet.

The sun is so intensely brilliant that to gaze at it is to go blind. But the moon, coincidentally the same size in the sky as the sun, shines with harmless reflected light that invites us to gaze and gaze—to become lunatics.

What Galileo Saw

It is possible to observe many features of the moon without a telescope. One of the first things you should try is to track its daily motion against the background stars (described in Chapter 1). Since the moon travels around the earth (360 degrees) in 27.3 days, it will travel through about 13 degrees in 24 hours, or about half a degree (its diameter) every hour.

Galileo was the first person to look at the moon through a telescope; indeed, its mottled gray face was one of the first celestial objects on which he trained his new instrument in 1609.

What he saw conflicted with existing theories that the surface was glassy smooth; it was instead rough and mountainous. He closely studied the *terminator*, or the boundary separating day and night, and noted the shining tops of mountains. Using simple geometry, he calculated the height of some of the mountains based on the angle of the sun and the estimated length of shadows cast. Galileo overestimated the height of the lunar mountains he observed; but he did conclude rightly that their altitudes were comparable to Earthly peaks.

Noticing mountains and craters on the moon was important, because it helped Galileo conclude that the moon was fundamentally not all that different from the earth. It had mountains, valleys, and it even had what were called seas—in Latin, *maria,*

Star Words

The **terminator** is the boundary separating light from dark, the daytime from nighttime hemispheres of the moon (or other planetary or lunar body). The terminator marches across the lunar surface at a very slow rate, crossing the entire surface of the moon once every 27.3 days. Now that's a very long day.

Star Words

Maria (pronounced "MAH-ree-uh") is the plural of mare (pronounced "mar-ay"), Latin for "sea." Maria are dark gray plains on the lunar surface, which to earlier observers resembled bodies of water seen at a distance.

151

though there is no indication that Galileo or anyone else maintained after telescopic observations that the *maria* were water-filled oceans. Conten-ding that the moon re-sembled the earth in 1609 was not a small thing. This statement implied that there was nothing supernatural or special about the moon or perhaps the planets and the stars, either. Followed to its conclusion, the observation implied that there was per-haps nothing divine or extraordinary about the earth itself. The earth was a body in space, like the moon and the other planets.

Astronomer's Notebook

Go outside at the next full moon in the early evening (the full moon will always rise at sunset!). You should be able to easily see several of the "seas" without a telescope. The large gray area that takes up most of the left half of the moon consists of the Mare Imbrium and the Oceanus Procellarum. Far in the upper right corner is an oval-shaped re-gion, Mare Crisium. And just below Mare Crisium and a little to the left is Mare Tranquilitatis, or more familiarly, the Sea of Tranquility, where the Apollo 11 astronauts walked. Can you see any footprints?

What You Can See

Even if you don't have a telescope, there are some very interesting lunar observations you can make. Have you ever thought that the moon looks bigger when it's closer to the horizon? It's just an optical illusion, but you can test it out. The angular size of the moon is sur-prisingly small. A circular piece of paper just about 0.2 inches in diameter held at arms' length should cover the moon. At the next full moon, cut out a little disk of that size and prove to yourself that the moon stays the same size as it rises high into the sky.

The telescope through which Galileo Galilei made his remarkable lunar observations was a brand-new and very rare instrument in 1609; but you can easily sur-pass the quality of his observations with even a mod-est amateur instrument.

Why is it so exciting to point your telescope at the moon?

Because no other celestial object is so close to us. Being so close, the moon provides the most detailed images of an extraterrestrial geography that you will ever see through your own telescope.

When should you look at the moon?

The easy answer is: anytime the sky is reasonably clear. But if you're thinking that you should always wait until the moon is full, think again. When is the best time to view a rugged Earthly landscape at its most dramatic? When the sun is low, early in the morning or late in the afternoon, and the light rakes across the earth, so that shadows are cast long and all stands in bold relief.

The same holds true for the moon. When you can see the sunrise or sunset line (the terminator), and the moon is not so full as to be blindingly bright, *that* is when the topography of the moon will leap out at you most vividly. This characteristic means that you'll get some very satisfying viewing when the moon is at one of its crescent phases, and probably not at its full phase.

But most important, get out and set up your telescope (or gaze at the moon) as often as possible. Take the time to observe the moon through all of its phases.

When the moon is about three or four days "old," Mare Crisium and other vivid features—including the prominent craters Burckhardt and Geminus—become dramatically visible, assuming it's a clear night. You can also begin to see Mare Tranquilitatis, the Sea of Tranquility.

At day seven, the moon is at its first quarter. At this time, mountains and craters are most dramatically visible. Indeed, this is the optimum night for looking at lunar features in their most deeply shadowed relief.

As the moon enters its waxing gibbous beyond first quarter phase, its full, bright light is cheerful, but so bright that it actually becomes more difficult to make out sharp details on the lunar surface. An inexpensive "moon filter" or variable polarizing filter fitted to your telescope can help increase contrast on the bright lunar surface. As the moon verges on full, we do get great views of the eastern *maria*, the lunar plains.

Past day 14, the moon begins to wane as the sunset terminator moves slowly across the lunar landscape. At about day 22, the Apennine Mountains are clearly visible. It was these mountains that Galileo studied most intensely, attempting to judge their height by the shadows they cast.

During the late waning phase of the moon, moonrise comes later and later at night, as the moon gradually catches up with the sun in the sky. By the time the moon passes day 26, it is nothing but a thin crescent of light present in the predawn sky. The new moon follows, and as the moon overtakes the sun, the crescent reappears (on the other side of the moon at sunset), and it begins to wax again.

Here are some cold, hard facts about a cold, hard place. The moon is Earth's only natural satellite, and in fact a very large satellite for a planet as small as the earth. The planet Mercury is only slightly larger than the moon. The mean distance between the earth and moon, as it orbits our planet from west to east, is 239,900 miles (386,239 km). The moon is less than one-third the size of the earth, with a diameter of about 2,160 miles (3,476 km) at its equator. Moreover, it is much less massive and less dense than the earth—$\frac{1}{80}$ as massive, with a density of 3.34 g/cm^3, in contrast to 5.52 g/cm^3 for the earth. If the earth were the size of your head, the orbiting moon would be a tennis ball 30 feet away.

Because the moon is so much less massive than the earth, and about a third as big, its surface gravity is about one-sixth that of our planet. That's why the *Apollo* astronauts could skip and jump like they did, even wearing those heavy space suits. If you weigh 160 pounds on the earth's surface, you would weigh only 27 pounds on the moon. This apparent change would give you the feeling of having great strength, since your body's muscles are accustomed to lifting and carrying six times the load that burdens them on the moon. Of course, your mass—how much matter is in you—does not change. If your mass is 60 kilograms (kg) on the earth, it will still be 60 kg on the moon.

As we learned in Chapter 3 "The Unexplained Motions of the Heavens," the moon is in a *synchronous orbit* around the earth; that is, it rotates once on its axis every 27.3 days, which is the same time that it takes to complete one orbit around the earth. Thus synchronized, we see only one side of the moon (except for the tantalizing peek at the far side that libration affords).

Astro Byte

The moon is much smaller in the night sky than you might think. Cut a small circle (about 1/5 inch in diameter) from a piece of paper. Hold that paper out at arm's length. That is how large the moon is in the night sky. And don't let anyone tell you that a harvest moon (a full moon in September) is any larger than any other full moon. The moon may be lower in the sky in autumn (as is the sun), so it may look larger, but its size is unchanged.

Star Words

A celestial object is in **synchronous orbit** when its period of rotation is exactly equal to its average orbital period; the moon, in synchronous orbit of the earth, presents only one face to the earth.

It's a Moon!

No one can say with certainty how the moon was formed (we weren't there!), but four major theories have been advanced.

A Daughter?

The oldest of the four theories speculates that the moon was originally part of the earth, and was somehow spun off a rapidly spinning, partially molten, newly forming planet.

Once prevalent, this theory (sometimes referred to as the fission theory) has largely been rejected, because it does not explain how the proto-Earth could have been spinning with sufficient velocity to eject the material that became the moon. Moreover, it is highly unlikely that such an ejection would have put the moon into a stable Earth orbit.

A Sister?

Another theory holds that the moon formed separately near the earth from the same material that formed the earth. In effect, the earth and the moon formed as a double-planet system.

This theory seemed quite plausible until lunar rock samples were recovered, revealing that the moon differs from Earth not only in density, but in composition. If the two bodies had formed out of essentially the same stuff, why would their compositions be so different?

A Captive?

A third theory suggests that the moon was formed independently and far from the earth, but was later

captured by the earth's gravitational pull when it
came too close.

This theory can account for the differences in
composition between the earth and the moon, but
it does not explain how the earth could have gravitationally captured such a large moon. Indeed,
attempts to model this scenario with computer
simulations have failed. Moreover, while the theory accounts for some of the chemical *differences*
between the earth and moon, it does not explain
the many chemical *similarities* that also exist.

A Fender Bender?

The favored theory today combines elements of
the daughter theory and the capture theory in
something called an impact theory. Most astronomers now believe that a very large object,
roughly the size of Mars, collided with the earth

Astronomer's Notebook

Ever felt like you've had a long
day? Well, none of your days are
as long as a day on the moon.
Because the moon rotates once
every 27.3 days (as it orbits the
earth once), a day on the moon
is 27.3 Earth days long. But then
again, you've got a nice long
night to rest up.

when it was still molten and forming. Assuming the impact was a glancing one, it is
suggested that shrapnel from the earth and the remnant of the other planetesimal (a
planet in an early stage of formation) were ejected and then slowly coalesced into a
stable orbit that formed the moon.

This model is also popular because it can explain some unusual aspects of the earth
(the "tip" of its rotational axis, perhaps) and the moon. In the impact model, it is further theorized that most of the iron core of the Mars-sized object would have been
left behind on the earth, eventually to become part of the earth's core, while the material that would coalesce into the moon acquired little of this metallic core. This
model can explain why the earth and moon share similar mantles (outer layers), but
apparently differ in core composition.

Give and Take

Remember Newton's law of gravitation from Chapter 4, "Astronomy Reborn:
1543–1687"? Newton proposed that every object with mass exerts a gravitational pull
or force on every other object with mass in the universe. Well, the earth is much
more (80× more) massive than the moon, which is why the moon orbits us, and not
we it. (If you want to get technical, we both actually orbit an imaginary point called
the center of mass.) However, the moon *is* sufficiently massive to make the effects of
its gravitational field felt on the earth.

Anyone who lives near the ocean is familiar with tides. Coastal areas experience
2 high and 2 low tides within any 24-hour period. The difference between high and
low tides is variable, but, out in the open ocean, the difference is somewhat more

than 3 feet. If you've ever lifted a large bucket of water, you know how heavy water is. Imagine the forces required to raise the level of an entire ocean 3 or more feet!

What force can accomplish this?

The *tidal* force of gravity exerted by the moon on the earth and its oceans.

The moon and the earth mutually pull on each other; the earth's gravity keeping the moon in its orbit, the moon's gravity causing a small deformity in the earth's shape. This deformity results because the moon does not pull equally on all parts of the earth. It exerts more force on parts of the earth that are closer, and less force on parts of the earth that are farther away. Just as Newton told us: Gravitational forces depend on distance. These differential or *tidal forces* are the cause of the earth's slightly distorted shape—it's ovoid rather than a perfect sphere—and they also make the oceans flow to two locations on the earth: directly below the moon, and on the opposite side. This flow causes the oceans to be deeper at these two locations, which are known as the *tidal bulges*. The entire Earth is pulled by the moon into a somewhat elongated—football—shape, but the oceans, being less rigid than the earth, undergo the greatest degree of deformity.

Star Words

A **tidal bulge** is any deformation of a celestial body caused by the gravitational force of another celestial body. The moon creates an elongation of the earth and its oceans—a tidal bulge.

Interestingly, the side of the earth farthest from the moon at any given time also exhibits a tidal bulge. This is because the Earth experiences a stronger gravitaional pull than the ocean on top of it, and the Earth is "pulled away" from the ocean on that side. As the Earth rotates beneath the slower-moving moon, the forces exerted on the water cause high and low tides to move across the face of the earth.

The tides of largest range are the spring tides, which occur at new moon, when the moon and the sun are in the same direction, and at full moon, when they are in opposite directions. The tides of smallest range are the neap tides, which occur when the sun and the moon are at 90 degrees to one another in the sky. Tides affect us every day, of course, especially if you happen to be a sailor or a fisherman. But even if you live high and dry in Kansas or Nebraska, say, tides (and the moon) still affect you. Every day, the earth is spinning a little slower on its axis because of the moon.

The earth's rotation is slowing down at a rate that increases the length of a day by approximately 2 milliseconds ($\frac{2}{1,000}$ of a second) every century. Over millions of years, though, this slowing effect adds up. Five hundred million years ago, a day was a little over 21 hours long, and a year (1 orbit of the sun) was packed with 410 days. When a planetesimal plowed into the earth early in the history of the solar system, it was rotating once every 6 hours. (And you think there aren't enough hours in the day now!)

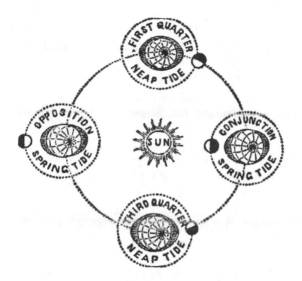

The tides illustrated: There are two high and two low tides each day at any given place, but they occur at times that change from day to day.

(Image from arttoday.com*)*

How does this happen? Well, let's think again of why tides occur. The moon's gravity causes two bulges to form in the earth's oceans, and the earth rotates (once every 24 hours) beneath that bulge. As the earth spins, friction between it and the oceans tends to pull the high tide ahead, so that the "bulge" actually leads the position of the moon overhead.

With the ocean's bulge thus slightly ahead of the moon's position, the moon's gravity exerts a force that tends to slow rotation. Eventually, the earth's rotation will slow sufficiently to become synchronized with the orbit of the moon around the earth. When that happens, the moon will always be above the same point on Earth, and the earth's rotation period will have slowed (billions of years from now) from its present 24 hours to 47 days.

But that is only half of the picture. The earth can't be slowing down without some-thing else speeding up (as a result of one of the fundamental conservation laws of physics). What's speeding up? *The moon.* And what does that mean? That it's spiraling away slowly, and getting smaller and smaller in the sky.

Green Cheese?

Even with the naked eye, the moon doesn't look particularly green. And Neil Arm-strong confirmed that the surface of the moon was more dusty than cheesy. On any night that the moon is visible, the large, dark *maria* are clearly visible. These are vast plains created by lava spread during a period of the moon's evolution marked by in-tense volcanic activity. The lighter areas visible to the naked eye are called highlands. Generally, the highlands represent the moon's surface layer, its *crust,* while the *maria* consist of much denser rock representative of the moon's lower layer, its *mantle.* The surface rock is fine-grained, as was made dramatically apparent by the image of the

first human footprint on the moon. The *mare* resemble terrestrial *basalt*, created by molten mantle material that, through volcanic activity, swelled through the crust.

The mass of the moon is insufficient for it to have held on to its atmosphere. As the sun heated up the molecules and atoms in whatever thin atmosphere the moon may have once had, they drifted away into space. With no atmosphere, the moon has no weather, no erosion—other than what is caused by asteroid impacts—and no life. While it was thought that the moon had absolutely no water, recent robotic lunar missions have shown that there may be water (in the form of ice) in the permanent shadows of the polar craters.

This footprint was left in the lunar dust by an Apollo 11 astronaut. Unless a stray meteoroid impact obliterates it, the print will last for millions of years on the waterless, windless moon.

(Image from NASA)

A Pocked Face

Look at the moon through even the most modest of telescopes—as Galileo did—and you are impressed first and foremost by the *craters* that pock its surface.

Most craters are the result of asteroid and meteoroid impacts. Only about a hundred craters have been identified on Earth, but the moon has thousands, great and small. Was the moon just unlucky? No. Many meteoroids that approach Earth burn up in our atmosphere before they strike ground. And the traces of those that do strike the ground are gradually covered by the effects of water and wind erosion as well as by plate tectonics. Without an atmosphere, the moon has been vulnerable to whatever comes its way, and preserves a nearly perfect record of every impact it has ever suffered.

Meteoroid collisions release terrific amounts of energy. Upon impact, heat is generated, melting and deforming the surface rock, while pushing rock up and out and creating an *ejecta blanket* of debris, including large boulders and dust. It is this ejected material that covers the lunar surface.

It is believed that the rate of meteoroid impact with the moon (and with other objects in the solar system) was once much higher than it is now. The rate dropped sharply about 3.9 billion years ago—at the end of the period in which it is believed that the planets of the solar system were formed—and, some time later, lunar volcanic activity filled the largest craters with lava, giving many of them a smooth-floored appearance.

And What's Inside?

Geologically, the moon is apparently as dead as it is biologically. Astronauts have left seismic instruments on the lunar surface, which have recorded only the slightest seismic activity, barely perceptible, in contrast to the exciting (and sometimes terribly destructive) seismic activity common on Earth and some other bodies in the solar system.

It is believed, then, that the interior of the moon is uniformly dense, poor in heavy elements (such as iron) but high in silicates. The *core* of the moon, about 250 miles (402 km) in diameter, may be partially molten. Around this core is probably an inner mantle, perhaps 300 miles (483 km) thick, consisting of semisolid rock, and around this layer, a solid outer mantle some 550 miles (885 km) thick. The lunar crust is of variable thickness, ranging from 40 to 90 miles (64 to 145 km) or so.

The moon is responsible for everything from the earth's tides to the length of our day, and perhaps the presence of seasons. Most astronomers think that the moon is with us today because of a gigantic collision early in the life of the solar system. The moon's gravity pulls tides across the earth's surface, and its presence has slowed the rotation of the earth from a frenetic 6 hours to our current 24. Think of that next time you see the moon shining peacefully over your head.

Star Words

Crater, the Latin word for "bowl," refers to the shape of depressions in the moon or other celestial objects created (mostly) by meteoroid impacts. An **ejecta blanket** is the debris displaced by a meteoroid impact.

Astro Byte

The lunar crust is thickest on the far side of the moon. In the moon's formative period, while it was in the process of solidifying and cooling (perhaps after a stupendous collision), the earth's gravity pulled the mantle more on the near side than on the far side, leaving more crust on the far side and increasing volcanic activity on the near side giving us the "maria."

The Least You Need to Know

➤ The features of the lunar surface are at their most dramatic when the moon is in one of its crescent phases; however, there are fascinating observations to make during all lunar phases.

➤ The earth's gravitational field holds the moon in orbit, but the moon's gravitation also profoundly influences the earth, creating ocean tides and slowing our days ever so slightly over the millenia.

➤ The moon is biologically dead and geologically inactive.

➤ There is abundant evidence that the moon was formed as a result of a collision between the earth and another planet-size object very early in the history of the solar system.

Chapter 11

Solar System Home Movie

In This Chapter

➤ How astronomers "read" the history of the solar system

➤ A theory of solar system origin

➤ The age of the solar system

➤ How the solar system was born and grew: the nebular hypothesis

➤ Nebular hypothesis revised: the condensation theory

➤ Why the planets differ from one another

➤ The death of the solar system

In the coming chapters of this section, we will explore the solar system from the familiar to the unfamiliar—from our own back yard, to the most distant reaches of the solar neighborhood. Before we begin this journey, though, we take a moment to think about where we came from, and where we are going; a little home movie, if you will.

One of the most difficult things about understanding the origins of our solar system, the sun, and the collection of planets that orbit it, is that it has been around for a very long time. The problem is akin to looking at a middle-aged man and being asked to describe the moment of his birth. Well, we might be able to watch other births happen, and assume that his birth was much the same. But no one was present to watch the birth of our solar system, and so uncovering its beginnings has required some serious sleuthing. Yet astronomers have accepted the challenge, and they believe they have made more than a promising beginning in understanding the history of our solar system. This chapter presents the prominent theories about how the solar system was born and developed—and how, in time, it will die.

Solar System History

Historians of human events generally enjoy two advantages over would-be historians of the solar system. Those who chronicle human history often have records, even eyewitness reports, and they have the availability of precedent events and subsequent events. For example, a historian of the American Civil War not only has a wealth of eyewitness accounts to draw on, but may also look to civil wars both before and after 1861 to 1865 to help explain the War between the States. Comparing and contrasting the American Civil War to the English civil wars of the seventeenth century or the Russian Civil War of the twentieth may help illuminate analysis and make explanations clearer.

The solar system historian lacks both witnesses and precedents. But she has several advantages as well. Human history is complicated by the infinite depths of the human mind. But if we understand the fundamental laws of physics, and make good observations of our solar system today, a recounting of the early history of the solar system should be within our grasp.

The Biggest Problem: We Weren't There

Of course, there was no one around to record the series of events that created the solar system. But there are a few fragments that survive from those early moments, like the years-old crumbs from behind the sofa, that give us clues to how the planets took shape around the youthful sun. The most important clues to the origin of the solar system are to be found not in the sun and planets, but in those untouched smaller fragments: the asteroids, meteoroids, and some of the planetary moons (including our own). These objects make up the incidental matter, the debris of the solar system, if you will.

This solar system debris, though rocky, is not mute. As the English Romantic poet William Wordsworth (1770–1850) wrote of hearing "sermons in stones," so modern astronomers have extracted eloquent wisdom of a different kind from meteorites as well as "moon rocks."

In Chapter 7, "Over the Rainbow,", we described how atoms are made up of three basic particles, protons and neutrons in the nucleus, orbited by electrons. Most elements exist in different atomic forms, which, while identical in their chemical properties, differ in

Star Words

Radioactive decay is the natural process whereby a specific atom or isotope is converted into another specific atom or isotope at a constant and known rate. When an atomic nucleus is overloaded with neutrons, it will be a "heavy" isotope. By measuring the ratios of isotopes of certain elements in meteorite samples, it is possible to determine the age of the sample.

the number of neutral particles (neutrons) in the nucleus. Deuterium and Tritium are famous radioactive isotopes of the more familiar hydrogen. For a single element, these atoms are called isotopes. Through a natural process called *radioactive decay,* a specific isotope of one atom is converted into another isotope at a constant and known rate, often over many millions of years, depending on the element and isotope involved. Using a device called a mass spectrometer, scientists can identify the "daughter" atoms formed from the "parent" atoms in a sample, such as a meteorite. If the decay rate of the elements in the sample is known, then the ratio of daughter atoms to parent atoms (called isotopic ratios), as revealed by the instrument, betrays the age of the sample.

The results of this dating process have been remarkably consistent. Most meteors and moon rocks (which are the only "bits" of the solar system other than the earth we have been able to study exhaustively) are from between 4.4 and 4.6 billion years old, which has led scientists to conclude with a high degree of confidence that the age of the solar system is about 4.6 billion years.

Astro Byte

A familiar example of how radioactive decay is used to date objects much younger than meteorites is carbon dating. Scientists use the fact that the ratio of two carbon isotopes (Carbon-14 and Carbon-12) has a specific value in living matter. Once the organism dies, the ratio changes at a known rate. As a result, the exact ratio of these two isotopes in something containing carbon (a piece of wood, a piece of parchment, the Shroud of Turin) will give its approximate age.

What Do We Really Know About the Solar System?

In a very real sense, then, we do have—in meteorites and moon rocks—"witnesses" to the creation of the solar system. These geological remnants are relatively unchanged from the time the solar system was born. But how do we make up for an absence of precedents from which to draw potentially illuminating analogies? Why can't we just go find another planetary system forming (around a star younger than the sun) and draw our analogies from it? Well, that has been one of the main goals of astronomers in the past decade or so. In fact, NASA has defined one of its primary missions in terms of this search, called the Origins program. We are just now starting to see the results of these searches. The *Hubble Space Telescope,* in particular, has given us tantalizing clues about the formation of planetary systems. Around the star Beta Pictoris, astonomers have imaged a disk of material larger than the orbit of the most distant planet in our solar system, Pluto. Are the inner reaches of this disk even now taking shape as planets around that star?

The truth is, even with the best instruments that we have today, we can still learn a lot more about how our solar system formed by looking closer to home. There are a number of fundamental things that we know about our solar system, and any explanation that we come up with must, at the very least, account for what we observe.

Here are some undeniable facts that the last 300 years of planetary exploration have given us:

➤ Most of the planets in the solar system rotate on their axis in the same direction as they orbit the sun (counterclockwise as seen from the North Pole of Earth), and their moons orbit around them in the same direction.

➤ The planets in the inner reaches of the solar system are rocky and bunched together, and those in the outer part are gaseous and widely spaced.

➤ Most of the planets (with the exception of Pluto) orbit the sun in elliptical paths that are very nearly circles.

➤ Except for the innermost planet (Mercury) and the outermost (Pluto), the planets orbit in approximately the same plane (near the ecliptic), and they all orbit in the same direction.

➤ Asteroids and comets are very old, and are located in particular places in the solar system. Comets are found in the Kuiper Belt and Oort Cloud, and asteroids in the asteroid belt between Mars and Jupiter.

In addition, it is clear that the asteroids we have examined are some of the oldest unchanged objects in the solar system, and that comets travel in highly elliptical orbits, originating in the far reaches of the solar system. The most important conclusion we can draw from these observations is that the solar system appears to be fundamentally orderly rather than random. It doesn't appear that the sun formed first, and then gradually captured its nine planets from surrounding space.

Star Words

The term **nebula** has several applications in astronomy and is used most generally to describe any fuzzy patch seen in the sky. Nebulae are often (though not always) vast clouds of dust and gas.

Although there are important exceptions, the "counterclockwise" (as viewed from above the North Pole of the Earth) aspect of so many properties in the solar system suggests that the planets fragmented and formed from a large rotating cloud of material. That the orbits of the remaining planets are very nearly circular suggests that the solar system has settled down, as it were. Any planets or planetesimals that were on highly elliptical orbits have been cleared out in the last 4.6 billion years. The inclined, markedly elliptical orbit of Pluto is one of the arguments for its being an escaped moon from an outer planet. The physical differences in planets that are related to their distances from the sun (Chapter 12, "Solar System Family Snapshot,") suggests that the sun influenced the formation of the planets; that is, the sun must have formed first.

From Contraction to Condensation

In 1755, the great German philosopher Immanuel Kant (1724–1804) theorized that the solar system had begun as a *nebula*—a cloud of dust and gas—that slowly rotated, gradually contracting until it became flattened into a spinning disk that variously co-alesced into the sun and planets.

Later, in 1796, the French astronomer and mathematician Pierre-Simon Laplace (1749–1827) suggested a similar hypothesis, though he thought the planets formed before the sun.

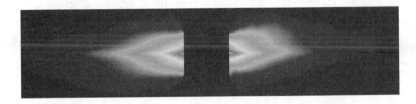

Beta Pictoris is believed to be a dusty disk surrounding a young star. The presence of this disk, imaged by the Hubble Space Telescope, *supports the nebular hypothesis.*

(Image from Al Schultz, CSC/STScI, and NASA)

Angular Momentum Explained

Most importantly, Laplace introduced conservation of *angular momentum* to the discussion of planetary formation. He demonstrated mathematically that the *solar nebula*—the gaseous mass that would become the solar system—would spin faster as it contracted. Anyone who has watched an ice skater spinning knows this is true. As a skater pulls in his arms, bringing his mass closer to his axis of rotation, he will spin faster. If he were to put his arms out at his side, his rotation would slow. Newton described how all objects with mass were mutually attracted. As the cloud of gas that eventually formed the solar system started to collapse, it would have to rotate faster and faster to conserve angular momentum. And, as the speed of rotation increased, the shape of the solar nebula would change, becoming the pancake-like disk Kant had first pictured. Think of that the next time you watch the local pizza maker throw dough in the air, making it spin, flatten, and strech all at once.

Laplace theorized that as the spinning disk contracted, it would form concentric rings, each of which would clump together into a "protoplanet"

Star Words

Angular momentum can be thought of as the rotating version of linear momentum. Any object that rotates possesses this property. If an object gets smaller, it must rotate faster to conserve angular momentum.

(a sort of embryonic planet), which ultimately developed into a mature planet. The center of the disk (in this picture) would coalesce into a hot, gaseous "protosun," which ultimately became the sun (we'll talk about that in Chapter 18, "Stellar Careers").

Pearls the Size of Worlds

Beginning in the 1940s, astronomers returned to the idea of the solar nebula to create a modification of it called the condensation theory.

There were critics of the nebular theory in the nineteenth century, among them James Clerk Maxwell, who had figured out the fundamentals of electromagnetic radiation. What Maxwell and the other critics of the Kant-Laplace theory didn't know about was interstellar dust. Microscopic dust grains—ice crystals and rocky matter—formed in the cooling atmosphere of dying stars, then grew by attracting additional atoms and molecules of various gases. These dust grains served two purposes in the formation of planets:

1. The presence of grains hastened the collapse of the nebular cloud by promoting the radiation of heat from it. This radiation of heat cooled the cloud, accelerating its collapse.

2. Each grain acted as a condensation nucleus, like the grain of sand in an oyster that eventually becomes a pearl. These grains eventually grew into pearls the size of worlds. In effect, these grains were planetary seeds.

Birth of the Planets

Let's put the nebular theory and the condensation theory together, as most current astronomers do.

Here is a possible portrait of the formation of our solar system: A cloud of interstellar dust, measuring about a light-year across, begins to contract, rotating more rapidly the more it contracts. With the accelerating rotation comes a flattening of the cloud into a pancake-like disk, perhaps 100 A.U. across—100 times the current distance between the earth and the sun.

The original gases and dust grains that had formed the nebular cloud have contracted into condensation nuclei, which begin to attract additional matter, forming clumps that rotate within the disk.

The clumps encounter other clumps and more matter, growing larger by accretion. Accretion is the gradual accumulation of mass, and usually refers to the building up of larger masses from smaller ones through the mutual gravitational attraction of matter.

Close Encounter

Startling new infrared and radio telescope evidence supporting the condensation theory was reported in April 1998 by astronomers working at the Keck Observatory, Mauna Kea, Hawaii, and at Cerro Tololo Inter-American Observatory in Chile. Studying a star known as HR4796A, 220 light-years from the earth, the astronomers discovered a vast dust disk forming around it. A doughnut-like hole, slightly larger than the distance between the sun and Pluto, surrounds the star, and the disk itself extends more than twice the distance of the doughnut hole.

While astronomers did not detect any planets in this very distant object, they believe the hole in the disk may be caused by the gravitational force of one or more inner planets. In effect, then, astronomers believe they are seeing a distant planetary system in the making.

In 1984, a disk was observed around another star, Beta Pictoris, but astronomers are even more excited about the 1998 discovery, because they are more certain of the young age of HR4796A than they are of the age of Beta Pictoris. At a mere 10 million years old, HR4796A is believed to be the right age for planetary formation.

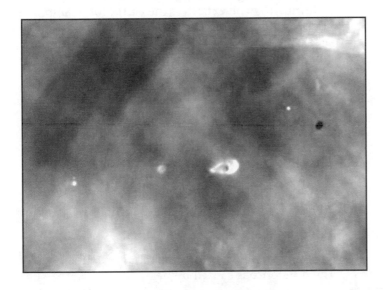

Further evidence for proto-planetary disks around young, low-mass stars was provided by the Hubble Space Telescope images of objects called proplyds, short for proto-planetary disks. This image shows a number of proplyds in the Orion nebula.

(Image from CSC/STScI and NASA)

Star Words

Planetesimals are embryonic planets in an early formative stage. Planetesimals, which are probably the size of small moons, develop into **protoplanets,** immature but full-scale planets. It is the protoplanets that go on to develop into mature planets.

Astro Byte

Astronomers estimate that the evolution from a collection of planetesimals to 9 protoplanets, many protomoons, and a protosolar mass at the center of it all consumed about 100 million years. After an additional billion years, it is believed that the leftover materials had assumed their present orbits in the asteroid belt, the Kuiper Belt, and the Oort Cloud. The high temperatures close to the protosun drove most of the icy material into the outer solar system where (with the exception of periodic cometary shows) it remains to this day.

Accretion and Fragmentation

The preplanetary clumps grew by accretion from objects that might be imagined to be the size of baseballs and basketballs to planetesimals, embryonic protoplanets several hundred miles across. The early solar system must have consisted of millions of planetesimals.

While smaller than mature planets, the planetesimals were large enough to have sufficiently powerful gravitational forces to affect each other. The result was near misses and collisions that merged planetesimals into bigger objects, but also fragmentation, as collisions resulted in chunks of some planetesimals breaking off. As we saw in the last chapter, the formation of the moon likely happened at this point in the history of the solar system.

The larger planetesimals, with their proportionately stronger gravitational fields, captured the lion's share of the fragments, growing yet larger, while the smaller planetesimals joined other planets or were "tossed out." A certain number of fragments escaped capture to become asteroids and comets.

Unlike the planets, whose atmospheres and internal geological activity (volcanism and tectonics) would continue to evolve matter (the earth, for example, has rocks and minerals that vary greatly in age), asteroids and comets remained geologically static, dead; therefore, their matter, unchanging, marks well the date of solar system birth.

Whipping Up the Recipe

While there is substantial variety among the nine planets, they tend to fall into two broad categories: the large gaseous outer planets, known as the jovians, and the smaller rocky inner planets, the terrestrials (these groups are fully described in the next chapter).

Why this particular differentiation?

As with just about any recipe in any kitchen, part of the difference is caused by heat.

Out of the Frying Pan

As the solar nebula contracted and flattened into its pancake-like shape, gravitational energy was released in the form of heat, increasing its temperature. Due to the inverse-square law of gravitational attraction, matter piled up mostly at the center of the collapsing cloud. The density of matter and the temperature were highest near the center of the system, closest to the protosun, and gradually dropped farther out into the disk.

At the very center of the nascent solar system, where heat and density were greatest, the solar mass coalesced. In this very hot region, the carefully assembled interstellar dust was pulled apart into its constituent atoms, while the dust in the outer regions of the disk remained intact. Once the gravitational collapse from a cloud to a disk was complete, the temperatures began to fall again, and new dust grains condensed out of the vaporized material toward the center of the solar system. This vaporization and recondensation process was an important step in the formation of the solar system, because it chemically differentiated the dust grains that would go on to form the planets. These grains originally had a uniform composition. In the regions nearest the protosun—where temperatures were highest—metallic grains formed, because metals survived the early heat. Moving farther out, silicates (rocky material), which could not survive intact close to the protosun, were condensed from the vapor. Farther out still, there were water ice grains, and, even farther, ammonia ice grains. What is fascinating to realize is that the heat of the protosun depleted the inner solar system (which is home to the earth) of water ice and organic carbon compounds. These molecules, as we will see, survived in the outer solar system and later rained onto the surfaces of the inner planets, making one of them habitable.

The composition of the surviving dust grains determined the type of planet that would form. Farthest from the sun, the most common substances in the preplanetary dust grains were water vapor, ammonia, and methane, in addition to the elements hydrogen, helium, carbon, nitrogen, and oxygen—which were distributed throughout the solar system. The jovian planets, therefore, formed around mostly icy material. And in the cooler temperatures farthest from the protosolar mass, greater amounts of material were able to condense, so the outer planets tended to be very massive. Their mass was such that, by gravitational force, they accreted hydrogen-rich nebular gases in addition to dust grains. Hydrogen and helium piled onto the outer planets, causing them to contract and heat up. Their central temperatures rose, but never high enough to trigger fusion, the process that produces a star's enormous energy. Thus the jovian worlds are huge, but also gaseous.

Into the Fire

Closer to the protosun, in the hottest regions of the forming solar system, it was the heaviest elements, not ices and gases, that survived to form the planets. Thus the terrestrial planets are rich in the elements silicon, iron, magnesium, and aluminum. The dust grains and then planetesimals from which these planets were formed were rocky rather than icy. It is fortunate that water ice and organic compounds later rained down on the early Earth, or the present-day planet would be as lifeless as the moon.

Close Encounter

If the central area of the forming solar system was too hot for light elements and gases to hang around, where did the earth's abundant volatile matter—the oxygen, nitrogen, water, and other elements and molecules so essential to life—come from?

One theory is that comets, which are icy fragments that formed in the outer solar system, were deflected out of their orbits by the intense gravitational force of the giant jovian planets. Bombarded by icy meteoroids after the planet had coalesced and cooled, the earth was thus resupplied with water and other essential elements.

Do the Pieces Fit?

By combining the nebular and condensation theories, we have arrived at an explanation that appears to address the major constraints that we listed at the beginning of the chapter. Does that mean that this theory is "right"?

Perhaps.

But like any model, it is subject to future observations that might cause us to reject or revise it. Let's revisit some of the constraints that we outlined.

➤ A rotating cloud of gas, collapsing gravitationally, can account for the "counter-clockwise" (as seen from the North Pole) orbit of the planets, rotation of the sun, and rotational orientation of their moons. What we are seeing in all of these is the direction of rotation of that original solar nebula.

➤ The rocky nature of the inner solar system, and the gaseous nature of the outer solar system follow directly from the temperature of these regions as the dust grains were formed. Only the heaviest materials (metals) survived intact close to the sun, whereas more fragile molecules (like water) survived in the outer reaches.

➤ The planets are all found close to the ecliptic because, as the solar nebula contracted gravitationally, it naturally flattened. This flattened disk was where the planets most likely formed.

➤ The existence and location of asteroids, comets, and other debris is a natural by-product of the accretion and early gravitational interaction process.

Yet, as expressed here, the condensation theory does not account for absolutely everything we observe in the solar system. Various apparent anomalies and irregularities, which we shall consider when we discuss the planets in detail in Chapters 13, 14, and 15, exist. Do they threaten to topple the condensation theory?

Probably not.

For the theory allows for an element of randomness, primarily in the form of close encounters and collisions among the planetesimals and protoplanets, which probably influenced certain variations we see in the orbital motions and orientation of some of the planets. As we saw in the last chapter, it is very likely that our own moon is the remnant of a catastrophic collision between the earth and a planetesimal that was the size of Mars. That collision also likely explains the anomalous tip of the earth's rotational axis, and thus the seasons that grace our planet.

Ashes to Ashes, Dust to Dust

In the chapters of this book's final section, we will consider questions of time and eternity as they relate to the universe. But as to the solar system, we know that it was born about 4.6 billion years ago, and that it will die when its source of energy (the sun) dies of old age.

Just as the specifics of the formation of the solar system depended on the formation of the sun, so its death will be intimately related to the future of our parent star. The evolution of the sun will presumably follow the same path of other stars of its size and mass (see Chapters 18, 19, and 20), which means that the sun will eventually consume the store of hydrogen fuel at its core. As this core fuel wanes, the sun will start to burn fuel in its outer layers, grow brighter, and its outer shell will expand. It will become a red giant (see Chapter 17), with its outer layers extending perhaps as far as the orbit of Venus. When the sun puffs up into a red giant, Mercury will slow in its orbit, and probably fall into the sun. Venus and the earth will certainly be transformed, their atmospheres (and, in the case of Earth, also water) being driven away by the intense heat of the swelling sun. Venus and Earth will return to their infant state, dry and lifeless.

Some recent models of solar evolution predict that the sun will slowly grow to this state sooner, giving us only another billion or so years before the earth becomes uninhabitable.

But don't fret. All of this is another one to five billion years away. The sun is in its midlife now, and, we hope, will avoid any crisis. The sun will then eject its outer layers (to become a planetary nebula), leaving behind a burned-out star called a white dwarf. A white dwarf does not have sufficient mass to continue fusing elements (as we will see some stars do in Chapter 19). It will slowly cool, radiating its internal heat into space, and eventually become a black dwarf—a strange object composed mostly of oxygen and carbon, the size of a planet with the mass of a star. Let's hope humanity has pushed on by then!

The planetary nebula NGC 1514 was discovered in 1790 by William Herschel. Planetary nebulae represent a brief phase late in the life of a low-mass star, when the outer layers of the star are ejected.

(Image from Tom Wickman)

The Least You Need to Know

➤ Based on studies of meteorites, astronomers believe the solar system is 4.6 billion years old.

➤ The philosopher Kant and the mathematician Laplace proposed, in the eighteenth century, the nebular hypothesis, a theory that the solar system was formed from a spinning cloud of interstellar dust.

➤ The early solar system was filled with dust grains that were pulled together by gravity to form planetesimals. The composition of the dust grains and thus the planetesimals depended on the distance from the sun.

➤ A gravitationally contracting nebula in which dust grains condense and collide to form planets can explain most of the patterns observed in our solar system, and even account for some of its irregularities.

Solar System Family Snapshot

In This Chapter

➤ A solar system inventory

➤ Distances within the solar system

➤ Introduction to the sun

➤ Planetary stats

➤ Terrestrial and jovian planets

➤ All about asteroids

➤ Comet hunting

➤ Meteors, meteoroids, and meteorites

A snapshot freezes an instant in time. When we think about our solar system, we usually assume that it has always been much as it is now, and always will be. But what we know of the solar system (4,000 years of accumulated knowledge) is only a mere snapshot in comparison to its 4.6 billion–year age. It took humankind millennia to reach the conclusion that our planet is part of a *solar* system, one of many planets spinning on its axis orbiting the sun. These were centuries of wrestling with the earth-centered planetary system first of Aristotle, then of Ptolemy, trying to make the expected planetary orbits coincide with actual observation. This knowledge arose in some sense as a side product of the real initial goals: to be able to predict the motion of the planets and stars for the purpose of creating calendars and (in some cases) as a means of for-tune-telling. However, even the earliest astronomers (of whom we know) wanted to do more than predict the planets' motions. They wanted to know what was "really" going on. When Copernicus, Galileo, Tycho Brahe, and Kepler finally succeeded in doing this quite well in the sixteenth and seventeenth centuries, it was a momentous time for astronomy and human understanding.

Understanding how the planets move is important, of course, but our understanding of the solar system hardly ends with that. In the last few decades of the twentieth century and now into the twenty-first, astronomers have learned more about the solar system than in all the 400 years since planetary motions were pretty well nailed down. As this chapter will show, the planetary neighborhood is a very interesting place, and our own world, the earth, is unique among the planets as a home to life.

Astro Byte

A few asteroids are so large that some astronomers call them minor planets or planetoids. Ceres, the largest known minor planet, has a diameter of about 581 miles (935 km), and Pallas measures 332 miles (535 km) across.

Star Words

An **asteroid** is one of thousands of small members of the solar system, which orbit the sun. Most asteroids are found between Mars and Jupiter in the asteroid belt.

A Beautiful Day in the Neighborhood: Let's Take a Stroll

Our solar system is centered on a single star, the sun. We have recently come to appreciate that about 50 percent of all stars form in binary systems (containing two stars), so our sun is a bit lonely as stars go. In orbit around the sun are nine planets (in order of distance from the sun): Mercury, Venus, Earth, Mars, Jupiter, Saturn, Uranus, Neptune, and Pluto. Around some of these planets orbit moons—more than 70, at latest count. By the 1990s, astronomers had observed more than 6,000 large *asteroids,* of which approximately 5,000 have been assigned catalog numbers. (Such an assignment is made as soon as accurate orbital data is recorded.) Most asteroids are rather small; it is estimated that there are 1 million with diameters greater than 1 km (or about ⅗ of a mile). Some, perhaps 250, have diameters of at least 62 miles (100 km). About 30 have diameters of more than 124 miles (200 km). All of these planets and asteroids are the debris from the formation of the sun. They coalesced slowly (as we saw in the preceding chapter) through the mutual attraction of gravity.

In addition, the solar system contains a great many comets and billions of smaller, rock-size meteoroids.

Some Points of Interest

The orbits of the planets lie nearly in the same plane, except for Mercury and Pluto, which deviate from this plane by 7 degrees and 17 degrees, respectively. Between the orbit of Mars and Jupiter is a concentration of asteroids known as the asteroid belt. Most of the solar system's asteroids are here.

The orbits of the planets are not equally spaced, tending (very roughly) to double between adjacent orbits as we move away from the sun.

To say that the distances between planets and the sun are very great is an understatement. Interplanetary distances are so great that it becomes awkward to speak in terms of miles or kilometers. For that reason, astronomers have agreed on something called an *astronomical unit* (*A.U.*), which is the average distance between the earth and the sun—that is, 149,603,500 kilometers or 92,754,170 miles.

Let's use these units to gauge the size of the solar system. From the sun to the average distance of the outermost planet, Pluto, is 40 A.U. (3,710,166,800 miles, or almost 6 billion km). At just about a million times the radius of the earth, that's quite a distance. Think of it this way: If the earth were a golfball, Pluto would be a chickpea about 8 miles away, Jupiter would be a basketball about 1 mile away, and the sun would go floor-to-ceiling in a 10-foot room and be less than a quarter-mile away. However, compared to, say, the distance from the earth to the nearest star (after the sun), even Pluto is a near neighbor. Forty A.U. is less than $\frac{1}{1000}$ of a *light-year*, the distance light travels in one year: almost 6 trillion miles. Alpha Centauri, the nearest star system to our sun, is about 4.3 light-years from us (more than 25 trillion miles). On our golf ball scale, Alpha Centauri would be about 55,000 miles away.

Star Words

An **astronomical unit** (**A.U.**) is a unit of measurement equivalent to the average distance from the earth to the Sun (149,603,500 kilometers or 92,754,170 miles). The most distant planet (usually Pluto) is located on average 40 A.U. from the Sun, and Mercury is the closest at 0.4 A.U.

Star Words

A **light–year,** an important astronomical unit of distance (not of time!), is the distance light travels in one Earth year: 9.5 trillion km or about 6 trillion miles. The closest star, Alpha Centauri, is 4.3 light-years away.

More or Less at the Center of It All

Near the center of the solar system—more accurately, at one focus of the elliptical orbits of the planets (see Chapter 4, "Astronomy Reborn: 1543–1687")—is the sun. The sun is most of the solar system, containing more than 99.9 percent of the matter that makes up the solar system. All of Chapter 16, "Our Star," is devoted to the sun, but be aware now just how massive an object it is. Jupiter, the largest planet in the solar system, is over 300 times the mass of the earth. And the sun, in turn, is more than a *thousand* times more massive than Jupiter (and about 300,000 times more massive than the earth).

Planetary Report Card

Let's make a survey of the planets. Here's what we'll be measuring and comparing in the table that follows:

➤ **Semi-major axis of orbit.** You'll recall from Chapter 4 that the planets orbit the sun not in perfectly circular paths, but elliptical ones. The semi-major axis of an ellipse is the distance from the center of the ellipse to its farthest point. This distance does not exactly correspond to the distance from the sun to the farthest point of a planet's orbit, since the sun is not at the center of the ellipse, but at one of the ellipse's two foci. We will express this number in A.U.

Star Words

The **terrestrial planets** are the rocky planets closest to the sun: Mercury, Venus, Earth, and Mars. The **jovian planets** are the gaseous planets farthest from the sun: Jupiter, Saturn, Uranus, and Neptune. Pluto is classed as neither terrestrial nor jovian.

➤ **Sidereal period.** The time it takes a planet to complete one orbit around the sun, usually expressed in Earth years.

➤ **Mass.** The quantity of matter a planet contains. The mass of the earth is 5.977×10^{24} kg. We will assign the earth's mass the value of 1.0 and compare the masses of the other planets to it.

➤ **Radius.** At the equator, the radius of the earth is slightly less than 6,400 km (3,963 miles). We will assign the radius of the earth a value of 1.0 and compare the radii of other planets to it.

➤ **Number of known moons.** Self-explanatory— an ever-changing number for the outer planets.

➤ **Average density.** This value is expressed in kilograms of mass per cubic meter. The substance of the inner planets is dense and tightly packed; in the outer planets, the densities are typically lower.

Planet	Semi-Major Axis of Orbit (in A.U.)	Sidereal Period (in Years)	Mass (in Earth Masses)	Radius (in Earth Masses)	Moons	Density (kg/m³)
Mercury	0.39	0.24	0.055	0.38	0	5,400
Venus	0.72	0.62	0.81	0.95	0	5,200
Earth	1.0	1.0	1.0	1.0	1	5,500
Mars	1.5	1.9	0.11	0.53	2	3,900
Jupiter	5.2	11.9	318	11.2	16	1,300
Saturn	9.5	29.5	95	9.5	20	700
Uranus	19.2	84	15	4.0	15	1,200
Neptune	30.1	165	17	3.9	8	1,700
Pluto	39.5	249	0.003	0.2	1	2,300

The Inner and Outer Circles

Astronomers think of the planets as falling into two broad categories—with one planet left over. The four planets (including the earth) closest to the sun are termed the *terrestrial planets*. The four farthest from the sun (not counting Pluto) are the *jovian planets*. And Pluto, usually the farthest out of all, is in an unnamed class by itself. Its location is jovian, while its size and composition put it more in a class with the moons of the jovian planets. Some astronomers prefer to think of it as the largest Kuiper Belt object rather than the smallest (and hardest-to-categorize) planet.

Snapshot of the Terrestrial Planets

Mercury, Venus, Mars, and Earth are called the terrestrial planets because they all possess certain Earth-like (terrestrial) properties. These include proximity to the sun (within 1.5 A.U), relatively closely spaced orbits, relatively small masses, relatively small radii, and high density (rocky and solid-surfaced). Compared to the larger, more distant jovian planets, the terrestrials rotate more slowly, possess weak magnetic fields, lack rings, and have few moons or none. In fact, within the terrestrial "club" the earth's large moon is unique. The moon is only slightly smaller than the planet Mercury and larger than Jupiter's moon Europa! As we have seen, the moon's large size is one clue to its origin. We'll take a closer look at the terrestrial planets in Chapter 13, "So Close and Yet So Far: The Inner Planets."

Snapshot of the Jovian Planets

The jovians are far from the sun and travel in widely spaced orbits. They are massive planets with large radii, yet they are of low density with predominantly gaseous makeup and no solid surface. In contrast to the terrestrial planets, they rotate faster, possess strong magnetic fields, have rings, and are orbited by many large moons. We'll discuss the jovian planets in Chapter 14, "Great Balls of Gas! The Outer Planets."

The outermost jovian planet is Neptune, and beyond it is Pluto.

Serving Up the Leftovers

What's the oldest stuff in your refrigerator (aside from that rubbery celery you bought but never ate)? Leftovers! The same is true in the solar system. The fragmentary leftovers of the formation of the sun and planets are some of the oldest objects in the solar system. For a long time, few scientists paid much attention to this debris or knew much about it. More recently, however, astronomers have come to realize that many significant clues to the origin and early evolution of the solar system are to be found not in the planets, but in the smaller bodies, the planetary moons and solar system debris. For the most part, the planets are very active places. Atmospheres have produced erosion, and internal geological activity has erased ancient surfaces. On the earth, weather, water, and tectonic motion have long since "recycled" the earth's original surface.

So studying the planets reveals relatively little about the origins of the solar system. However, on moons and asteroids, atmospheres are sparse or nonexistent, and geological activity is minimal or absent. The result? Many of these bodies have changed little since the solar system was born. They are, in effect, cosmic leftovers.

Star Words

A **retrograde** orbit is one that is backwards or contrary to the direction of the planets. An **ellipse** is called **eccentric** when it is noncircular. An ellipse with an eccentricity of 0 is a circle, and an ellipse with an eccentricity of close to 1 would be very oblong. Even planets with the most eccentric orbits (Pluto and Mercury) have eccentricities of only about 0.2.

The Asteroid Belt

Astronomers have noted and cataloged more than 6,000 asteroids with regular orbits, most of them concentrated in the asteroid belt, between the orbits of Mars and Jupiter. So far, every asteroid that has been noted orbits in the same direction as the earth and other planets—except one, whose orbit is *retrograde* (backward, or contrary to the direction of the planets). Although the asteroids move in the same direction—and pretty much on the same plane—as the planets, the shape of their orbits is different. Many asteroid orbits are more *eccentric* (the ellipse is more exaggerated and oblong) than those of the planets.

Landing on Eros—The Love Boat

In early 2001, an asteroid-exploring probe orbited and finally landed on the surface of Eros. As it (slowly) crashed to the asteroid's surface, it sent back tantalizing close-up images of the surface.

Rocks and Hard Places

Asteroids are composed of stony as well as metallic materials—mostly iron—and are basically tiny planets without atmospheres. Some asteroids have a good deal of carbon in their composition as well. These, called carbonaceous chondrites, are thought to represent the very first materials that came together to form the objects of the solar system. Carbonaceous chondrites are truly the solar system's fossils, having avoided change for billions upon billions of years.

Earlier astronomers surmised that asteroids were fragments resulting from various meteoric collisions. While some of the smaller meteoroids were likely produced this way, the major asteroids probably came into being at the time of the formation of the solar system as a whole. Theoretical studies show that no planet could have formed at the radius of the asteroid belt (at about 3 A.U. from the sun). The region between Mars and Jupiter is dominated by the gravitational influence of the giant planet Jupiter. These forces stirred up the potential planet-forming material, causing it to collide and break up instead of coming together to create a planet-sized object.

The smaller asteroids come in a wide variety of shapes, ranging from nearly spherical, to slab-like, to highly irregular.

During 1993–1994, the Galileo probe (see Chapter 9, "Space Race: From *Sputnik* to the International Space Station") passed through the asteroid belt on its way to Jupiter and took pictures of an asteroid orbited by its own miniature moon. Potato-shaped, the asteroid was named Ida, and is about 35 miles (56 km) long, orbited at a distance of roughly 60 miles (97 km) by a rock less than 1 mile in diameter. This little moon is the smallest known natural satellite in the solar system.

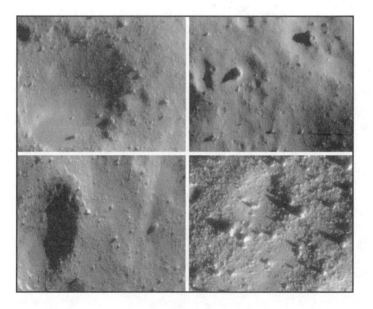

These images of Eros show the closest views of an asteroid ever seen. The top images are 550 m across, and the bottom images are 230 m across. The surface is dominated by small boulders.

(Image from NASA)

Impact? The Earth-Crossing Asteroids

Most of the asteroids in the asteroid belt remain there, but some have highly eccentric orbits that take them out of the asteroid belt and across the orbital path of the earth (as well as the paths of other terrestrial planets).

Star Words

Apollo asteroids are those with sufficiently eccentric orbits to cross paths with the earth (and other terrestrial planets).

Astro Byte

One widely accepted theory blames an asteroid impact for the extinction of the dinosaurs about 65 million years ago. Some believe that the impact of an asteroid only some 6 miles (10 km) in diameter just north of the Yucatan Peninsula of Mexico created a massive dust cloud that killed many plant species essential to the dinosaurs' survival. The impact also appears to have triggered worldwide wildfires that decimated the earth's ecosystem.

Nearly 100 of these so-called *Apollo asteroids* have been identified so far, and a number of astronomers advise funding efforts to identify and track even more, because the potential for a collision with Earth is real. With advance warning, scientists believe, missiles with thermonuclear warheads could be exploded near an incoming asteroid, sufficiently altering its course to make it avoid the earth, or shattering it into a large number of smaller asteroids. You're local movie theater or video store is a good source to study Hollywood's take on these nightmare scenarios, but they are a very real threat. Project NEAT (Near Earth Asteroid Tracking) is funded by NASA. For more information see neat.jpl.nasa.gov.

It is believed that a few asteroids of more than a half-mile diameter might collide with the earth in the course of a million years. Such impacts would be disastrous, each the equivalent of the detonation of several hydrogen bombs. Not only would a great crater, some eight miles across, be formed, but an Earth-enveloping dust cloud would darken the skies. It is thought that the great extinction of dinosaurs 65 million years ago was due to such an impact. Were the impact to occur in the ocean, tidal waves and massive flooding would result.

Earth impacts of smaller objects are not uncommon, but on June 30, 1908, a larger object—apparently the icy nucleus of a very small comet—fell in the sparsely inhabited Tunguska region of Siberia. The falling object outshone the sun, and its explosive impact was felt at a distance of more than six hundred miles. A very wide area of forest was obliterated—quite literally flattened. Pictures from the time show miles of forest with trees stripped and lying on their sides, pointing away from the impact site.

Anatomy of a Comet

The word *comet* derives from the Greek word *kome,* meaning "hair." The name describes the blurry, diaphanous appearance of a comet's long tail.

But the tail is only part of the anatomy of a comet, and it is not even a permanent part, forming only as the comet nears the sun. For most of the comet's orbit, only its main, solid body—its nucleus—exists. It is a relatively small (a few miles in diameter) mass of irregular shape made up of ice and something like soot, consisting of the same hydrocarbons and silicates that we find in asteroids.

The orbit of the typical comet is extremely eccentric (elongated), so that most comets (called long-period comets) travel even beyond Pluto and may take millions of years to complete a single orbit. So-called "short-period" comets don't venture beyond Pluto and, therefore, have much shorter orbital periods.

Star Words

A **comet** is a "dirty snowball," a small celestial body composed mainly of ice and dust, which has a highly eccentric orbit around the sun. As it approaches the sun, some of its material is vaporized and ionized to create an atmosphere ("coma") and long tail.

Comet Hale-Bopp, photographed by Gordon Garradd in Loomberah, NSW, Australia, February 21, 1998.

(Image from Gordon Garradd)

As a comet approaches the sun, the dust on its surface becomes hotter, and the ice below the crusty surface of the nucleus sublimates—immediately changes to a gas without first becoming liquid. The gas leaves the comet, carrying with it some of the dust. The gas molecules absorb solar radiation, then reradiate it at another wavelength while the dust acts to scatter the sunlight. The effect of this is the creation of a *coma,* a spherical envelope of gas and dust (perhaps 60,000 miles across) surrounding the nucleus and a long *tail* consisting of gases and more dust particles.

Star Words

Solar wind is the continuous stream of radiation and matter that escapes from the sun. Its effects may be seen in how it blows the tails of a comet approaching the sun.

A Tale of Two Tails

Most comets actually have two tails. The dust tail is usually broader and more diffuse than the ion tail, which is more linear. The ion tail is made up of ionized atoms—that is, atoms that have lost one or more electrons and that, therefore, are now electrically charged. Both the dust tail and the ion tail point away from the sun. But the dust tail is usually seen to have a curved shape that trails the direction of motion of the comet. Careful telescopic or binocular observations of nearby comets can reveal both of these tails.

What we cannot see optically is the vast hydrogen envelope that surrounds the coma and the tail. It is invisible to optical observations.

Common sense tells us that the tail would stream behind the fast-moving nucleus of the comet. This is not the case, however. The ion tail (far from the sun) or tails (the dust tail appears as the comet gets close to the sun) point away from the sun, regardless of the direction of the comet's travel. Indeed, as the comet rounds the sun and begins to leave the sun's proximity, the tail actually *leads* the nucleus and coma. This is because the tail is "blown" like a wind sock by the *solar wind,* an invisible stream of matter and radiation that continually escapes from the sun. It was by observing the behavior of comet tails that astronomers discovered the existence of the solar wind.

"Mommy, Where Do Comets Come From?"

The solar system has two cometary reservoirs, both named after the Dutch astronomers who discovered them. The nearer reservoir is called the Kuiper Belt. The short-period comets, those whose orbital period is less than 200 years, are believed to come from this region, which extends from the orbit of Pluto out to several 100 A.U. Comets from this region orbit peacefully unless some gravitational influence sends one into an eccentric orbit that takes it outside of the belt.

Long-period comets, it is believed, originate in the Oort Cloud, a vast area (some 50,000–100,000 A.U. in radius) surrounding the solar system and consisting of comets orbiting in various planes. Oort comets are distributed in a spherical cloud instead of a disk.

Close Encounter

Comets do not randomly occur, but are regular, orbiting members of the solar system. The most famous comet of all is Halley's Comet, named after the British astronomer Edmund Halley (1656–1742), who published a book in 1705, showing by mathematical calculation that comets observed in 1531, 1607, and 1682 were actually a single comet. Halley predicted the comet would return in 1758, and, when it did, it was named in his honor.

Subsequent calculations show that Halley's Comet, which appears every 76 years, had been noted as early as 240 B.C.E. and was always a source of great wonder and even fear. (In many cultures and at various times, comets have been seen as omens of great good or great evil—usually evil.)

More is known about Halley's Comet than any other, since it has been studied by earthbound astronomers as well as by unmanned spacecraft. Its tail is millions of miles long, and the comet travels at about 80,000 miles per hour (129,000 km/h). Its nucleus is oblong, about 9×5 miles (15×8 km), and it consists primarily of ice and rock dust covered by a porous black crust. Halley's Comet most recently visited the earth in 1986, and will not return until 2062.

Edmund Halley (1656–1742), who, in 1705, showed by mathematical calculation that comets observed in 1531, 1607, and 1682 were in fact a single returning comet. Halley predicted it would return again in 1758, and, after it did, it was named in his honor.

(Image from arttoday.com)

183

The Oort Cloud is at such a great distance from the sun, that it extends about ⅓ of the distance to the nearest star. We don't see the vast majority of these comets, because their orbital paths, though still bound by the sun's gravitational pull, never approach the perimeter of the solar system. However, it is believed that the gravitational field of a passing star from time to time deflects a comet out of its orbit within the Oort Cloud, sending it on a path to the inner solar system, perhaps sealing our fate.

Once a short-period or long-period comet is kicked out of its Kuiper Belt or Oort Cloud home, it assumes its eccentric orbit indefinitely. That is, it can't go home again. A comet will, each time it passes close to the sun, lose a bit of its mass as it is boiled away. A typical comet loses about ¹⁄₁₀₀ of its mass each time it passes the sun, and so, after 100 passages, will typically fragment and continue to orbit or coalesce with the sun as a collection of debris. As the earth passes through these orbital paths, we experience meteor showers.

A-Hunting We Will Go

Visitations by major comets, such as Comet Hyakutake in 1966 and Hale-Bopp in 1997, are newsworthy events. Turn on the television or read a newspaper, and you'll be told where to look and when. But most comets don't make the front pages. For the latest comet news, check out the NASA comet home page at encke.jpl.nasa.gov.

Sky and Telescope magazine also publishes comet information (see Appendix E, "Sources for Astronomers").

Astronomer's Notebook

What are your chances of finding a new comet? David H. Levy, discoverer of several comets and co-discoverer of the celebrated Comet Shoemaker-Levy, reports that, on average, observers hunt about 400 hours for their first comet and 200 for each additional one. Apparently there is a bulk discount for comet discoveries.

Of course, you don't have to limit yourself to looking for comets whose presence or approach is already known. You can head out with your trusty telescope and hunt for new ones.

Comet hunting can be done with or without a telescope, but a good telescope greatly increases your chances of finding a new comet. Remember that telescopes catch more light than our eyes, and most comets are discovered as a tiny, wispy smudge. The coma will not appear much different from a star, but you should see a gradual, not sharply defined, tail attached to it. The tail may be a short, broad wedge or a long ion streamer.

The following tips will increase your chances of finding a comet:

➤ Set up your telescope in a rural area, away from city lights. Choose a moonless night so that the skies are as dark as possible. You will be looking for a faint object.

➤ According to David H. Levy, just before dawn, two days before or five days after the new moon, is an ideal time to search.

➤ Comets can be seen in any part of the sky, but they are brightest when they approach within 90 degrees of the sun. You might concentrate on this part of the sky. That is, at sunset you could look from directly overhead to the western horizon.

➤ Gradually and methodically sweep the sky with your telescope. Stake out perhaps 40 degrees of sky and sweep in one direction (either from east to west or west to east).

➤ Remember one thing. Discovering a comet requires you to see something unusual or different in the sky. For this reason, you would do well to spend time becoming familiar with the sky, the constellations, and your telescope, so that you will be better able to recognize when something is not quite right.

Close Encounter

If you are really serious about comet hunting, you'll need to consult a very good star atlas or star catalog (see Appendix E) in order to rule out known objects. The Messier Catalog was originally compiled in the eighteenth century as a list of objects commonly mistaken for comets (see Appendix D, "The Messier Catalog"). You should also check the Web sites noted earlier to see if your comet has already been reported. Finally, if you are convinced that you have found something new, you may report the find to the Central Bureau of Astronomical Telegrams, which can be reached by telegram or telex: 710-320-6842 ASTROGRAM CAM. Write to the Bureau at 60 Garden Street, Cambridge, MA 02138, for information on what to include in a reporting telegram.

Catch a Falling Star

While sighting a comet—especially a hitherto unrecorded one—is certainly a momentous event, few astronomical phenomena are more thrilling than the sight of a meteor. Best of all, such sightings are common and require no telescope.

Meteors, Meteoroids, and Meteorites

Meteors are commonly called shooting stars, although they have nothing to do with stars at all. A *meteor* is a streak of light in the sky resulting from the ionization of a narrow channel in the Earth's upper atmosphere. The heat generated by friction with air molecules ionizes a pathway behind the piece of debris.

While smaller meteoroids (often called *micrometeoroids*) are typically the rocky fragments left over from a broken-up comet, the meteor phenomenon is very different from a comet. A meteor sighting is a momentary event. The meteor streaks across a part of the sky. As we have seen, a comet does not streak rapidly and may, in fact, be visible for many months because of its great distance from the earth. A meteor is an atmospheric event, whereas a comet is typically many A.U. distant from the earth.

Meteor is the term for the sight of the streak of light caused by a *meteoroid*—which is the term for the actual rocky object that enters the atmosphere. Most meteoroids are completely burned up in our atmosphere, but a few do get through to strike the earth. Any fragments recovered are called *meteorites*.

While most of the meteors we see are caused by small meteoroids associated with comet fragments (about the size of a pea), larger meteoroids, more than an inch or so, are probably asteroid fragments that have strayed from their orbit in the asteroid belt. Such fragments enter the earth's atmosphere at supersonic speeds of several miles per second and often generate sonic booms. If you see a very bright meteor—the brightness of the planet Venus or even brighter—it is one of these so-called *fireballs*. It is estimated that about 100 tons of meteoric material fall on the earth each day.

Star Words

Meteor is the term for what you see streaking across the night sky—a shooting star. **Meteoroid** is the object itself, a rocky thing that is typically a fragment shed from a comet or an asteroid. A **micrometeoroid** is a very small meteoroid. Those few meteoroids that are not consumed in the earth's atmosphere reach the ground as **meteorites.** Many science museums have collections of meteorites.

Star Words

A **fireball** results from a substantial meteoroid (measuring perhaps an inch or so in diameter), which creates a spectacular display when it enters the atmosphere. Meteorites must be rather large (larger than your head) to avoid burning up entirely before hitting the surface of the earth.

News from NEAT

Very rarely—perhaps once in several hundred thousand years—a large meteoroid strikes the earth, causing great damage, as discussed in "Impact? The Earth-Crossing Asteroids," earlier in the chapter.

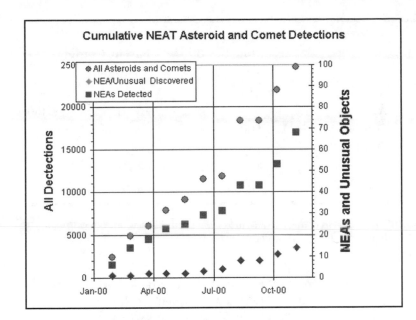

Cumulative NEAT Asteroid and Comet Detections

Counts of near-earth asteroids from Project NEAT (Near Earth Asteroid Tracking).

(Table from NASA/NEAT)

April Showers (or the Lyrids)

Whenever a comet makes its nearest approach to the sun, some pieces break off from its nucleus. The larger fragments take up orbits near the parent comet, but some fall behind, so that the comet's path is eventually filled with these tiny micrometeoroids. Periodically, the earth's orbit intersects with a cluster of such micrometeoroids, resulting in a *meteor shower* as the fragments burn up in our upper atmosphere.

Meteor showers associated with certain comets occur with high regularity. They are known by the constellation from which their streaks appear to radiate. The following table lists the most common and prominent showers. The shower names are genitive forms of the constellation name; for example, the Perseid shower comes from the direction of the constellation Perseus, the Lyrids from Lyra. The dates listed are those of maximum expected activity, and you can judge the intensity of the shower by the estimated hourly count. The table also lists the parent comet, when known.

Star Words

When the earth's orbit intersects the debris that litter the path of a comet, we see a **meteor shower,** a period when we see more meteors than average.

Name of Shower	Maximum Activity	Estimated Hourly Count	Parent Comet
Quadrantid	January 3	50	unknown
Beta Taurid	June 30	25	Encke
Perseid	August 12	50+	1862III (Swift-Tuttle)
Draconid	October 8–9	500+	Giacobini-Zimmer
Orionid	October 20	25	Halley
Leonid	November 16–17	10*	1866I (Tuttle)
Geminid	December 11–17	50–75	3200 Phaeton

Every 33 years, the earth's orbit intersects the densest part of the Leonid debris path, resulting in the potential for a meteor infall rate of 1,000 a minute! Such an intersection occurred in 1999, and will happen again in 2032.

You'll recall from Chapter 8, "Seeing in the Dark," that you can detect meteor showers on your FM radio or even on unused VHF television frequencies. But if it's clear outside, we suggest that you take your radio outside, and as you listen for distant radio stations to pop up, look up at the skies and watch as well. It might be hard to believe that most of those streaks of light are following meteoroids no larger than a pea. But be thankful that they are!

The Least You Need to Know

➤ Eight of the nine planets of the solar system are divided into the rocky terrestrial planets (those nearest the sun: Mercury, Venus, Earth, and Mars) and gaseous jovian planets (those farthest from the sun: Jupiter, Saturn, Uranus, and Neptune). Pluto, usually the most distant planet from the sun, is not categorized, though it has a lot in common with the moons of the outer planets.

➤ While the sun and planets are certainly the major objects in the solar system, astronomers also pay close attention to the minor bodies—asteroids, comets, meteors, and planetary moons—that tell us a lot about the origin of the solar system.

➤ While most asteroids are restricted to highly predictable orbits, a few cross the earth's orbital path, posing a potentially catastrophic hazard.

➤ Both comets and meteors present ample opportunity for exciting amateur observation.

So Close and Yet So Far: The Inner Planets

Our two closest neighbors in the solar system, Mars and Venus, are a constant reminder of how easily things could have turned out differently here on Earth. Venus is so hot and forbidding that it might be a good place to film Dante's *Inferno,* and while pictures of Mars may resemble the American Southwest, it has an atmosphere so cold and thin that it's hardly there. Putting a planet a little closer to the sun or a little farther away can truly make all the difference in the world. Equally amazing is that all three planets fall within what is called the "habitable zone" of the sun (see Chapter 24, "Table for One?"). The habitable zone is the distance range from the sun within which water could be liquid on the planet's surface. But only one, Earth, has abundant liquid water. Recent imaging of the surface of Mars indicates that water may still briefly exist in liquid form on its surface.

In this chapter, we will take a closer look at the four rocky planets that are closest to the sun. Astronomers call these four planets terrestrial, meaning Earth-like. To be sure, Mercury, Venus, and Mars resemble Earth more than they do the jovian planets—Jupiter, Saturn, Uranus, and Neptune—but, as we'll see, these four worlds are as different from each other as one could imagine. And the earth is the only one of the four that has sustained life.

Why is that? What is it about the earth that allowed life to thrive here and not on the other three planets in the inner solar system? Let's start looking for some answers.

The Terrestrial Roster

The terrestrial planets are Mercury, Venus, Earth, and Mars. Except for Earth, all are named after Roman gods. Mercury, the winged-foot messenger of the gods, is an apt name for the planet closest to the sun; its sidereal period is a mere 88 Earth days, and its average orbital speed (30 miles per second or 48 km/s) is the fastest of all the planets. Mercury orbits the sun in less than a college semester, or about four times for each Earth orbit.

Venus, named for the Roman goddess of love and fertility, is (to observers on Earth) the brightest of the planets, and, even to the naked eye, quite beautiful to behold. Its atmosphere, we shall see, is not so loving. The planet is completely enveloped by carbon dioxide and thick clouds that consist mostly of sulfuric acid.

The name of the bloody Roman war god, Mars, suits the orange-red face of our nearest planetary neighbor—the planet that has most intrigued observers and that seems, at first glance, the least alien of all our fellow travelers around the sun.

We looked at some vital statistics of the planets in Chapter 12, "Solar System Family Snapshot." Here are some more numbers, specifically for the terrestrial planets. Notice that the presence of an atmosphere (on Venus and Earth) causes there to be much less variation in surface temperature.

Planet	Mass in Kilograms	Radius in Miles (and km)	Surface Gravity (Relative to Earth)	Rotation Period in Solar Days	Surface Temperature in K
Mercury	3.3×10^{23}	1,488 (2,400)	0.4	59	100–700
Venus	4.9×10^{24}	3,782 (6,100)	0.9	–243*	730
Earth	6.0×10^{24}	3,968 (6,400)	1.0	1.0	290
Mars	6.4×10^{23}	2,108 (3,400)	0.4	1.0	100–250

The reason for the minus sign is that the rotation of Venus is retrograde; that is, the planet rotates on its axis in the opposite direction from the other planets. Viewed from above the North Pole of the earth, all of the planets except Venus rotate counterclockwise. That means that on Venus, the sun would rise (if you could see it through the thick cloud cover) in the west.

If you recall, when we discussed the formation of the solar system, we mentioned a few observational facts that "constrained" our models of formation. A few rules of planetary motions are immediately apparent. All four terrestrial planets orbit the sun in the same direction. All except Venus rotate on their axes in the same direction as they orbit the sun. The orbital paths of the inner four planets are *nearly* circular. And the planets all orbit the sun in roughly the same plane.

But the solar system is a dynamic and real system, not a theoretical construct, and there are interesting exceptions to these rules. The exceptions can give us insight into the formation of the solar system.

Close Encounter

Most of us in the United States are accustomed to the Fahrenheit temperature scale. The rest of the world uses the Celsius (Centigrade) scale. Astronomers, like most scientists, measure temperature on the Kelvin scale. Throughout this book, we have expressed distance in the units familiar to most of our readers: miles (with kilometers or meters given parenthetically). For mass we give all values in kilograms. The Kelvin scale for temperature is conventional and very useful in astronomy; let us explain.

The Fahrenheit scale is really quite arbitrary, because its zero point is based on the temperature at which alcohol freezes. What's fundamental about that? Worse, it puts at peculiar points the benchmarks that most of us *do* care about. For example, water freezes at 32° F at atmospheric pressure and boils at 212° F. The Celsius scale is somewhat less arbitrary, because it is laid out such that water freezes at 0° C and boils at 100° C at atmospheric pressure. But the atmospheric pressure of the earth is by no means a universal quantity, so astronomers and others looked for more fundamental benchmarks.

The Kelvin scale is least arbitrary of all. It forces us to ask a fundamental question: What is heat?

The atoms and molecules in any matter are in constant random motion, which represents thermal energy. As long as there is atomic or molecular motion, there is heat (even in objects that, to the human senses, feel very cold). We know of no matter in the universe whose atoms and molecules are entirely motionless, but, in theory, such an absolute zero point does exist. The Kelvin scale begins at that theoretical absolute zero, the point at which there is no atomic or molecular motion. On the Fahrenheit scale, that temperature is −459 degrees. On the Celsius scale, it is −273 degrees. On the Kelvin scale, it is merely 0 degrees. Thus, in the Kelvin scale there are no negative numbers, because absolute zero is—well—absolute. After all, atoms and molecules cannot get any *more* motionless than motionless, and being absolutely still would cause the constituents of atoms to violate some fundamental physical principles.

Note that water freezes at 273 K and boils at 373 K. If you want to convert Kelvin temperatures to Celsius temperatures, subtract 273 (to be precise, 273.15) from the Kelvin temperature. If you *really* want to, you can then convert the Celsius temperature to Fahrenheit by multiplying the Celsius reading by 9/5, then adding 32.

Mercury: The Moon's Twin

In many ways, Mercury has more in common with the lifeless moon of our own planet than with the other terrestrial planets. Its face is scarred with ancient craters, the result of massive bombardment that occurred early in the solar system's history. These craters remain untouched because Mercury has no water, erosion, or atmosphere to erase them. The closest planet to the sun—with an average distance of 960,000 miles (1,546,000 km)—Mercury is difficult to observe from the earth, and can only be viewed near sunrise or sunset.

Its surface, revealed in detail for the first time in images transmitted by such unmanned probes as *Mariner 10* (in the 1970s), is pocked with moonlike craters. *Mariner 10* also discovered a weak but detectable magnetic field around Mercury. As a result, astronomers concluded that the planet must have a core rich in molten iron. This contention is consistent with the planet's position closest to the center of the solar system, where most of the preplanetary matter—the seeds that formed the planets—would have been metallic in composition (see Chapter 11, "Solar System Home Movie").

Astro Byte

Mercury's temperature range of 600 K (it varies from 100 K to 700 K) is the greatest of all the planets. Although Mercury is closest to the sun, Venus's surface temperature is higher. A planet's surface temperature and its mass determine how much (if any) of its atmosphere it will be able to hold on to.

Eighteen-image mosaic of Mercury taken by Mariner 10 *during its approach on March 29, 1974. The space craft was about 124,000 miles (200,000 km) above the planet. Note how closely the surface of Mercury resembles the Moon's.*

(Image from NASA)

Lashed to the Sun

In the days before space-based telescopes and probes, earthbound astronomers did the best they could to gauge the rotation of Mercury. The nineteenth-century astronomer Giovanni Schiaparelli observed the movement of what few, indistinct surface features he could discern and concluded that, unlike any other planet's, Mercury's rotation was synchronous with its orbit around the sun.

Synchronous orbit means that Mercury always keeps one face toward the sun, and the other away from it, much as the moon always presents the same face to the earth.

Close Encounter

If Mercury was difficult for a professional astronomer like Schiaparelli to observe, it is even more challenging for the amateur. It is never farther than 28 degrees from the sun (due to its small orbital radius) and always seen very low in the sky, either in the west just after sunset or in the east, just before sunrise. Because it is visible only close to the horizon, obstacles and atmospheric conditions (light pollution, smog, and turbulence) may often make it impossible to see. Like the moon (and, as we saw in Chapter 2, Venus), Mercury exhibits phases as different fractions of its face are seen to be illuminated by the sun. The best time to see Mercury is at its crescent phase, because it appears largest in the sky at this time. The reason for the variation in size with phase is that when the planet is on the near side of the sun (at a distance of approximately 0.6 A.U. from us), it is backlit and closer and thus appears large. When it is on the far side of the sun, it is fully illuminated (full), is 1.4 A.U. away, and appears smaller. To get a good look at Mercury, you need a telescope, preferably fitted with an eyepiece that offers about 150× magnification.

It is also possible to see Mercury in the daytime, but this can be dangerous. Because the planet is so close to the sun, there is a real danger that you might accidentally focus on the sun. Doing so for even a moment can permanently damage your eyesight! If you want to look for Mercury during the day, you should consult a good ephemerides guide (see Chapter 17 and Appendix E) and use a telescope fitted with setting circles (see your telescope's instruction manual and Chapter 17) to locate the planet precisely. For added safety, always keep a solar filter on the telescope until you have precisely located the planet.

Better yet: Restrict your viewing of Mercury to just before sunrise or shortly after sunset.

Technology marches on. In 1965, by means of radar imaging, unavailable to Schiaparelli in the nineteenth century, astronomers discovered that Mercury's rotation period was not 88 days, but only 59 days. This discovery implied that Mercury's rotation was not precisely synchronous with its orbit, but that it rotated three times around its axis every two orbits of the sun.

"I Can't Breathe!"

Like the earth's moon, Mercury possesses insufficient mass to hold—by gravitation—an atmosphere for very long. In the same way that mass attracting mass built up planetesimals, so the early planets built up atmospheres by hanging on to them with their gravitational pull. If an atmosphere was ever associated with Mercury, the heating of the sun and the planet's small mass helped it to escape long ago. Without an atmosphere to speak of, the planet is vulnerable to bombardment by meteoroids, x-rays, and ultraviolet radiation, as well as extremes of heat and cold. In sunlight, the planet heats to 700 K. In darkness, with no atmosphere to retain heat, it cools to 100 K.

Radar mapping carried out by the Magellan *probe made this image of volcanic domes in the Eistla region of Venus in 1991.*

(Image from NASA)

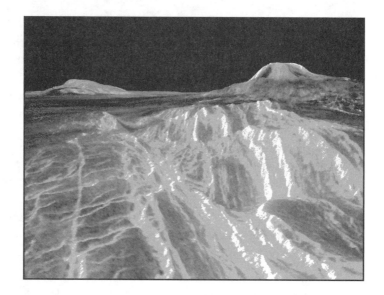

Despite the absence of atmosphere, regions at the poles of Mercury may remain permanently in shadow, with temperatures as low as 125 K. These regions, and similar regions on the earth's moon, may have retained some water ice.

Forecast for Venus: "Hot, Overcast, and Dense"

Venus's thick atmosphere and its proximity to the sun are a cruel combination. The planet absorbs more of the sun's energy (being closer to the sun than the earth) and because of its heavy cloud cover, is unable to radiate away much of the heat. Even before astronomers saw pictures of the planet's surface, they knew that it would not be a welcoming place.

Until the advent of radar imaging aboard space probes such as *Pioneer Venus* (in the late 1970s) and *Magellan* (in the mid-1990s), the surface of Venus was a shrouded mystery. Optical photons bounce off the upper clouds of the planet, and all we can see with even the best optical telescopes is the planet's swirling upper atmosphere. Modern radio imaging techniques (which involve bouncing radio signals off the surface) have revealed a Venusian surface of rolling plains punctuated by a pair of raised land masses that resemble the earth's continents. Venus has no coastlines, all of it's surface water having long ago evaporated in the ghastly heat. These land masses, called Ishtar Terra and Aphrodite Terra, are plateaus in a harsh waterless world.

The Venusian landscape sports some low mountains and volcanoes. Volcanic activity on the surface has produced *calderas* (volcanic craters) and *coronae,* which are vast, rough, circular areas created by titanic volcanic upwellings of the mantle.

Venus is surely lifeless biologically, but geologically it is very active. Volcanic activity is ongoing, and many astronomers believe that the significant, but fluctuating, level of sulfur dioxide above the Venusian cloud cover is the result of volcanic eruptions. Probes sent to Venus thus far have not detected a magnetosphere; however, astronomers still believe that the planet has an iron-rich core. Scientists reason that the core of Venus might simply rotate too slowly to generate a detectable magnetic field.

Star Words

Calderas are craters produced not by meteoroid impact, but by volcanic activity. **Coronae** are another effect of volcanic activity: large upwellings in the mantle, which take the form of concentric fissures, or stretch marks in the planet's surface.

The Sun Sets on Venus (in the East)

As we've seen, Mercury's peculiar rotational pattern can be explained by its proximity to the sun. But no such gravitational explanation is available for the peculiar behavior of Venus. If at 59 days, Mercury rotates on its axis slowly, Venus is even more sluggish, consuming 243 Earth days to accomplish a single spin.

What's more, it spins backwards! That is, viewed from a perspective above the earth's North Pole, all of the planets (terrestrial and jovian) spin counterclockwise—except for Venus, which spins clockwise.

Nobody knows why for sure, but we can guess that the rotational peculiarities of Venus were caused by some random event that occurred during the formation of the solar system—a collision or close encounter with another planetesimal, perhaps. A violent collision, like the one that formed the earth's moon, might have started Venus on its slow backward spin.

Close Encounter

Except for the moon, Venus is the brightest object in the night sky and is, therefore, quite easy to find. Just look for a bright, steadily shining (not twinkling) object in the western sky at or before sunset, or in the eastern sky at or before sunrise. It is so bright, in fact, that training your telescope on it during a very dark night might prove disappointing. If you make your observations in the evening, before full darkness, or in the early morning twilight, you stand a better chance of making out some atmospheric features. Unfortunately, you can't see any surface features because of the planet's thick atmosphere. The most interesting phenomena to observe are the phases, which resemble those of the moon. As with Mercury, the planet appears largest during its crescent phase (when it is backlit by the sun, and closest to the earth). When Galileo saw the variations in the size of the disk of Venus, he realized that it confirmed that the planets moved around the sun, and not around the earth.

Don't totally give up on Venus in a dark sky. If you observe the planet in its thinnest crescent phase, you may (as with observations of the crescent moon) see the rest of the shadowed disk.

Often, Venus can also be observed during the daytime, even at low magnification. As with observing Mercury in daylight, be very careful to avoid pointing your telescope or binoculars at the sun. To do so, even for a moment, will cause permanent damage to your eyesight.

Venusian Atmosphere

Chemically, the atmosphere of Venus consists mostly of carbon dioxide (96.5 percent). The remainder is mostly nitrogen. These are organic gases, which might lead one to jump to the conclusion that life—*some* form of life—may exist on Venus. Indeed, during the 1930s, spectroscopic studies of Venus revealed the temperature of the planet's upper atmosphere to be about 240 K—close to the earth's *surface* temperature of 290 K. Some speculated that the environment of Venus might be a dense jungle.

In the 1950s, radio astronomy was used for the first time to penetrate the dense cloud layer that envelops Venus. It turned out that surface temperatures were not 240 K, but were closer to 600 K. Those temperatures are incompatible with any form of life we know. But the outlook got only worse. Spacecraft probes soon revealed that the dense atmosphere of Venus creates high surface pressure—the crushing equivalent of 90 Earth atmospheres—and that surface temperatures actually top 730 K.

And what about those clouds?

On Earth, clouds are composed of water vapor. But Venus shows little sign of water. Its clouds consist of sulfuric acid droplets.

Venus global surface view, from the Magellan *probe. This view was hidden from astronomers until the advent of high-resolution radar mapping.*

(Image from NASA)

The Earth: Just Right

In our march through the terrestrial planets, the next logical stop would be Earth. We have mentioned some of the unique aspects of our home planet in earlier chapters, and will mention more in the course of the book. In particular, we will look at the earth as a home to life when we discuss the search for life elsewhere in the Milky Way.

But let's take a brief moment to think of the earth as just another one of the terrestrial planets. The earth is almost the same size as Venus, and has a rotational period and inclination on its axis almost identical to Mars. How is it, then, that the earth is apparently the only one of these three planets to support life?

As in real estate, it comes down to three things: location, location, and location. The earth is far enough from the sun that it has not experienced the runaway greenhouse effect of Venus. It is close enough to the sun to maintain a surface temperature that allows for liquid water, and massive enough to hold onto its atmosphere. The molten rock in the mantle layer above its core keeps the crust of the earth in motion (called plate tectonics), and the rotation of this charged material has generated a magnetic field that absorbs and holds on to charged particles that escape from the sun in the solar wind.

These conditions have created an environment in which life has gotten a foothold and flourished. And life has acquired enough diversity that the occasional setback (like the asteroid that struck the earth some 65 million years ago) may change the course of evolution of life on the planet, but has not yet wiped it out.

Our home planet is truly remarkable, and remarkably balanced. The more we learn about our terrestrial neighbors, the more we should appreciate the delicate balance that supports life on Earth.

Mars: "That Looks Like New Mexico!"

Those of us who were glued to our television sets when NASA shared images of the Martian surface produced by the *Mars Pathfinder* probe were struck by the resemblance of the landscape to the earth. Even the vivid red coloring of the rocky soil seems familiar to anyone who has been to parts of Australia or even the state of Georgia—though the general landscape, apart from its color, more closely resembles desert New Mexico.

In contrast to Mercury and Venus, which are barely inclined on their axes (in fact, their axes are almost perpendicular to their orbital planes), Mars is inclined at an angle of 25.2 degrees—quite close to the earth's inclination of 23.5 degrees.

And that's only one similarity. While Mercury and Venus move in ways very different from the earth, Mars moves through space in ways that should seem quite familiar to us. It rotates on its axis once in every 24.6 hours—a little more than an Earth day—and because it is inclined much as the earth is, it also experiences familiar seasonal cycles.

The peculiarities of Mercury and Venus make Mars look more similar to the earth than it really is. Generations have looked to the red planet as a kind of solar system brother, partly believing, partly wishing, partly fearing that life might be found there. But the fact is that life as it exists on Earth cannot exist on the other terrestrial planets.

Martian Weather Report: Cold and Thin Skies

While the atmosphere of Venus is very thick, that of Mars is very thin. It consists of about 95 percent carbon dioxide, 3 percent nitrogen, 2 percent argon, and trace amounts of oxygen, carbon monoxide, and water vapor. While our imaginations might tend to paint Mars as a hot desert planet, it is actually a very cold, very dry, place—some 50 K (on average) colder than the earth.

Section of a geometrically improved, color-enhanced version of a 360-degree pan taken by the Imager for Mars Pathfinder *during 3 Martian days in 1997. Note Sojourner rover vehicle and its tracks.*

(Image from NASA)

The Martian Chronicles

Percival Lowell was born in Boston in 1855, son of one of New England's wealthiest and most distinguished families. His early career was absorbed in literature (his sister Amy Lowell became a famous poet) and Far Eastern travel. He became a diplomat, serving as counselor and foreign secretary to the Korean Special Mission to the United States. But in the 1890s, he read a translation of an 1877 book by Giovanni Schiaparelli, the same Italian astronomer who had concluded that Mercury's rotation was synchronized with its orbit. Reporting his observations of the surface of Mars, Schiaparelli mentioned having discovered *canali*. The word, which means nothing more than "channels" in Italian, was mistranslated as "canals" in what Lowell read, and the budding astronomer, already charmed by exotic places, set off in quest of the *most* exotic of all: Mars—and whatever race of beings had excavated canals upon it.

Lowell dedicated his considerable family fortune to the study of the planet Mars. He built a private observatory in Flagstaff, Arizona, and, after years of

Astro Byte

Atmospheric pressure on Mars is only about $1/150$ that of the earth. On Venus the atmospheric surface pressure is about 100 times greater. Again: Earth is "just right" for us.

observation, published *Mars and Its Canals* in 1906. Noting that the canal network underwent seasonal changes, growing darker in the summer, Lowell theorized that technologically sophisticated beings had created the canals to transport crop-irrigation water from the Martian polar ice caps. In 1924, astronomers searched for radio signals from the planet (using a technique that anticipated the current search for radio signals from the universe), but to no avail. Yet the idea of intelligent life on Mars was so ingrained in the public imagination that, on October 30, 1938, Orson Welles's celebrated radio adaptation of H. G. Wells's 1898 science fiction novel about an invasion from Mars, *War of the Worlds,* triggered national panic.

A variety of space probes have now yielded very high resolution images of Mars, revealing the apparent canals as natural features, such as craters and canyons. While it is true that Mars undergoes seasonal changes, the ice caps consist of a combination of frozen carbon dioxide and water.

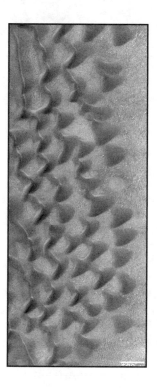

Sand Dunes in Proctor Crater, Mars. The Mars Global Surveyor *has revealed the surface of Mars in unprecedented detail.*

(Image from NASA)

Why Mars Is Red

If we feel any disappointment at the loss of our cherished Martian canals, at least we can still enjoy the image of the "angry red planet." Yet the source of the reddish hue is not the bloody spirit of the Roman god of war, but simple iron ore. The Martian surface contains large amounts of iron oxide, red and rusting. As *Viking 1* and *Mars Pathfinder* images revealed, even the Martian sky takes on a rust-pink tinge during seasonal dust storms.

The dust is blown about by winds that kick up in the Martian summer. These winds play a prominent role on Mars, forming vast dunes and streaking craters. An especially large dune is found around the north polar cap.

Volcanoes, Craters, and a "Grand Canyon"

The *Mariner* series of planetary probes launched in the 1960s and 1970s revealed a startling difference between the southern and northern hemispheres of Mars. The southern hemisphere is far more cratered than the northern hemisphere, which is covered with wind-blown material as well as volcanic lava. There have even been recent proposals that the smooth northern hemisphere hides a frozen ocean.

Volcanoes and lava plains from ancient volcanic activity abound on Mars. Because the planet's surface gravity is low (0.38 that of the earth), the volcanoes can rise to spectacular heights. Like Venus, Mars lacks a strong magnetic field, but, in contrast to Venus, it rotates rapidly; therefore, astronomers conclude that the core of Mars is nonmetallic, nonliquid, or both. Astronomers believe that the core of the smaller Mars has cooled and is likely solid, consisting largely of iron sulfide.

Unlike the earth, Mars failed to develop much *tectonic activity* (instability of the crust), probably because its smaller size meant that the outer layers of the planet cooled rapidly. Instead, volcanic activity was probably quite intense some 2 billion years ago.

Astro Byte

Olympus Mons, found on Mars, is the largest known volcano in the solar system. It is 340 miles (544 km) in diameter and almost 17 miles (27 km) high.

This Mars Orbiter Laser Altimeter *(MOLA) image of the Martian surface shows the famous Valles Marineris region as well as the huge Olympus Mons and Tharsis Bulge regions. The vertical accuracy is 5 meters.*

(Image from NASA and MOLA Science Team)

201

Close Encounter

While you can see Mars on many nights, it is best seen for a span of a few months every 26 months when it is in *opposition:* closest to the earth *and* on the opposite side of the earth from the sun (that is, when the earth comes between it and the sun). Consult one of the annual guides listed in Appendix E for the dates of the next several oppositions. When Mars is at opposition, you can make out remarkable detail, even with a modest telescope, including at least one of the polar caps as well as dark areas (*maria* or "seas") and contrastingly bright "desert" stretches. Colored filters can help enhance certain surface features such as the polar caps.

Mars is the only planet in the solar system that lets us see its surface in detail; however, even when close during opposition, it is possible that Martian dust storms will obscure its surface.

Also impressive are Martian canyons, including Valles Marineris, the "Mariner Valley," which runs some 2,500 miles (4,025 km) along the Martian equator and is as much as 75 miles (120 km) wide and, in some places, more than four miles (6.5 km) deep. The Valles Marineris is not a canyon in the earthly sense, since it was not cut by flowing water, but is a geological fault feature.

Water, Water Anywhere?

Clearly visible on images produced by Martian probes are runoff and outflow channels, which are believed to be dry river beds, evidence that water once flowed as a liquid on Mars. Geological evidence dates the Martian highlands to four billion years ago, the time in which water was apparently sufficiently plentiful to cause widespread flooding. Recent theories suggest that at the time, Mars had a thicker atmosphere that allowed water to exist in a liquid state, even at its low surface temperatures.

The *Mars Global Surveyor* mission, which has established an orbit around the red planet and is transmitting early data back to the earth, has found further geological evidence for the presence of liquid and subsurface water. Such evidence has kept alive hopes that life may have existed—or may even yet exist, perhaps on a microbial level—on Mars.

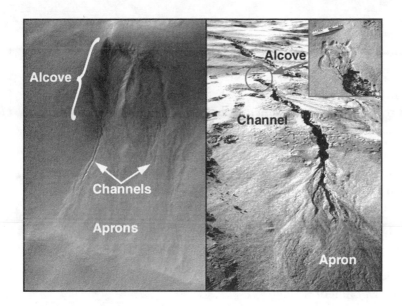

Comparison between similar geological features associated with water runoff on Mars (left) and the Earth (right).

(Image from NASA)

Martian Moons

Mars and Earth are the only terrestrial planets with moons. As we have said, the earth's moon is remarkably large, comparable in size to some of the moons of Jupiter. The moons of Mars, colorfully named Phobos (Fear) and Deimos (Panic), after the horses that drew the chariot of the Roman war god, were not discovered until 1877. They are rather unimpressive as moons go, resembling large asteroids. They are small and irregularly shaped (Phobos is 17.4 miles long × 12.4 miles [28 km × 20 km] wide, and Deimos is 10 miles × 6.2 miles [16 km × 10 km]). They are almost certainly asteroids that were gravitationally captured by the planet and fell into orbit around it.

Where to Next?

Mercury, Venus, Mars: our neighbors, the planets we know most about and, perhaps, feel the closest connection to. At that, however, they are still strange and inhospitable worlds. Strange? Inhospitable? Well, as Al Jolson was famous for saying, "You ain't seen nothin' yet." In the next chapter we visit the jovian planets, the gas giants of the outer solar system.

The Least You Need to Know

➤ The terrestrial planets are Mercury, Venus, Earth, and Mars.

➤ While the terrestrial planets share certain Earth-like qualities, they differ in significant ways that, among other things, make the existence of life on those planets either impossible or highly unlikely.

➤ Mercury and Venus display rotational peculiarities. In the case of Mercury, its rotation can be explained by its proximity to the sun; but the slow retrograde rotation of Venus can be explained only by the occurrence of some random event (probably a collision) early in the formation of the solar system.

➤ Of the terrestrial planets, only the earth has an atmosphere and environment conducive to life.

➤ While Mars may look and seem familiar, its cold surface whipped by dust storms is a harsh environment.

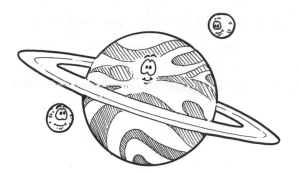

Great Balls of Gas! The Outer Planets

If you've ever been outside late at night looking to the south, chances are you've already seen the largest planet in the solar system, Jupiter. You may have thought that it was just a bright star, and that is exactly what the ancients thought, except that they realized it moved in a way unlike the other stars. Imagine Galileo's surprise, then, in 1610, when he pointed a telescope at the planet and saw its surface and four smaller bodies orbiting it. His discovery would cause a good deal of upheaval in the way humans viewed themselves in the universe, and Galileo himself would end up in trouble with the Church. All this because of that wandering star in the sky.

All of the planets are found near an imaginary arc across the sky that we call the *ecliptic*. Long before astronomers knew that the *terrestrial* planets shared common features, they knew that two of the "wanderers" that they watched were different.

While Mercury and Venus never strayed far from the sun, and Mars moved in a fairly rapid path across the sky, Jupiter and Saturn moved ponderously, majestically across the stellar ocean. In that motion, we had a clue that the outer planets—those farthest from the sun—were unique long before we had telescopes. Mercury, Venus, and Mars may seem inhospitable, forbidding, and downright deadly, but our sister terrestrial planets have more in common with the earth than the giants of the solar system's farthest reaches. The jovian planets are truly *other*-worldly, many times larger and more massive than the earth, yet less dense: They are balls of gas that coalesced around a dense core, accompanied by multiple moons and even rings.

In recent years, thanks to the *Hubble Space Telescope* and planetary probes such as the *Voyagers* and *Galileo* (see Chapter 9, "Space Race: From *Sputnik* to the International Space Station"), the jovian planets and their moons have given up some of their mysteries. In this chapter we explore these worlds. Their moons are so numerous and interesting that we will save a discussion of them until the following chapter.

The Jovian Line-Up

Jupiter: the supreme god in Roman mythology. His name was transmuted into Jove in Middle English, from which the word *jovial* comes, because, during the astrologically obsessed Middle Ages, the influence of Jupiter was regarded as the source of human happiness. Another adjective, *jovian,* usually spelled with a lowercase initial, means "like Jupiter." If the terrestrial planets have Earthlike ("terrestrial") qualities, the jovian planets have much in common with Jupiter.

In addition to Jupiter, by far the largest planet in the solar system (about 300 times more massive than the earth and with a radius 11 times greater), the jovians include Saturn, Uranus, and Neptune.

Planetary Stats

The most immediately striking difference between the terrestrial and jovian worlds are in size and density. It is useful to recall our rough scale: If the earth is a golf ball 0.2 miles from the sun, then Jupiter is a basketball 1 mile away from the sun, and Pluto is a chickpea 8 miles away. At this scale, the sun's diameter would be as big as the height of a typical ceiling (almost 10 feet). While the jovian planets dwarf the terrestrials, they are much less dense. Let's sum up the jovians, compared to the earth:

Planet	Mass in Kilograms	Radius in Miles (and km)	Surface Magnetic Field (Relative to Earth)	"Surface" Gravity (Relative to Earth)	Rotation Period in Solar Days	Atmospheric Temperature in K
Earth	6.0×10^{24}	3,968 (6,380 km)	1.00	1.00	1.00	290
Jupiter	1.9×10^{27}	44,020 (71,400 km)	14	2.5	0.41	124
Saturn	5.7×10^{26}	37,200 (60,200 km)	0.7	1.1	0.43	97
Uranus	8.8×10^{25}	16,120 (25,600 km)	0.7	0.9	−0.72*	58
Neptune	1.0×10^{26}	15,500 (24,800 km)	0.4	1.2	0.67	59

The reason for the minus sign is that the rotation of Uranus is retrograde; like Venus, it rotates on its axis in the opposite direction from the other planets.

We have given a gravitational force and a temperature at the surface of the jovian planets, but as we'll see, they do not really have a surface in the sense that the terrestrial planets do. These numbers are the values for the outer radius of their swirling atmosphere. One surprise might be that the surface gravity of Saturn, Uranus, and Neptune is very close to what we have at the surface the earth. Gravitational force depends on two factors, you'll recall, mass and radius. Although these outer planets are much more massive, their radii are so large that the force of gravity at their "surfaces" is close to that of the smaller, less massive Earth.

Of the jovians, Jupiter and Saturn have the most in common with one another. Both are huge, with their bulk mainly hydrogen and helium. If you recall from our description of the early, developing solar system, the outer solar system (farther from the sun) contained more water and organic materials, and the huge mass and cooler temperatures of the outer planets meant that they were able to gravitationally hold on to the hydrogen and helium in their atmospheres.

The terrestrials consist mostly of rocky and metallic materials, and the jovian planets primarily of lighter elements. The density of a planet is determined by dividing its mass by its volume. While the outer planets are clearly much more massive

Astronomer's Notebook

Why are the jovian planets so big? Recall from Chapter 11 that the jovian planets formed from nebular material far from the sun, in regions that were relatively cool. The lower temperatures allowed water ice and other molecules to condense in the outer solar system sooner, when they were still gaseous in the region closer to the protosun. As a result, there was a larger well of solid material available in the outer solar system to start building up into planetesimals—the outer solar system got a head start. These larger planetesimals (and their stronger gravitational fields) held on to a larger mass of hydrogen and helium, and the gas giants were born.

(which, one might think, would make them more dense), they are much larger in radius, and so encompass a far greater volume. For that reason, the outer planets have (on average) a much lower density than the inner planets, as is clear from the following table:

Planet	Density (kg/m³)
Earth	5,500
Jupiter	1,330
Saturn	710
Uranus	1,240
Neptune	1,670

But what of Uranus and Neptune—distant, faint, and unknown to ancient astronomers?

While they are both much larger than the earth, they are less than half the diameter of Jupiter and Saturn; in our scale model, they would be about the size of a cantaloupe. Uranus and Neptune, though less massive than Saturn, are significantly more dense. Neptune is more dense than Jupiter as well, and Uranus approaches Jupiter in density. Take a look at the following "Astronomer's Notebook" sidebar to understand why this is so.

Consider Neptune. Remember, density is equal to the mass of an object divided by its volume. While the mass of Neptune is about 19 times smaller than that of Jupiter, its volume is 24 times smaller. Thus, we expect its density to be about $^{24}/_{19}$ or 1.3 times greater. While we cannot yet peer beneath the atmospheric surface of Uranus and Neptune, the higher densities of these two planets provide a valuable clue to what's inside.

Reflecting their genesis, all of the jovian planets have thick atmospheres of hydrogen and helium covering a core slightly larger than Earth or Venus. The rocky cores of all four of the jovian planets are believed to have similar radii, on the order of 4,300 to 6,200 miles (7,000 to 10,000 km); but this core represents a much smaller fraction of the full radius of Jupiter and Saturn than do the cores of smaller Uranus and Neptune—thus the higher average density of the latter two planets.

Astronomer's Notebook

To understand the difference between radius and volume, think of our scale system of the planets. The difference in radius between the Earth and Jupiter tells you how many golf balls you can place next to the basketball to equal its diameter. The difference in *volume* between Earth and Jupiter tells you how many golf balls you could crush into dust and then fit *inside* the basketball. Volume goes like radius to the power of 3. So if object A has twice the radius of object B, object A has 2 × 2 × 2 = 8 times B's volume!

The atmospheres of the jovians are ancient, probably little changed from what they were early in the creation of the solar system. With their strong gravitational fields and great mass, these planets have held onto their primordial atmospheric hydrogen and helium, whereas most of these elements long ago escaped from the less massive terrestrial planets, which have much weaker gravitational pull.

But here's where it gets really strange. On the earth, we have the sky (and atmosphere) above, and the solid ground below. In the case of the jovians, the gaseous atmosphere never really ends. It just becomes denser with depth, as layer upon layer of it presses down.

There is no "normal" solid surface to these planets! As the gases become more dense, they become liquid, which is presumably what lies at the core of the jovian worlds. When astronomers speak of the "rocky" cores of these planets, they are talking about chemical composition rather than physical state. Even on the earth, rock may be heated and pressed sufficiently to liquefy it (think of volcanic lava). Thus it is on the jovians: gas giants, whose atmospheres become increasingly dense, but never solid, surrounding a liquid core. In the case of Jupiter and Saturn, the pressures are so great that even the element hydrogen takes on a liquid metallic form.

Latecomers: Uranus and Neptune

Since ancient times, the inventory of the solar system was clear and seemingly complete: a sun and, in addition to Earth, five planets, Mercury, Venus, Mars, Jupiter, and Saturn. Then came along one of those scientific busybodies that the eighteenth century produced in abundance. Johann Daniel Titius, or Tietz (1729–1796), a Prussian born in what is now Poland, poked his curious nose into everything. He was a physicist, biologist, and astronomer who taught at the University of Wittenberg.

It occurred to him, in 1766, that the spacing of the planetary orbits from the sun followed a fairly regular mathematical sequence. He doubled a sequence of numbers beginning with 0 and 3, like this: 0, 3, 6, 12, 24, 48, and so on.

He added 4 to each number in the sequence, then divided each result by 10. Of the first seven answers Titius derived—0.4, 0.7, 1.0, 1.6, 2.8, 5.2, 10.0—six very closely approximated the relative distances from the sun, expressed in astronomical units (remember, an A.U. is the mean distance between the earth and the sun), of the six known planets.

Astro Byte

Caroline Herschel (1750–1848) was an important astronomer in her own right. She discovered eight comets and three nebulae, and reduced and published her brother William's observations of nebulae after his death. In 1835 she was the first woman to be elected as an honorary fellow of the Royal Astronomical Society. The election was "honorary" because, at the time, women were not allowed to be regular members of the RAS.

No one paid much attention to Titius's mathematical curiosity until another Prussian astronomer, Johann Bode (1747–1826), popularized the sequence in 1772.

Neat as it was, the sequence, which became known as the Titius-Bode Law or simply *Bode's Law,* is now generally thought to be nothing more than numerology. For one thing, there is no planet at 2.8 A.U. This gap would be filled later by the discovery of the asteroid belt at this location. While the rule gives a number that is close to Uranus, it breaks down for the positions of Neptune and Pluto. Since those planets had yet to be discovered, no one saw it as a problem.

Astro Byte

Herschel, the court astronomer, knew what side his bread was buttered on and wanted to name the new planet Georgium Sidus—George's Star—after his patron, King George III of England. Bode intervened to suggest Uranus—in Roman mythology, the father of Saturn.

Astronomer's Notebook

If you want to know the right ascension, declination, altitude, and azimuth of any of the planets as viewed from anywhere on the surface of the earth, point your Web browser to the planetary position calculator at http://.imagiware.com/astro/planets.cgi.

But what about the numbers beyond 10.0 A.U.? Did the Titius-Bode law predict other, as yet unknown, planets?

The people of our planet did not have to wait long for an answer. On March 13, 1781, the great British astronomer William Herschel, tirelessly mapping the skies with his sister Caroline, took note of what he believed to be a comet in the region of a star called H Geminorum. On August 31 of the same year, a mathematician named Lexell pegged the orbit of this "comet" at 16 A.U.: precisely the next vacant slot the Titius-Bode Law had predicted.

Herschel, with the aid of a telescope, had discovered the first new planet since ancient times.

Once the planet had been found, a number of astronomers began plotting its orbit. But something was wrong. Repeatedly, over the next half century, the planet's observed positions did not totally coincide with its mathematically predicted positions. By the early nineteenth century, a number of astronomers began speculating that the new planet's apparent violation of Newton's laws of motion had to be caused by the influence of some as yet undiscovered celestial body—that is, yet another planet. For the first time, Isaac Newton's work was used to identify the irregularity in a planet's orbit and to predict where another planet should be. All good scientific theories are able to make testable predictions, and here was a golden opportunity for Newton's theory of gravity.

On July 3, 1841, John Couch Adams (1819–1892), a Cambridge University student, wrote in his diary: "Formed a design in the beginning of this week of investigating, as soon as possible after taking my degree, the irregularities in the motion of Uranus ... in order to find out whether they may be attributed to the action of an undiscovered planet beyond it ...." True to his word, in 1845, he sent to James Challis, director of the Cambridge Observatory, his calculations on where the new planet, as yet undiscovered, could be found. Challis passed the information to another astronomer, George Airy, who didn't get around to doing anything with the figures for a year. By that time, working with calculations supplied by another astronomer (a Frenchman named Jean Joseph Leverrier), Johann Galle, of the Berlin Observatory, found the planet that would be called Neptune. The date was September 23, 1846.

Earthbound Views: Uranus and Neptune

It is possible for the amateur astronomer to see both Uranus and Neptune. In fact, if you know where to look, Uranus is visible, albeit very faintly, even to the naked eye, provided that the night is very dark, very clear, and you are far from sources of light pollution. To view Neptune, which is much fainter than Uranus, requires an advanced amateur telescope. You don't have to spend years sweeping the skies to find these dim and distant worlds. Consult one of the annual guides or Web sites listed this chapter or in Appendix E, "Sources for Astronomers," for accurate information on positions of these planets.

Astro Byte

In addition to Titania and Oberon, the moons of Uranus are Miranda, Ariel, and Umbriel. Titania and Oberon are slightly smaller than the planet Pluto and about half the diameter of the moon. All but Umbriel are named for Shakespearean characters: Oberon and Titania are the king and queen of the fairies in *A Midsummer Night's Dream*, while Miranda is the heroine of *The Tempest*, and Ariel a magical spirit in that play. Umbriel is a "melancholy sprite" in Alexander Pope's narrative poem *The Rape of the Lock* (1712–1714).

Voyager 2 *revealed many secrets of the outer planets, including the faint outermost ring around Uranus. This ring is a less impressive version of the famous rings of Saturn.*

(Image from NASA)

But what can you expect to see? Uranus will appear as a greenish disk, probably featureless—though it is not impossible, given a very good telescope and superb atmospheric conditions, to see atmospheric features and bright spots. It is even possible to see Titania and Oberon, the largest of the planet's five moons.

Even many advanced amateur astronomers have not seen Neptune. Blue in color, it is aptly named for the Roman god of the sea. If you locate the planet at all, it will be a featureless disk.

A Voyager 2 *photograph of Neptune.*

(Image from NASA)

Earthbound Views: Jupiter and Saturn

In contrast to Uranus and Neptune, Jupiter and Saturn make for easy viewing. On a good, dark night, even a quite modest telescope will reveal the planets' belts. The use of colored filters can enhance bands in Jupiter's atmosphere. Moreover, Jupiter rotates so fast (its day consumes a mere ten hours) that any details you see will perceptibly move across the planet's face if you observe long enough. Its rapid rotation also makes the planet appear noticeably *oblate* (elongated). It is even possible to observe the near moons (like Io) emerging from behind Jupiter as they orbit.

Although smaller and nearly twice as distant as Jupiter—and therefore appearing much smaller and dimmer than the larger planet—the sight of Saturn through a refractor of at least a 4-inch aperture or a reflector with at least a 6-inch aperture is thrilling. Expect to see the planetary disk and its belts and zones, as well as its celebrated rings (discussed later in this chapter). You may even catch a glimpse of the moons, including Titan, brightest and biggest of Saturn's *nine* moons (which we will discuss in the next chapter). Titan's atmospheric pressure is similar to Earth's, although its composition and temperature are different. Titan is slightly larger in diameter than the planet Mercury.

Close Encounter

The dark brownish strips across Jupiter are *belts*, the brighter strips are *zones*. Belts are dark, cooler regions, settling lower into the atmosphere as part of a convective cycle. Zones are regions of rising hot atmospheric gas. The bands are the result of regions of the atmosphere moving from high pressure to low pressure regions (much as they do on the earth). The rapid rotation of Jupiter confines this movement to narrow belts. The planet does have an atmospheric geography that can be mapped:

➤ The light-colored central band is the equatorial zone. It may appear white, orange, or yellow.

➤ The Great Red Spot, a hurricane that has been observed south of the equatorial zone since the invention of telescopes, has a diameter approximately twice that of the earth.

➤ On either side of the equatorial zone are dark bands called the north and south equatorial belts. At times, you may witness a south equatorial belt disturbance: an atmospheric storm.

➤ North and south of the equatorial belts are the *north* and *south temperate belts.*

➤ At the extreme northern and southern ends of Jupiter are the *polar regions,* which are sometimes barely perceptible and sometimes quite apparent dark areas.

The Cassini *spacecraft is now on its way to Saturn, scheduled to arrive in 2004. On its way, it swept past Jupiter and tried out its cameras. This image shows some of the most recent close-ups of Jupiter as seen by* Cassini *in the blue (optical), ultraviolet, and near-infrared.*

(*Image from NASA/JPL/ University of Arizona*)

The moons of the outer planets are worlds unto themselves. In this image, the moon Io hovers over the swirling gas surface of Jupiter. The shading of this image reveals that the Sun must be off to the left, 5.2 AU away.

(Image from NASA/JPL/ University of Arizona)

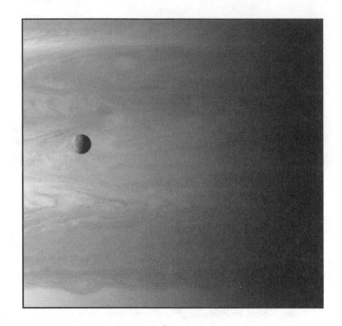

Views from the Voyagers *and* Galileo

During the 1970s and 1980s, two *Voyager* space probes (see Chapter 9) gave us unprecedented images of the jovian planets. *Voyager 1* visited Jupiter and Saturn, and *Voyager 2* added Uranus and Neptune to the list.

Close Encounter

Late in July 1994, *Galileo* was in orbit on the far side of Jupiter when more than 20 fragments of Comet Shoemaker-Levy 9 plunged into the atmosphere over a 6-day period. The fragments were the result of Jupiter's tidal forces that pulled a previously normal comet into a chain of smaller comets. Traveling at more than 40 miles per second (60 km/s), the fragments created a series of spectacular explosions in the planet's upper atmosphere, each with a force comparable to the detonation of a billion atomic bombs. So-called "black-eye" impact sites were created, the result of changes that the impacts produced in the upper atmosphere of the planet.

The *Voyager* missions also revealed volcanic activity on Io, one of Jupiter's moons. As for Saturn, a new, previously unknown system of rings emerged: several thousand ringlets. Ten additional moons were discovered orbiting Uranus, which also revealed the presence of a stronger magnetic field than had been predicted. And the Neptune flyby led to the discovery of three planetary rings as well as six previously unknown moons. The hitherto featureless blue face of the planet was now resolved into atmospheric bands, as well as giant cloud streaks. As a result of the *Voyager 2* flyby, the *magnetospheres* of Neptune and Uranus were detected. As with the *Van Allen belts* around the earth, the magnetospheres of these planets trap charged particles (protons and electrons) from the solar wind.

If only its namesake could have lived to see it. Launched in 1989, *Galileo* reached Jupiter in 1995 and began a complex 23-month orbital tour of the planet and its moons (see Chapter 15, "The Far End of the Block,") almost 400 years after the Italian astronomer first gazed on its colored bands and moons. Among the most extraordinary of *Galileo's* discoveries is a new ring of dust that has a retrograde (backward) orbit around Jupiter. About 700,000 miles (1,120,000 km) in diameter, this doughnut-shaped ring moves in the opposite direction of the rotating planet and its moons. Why does it move in this fashion? No one yet knows.

The *Cassini* space probe passed Jupiter in early 2001 and sent back images from its many cameras. It will reach Saturn in 2004. Stay tuned for more remarkable discoveries.

Rotation: A New Twist

With all the bands and surface features of the biggest jovian planets, you'd think it would be relatively easy to calculate rotation rates "by eye." Just look for a prominent surface feature and time how long it takes that feature to make one trip around.

Well, it's not so easy. Because these planets lack solid surfaces, different features on the surface actually rotate at differing rates! This *differential rotation* is not dramatic in the case of Jupiter,

Star Words

A **magnetosphere** is a zone of electrically charged particles trapped by a planet's magnetic field. The magnetosphere lies above the planet's atmosphere. The **Van Allen belts,** named for their discoverer, American physicist James A. Van Allen, are vast doughnut-shaped zones of highly energetic charged particles that are trapped in the magnetic field of the earth. The zones were discovered in 1958.

Astro Byte

The radio signals sent back from the outer solar system by *Voyager 2* were so faint that the Very Large Array in Socorro, New Mexico had to be equipped by NASA with special receivers to detect the signals.

whose equatorial region rotates only slightly faster than regions at higher latitudes. East-west winds move at about 190 miles per hour (300 km/h) in Jupiter's equatorial regions, and at a zippy 800 miles per hour (1,300 km/h) in the equatorial regions of Saturn. It turns out that the best way to clock the rotation rates of these planets is not to look at their atmospheres, but to measure something tied to the planets' cores. The periods of fluctuation in the radio emission (which arise from the planets' magnetic fields) are taken to be the "true" rotation rate.

While Neptune and Saturn are slightly tipped on their axes similar to the earth (30, 27, and 24 degrees, respectively), Jupiter's axis is nearly perpendicular to the plane of its orbit; the planet tilts from the perpendicular a mere 3 degrees.

Star Words

Differential rotation is a property of anything that rotates, and is not rigid. A spinning CD is a rigid rotator. A spinning piece of gelatin is not as rigid, and a spinning cloud of gas is even less so. The atmospheres of the outer planets and of the sun have equatorial regions that rotate at a different rate from the polar regions.

The true oddball in this respect is Uranus, which tilts 98 degrees, in effect lying on its side. The result of this peculiarity is that Uranus has the most extreme seasons in the solar system. While one pole experiences continuous daylight for 42 Earth years at a stretch, the other is plunged into an equal period of darkness.

It's interesting to note that if the earth were tipped on its axis like Uranus, a city like Atlanta would experience 70 days when the sun never rose, and 70 days when the sun never set. The North Pole would have 6 months of darkness, and 6 months of sunlight. On the vernal and autumnal equinoxes, day and night in Atlanta would still each last 12 hours.

Stormy Weather

Jupiter's spectacular surface features belie a turbulent atmosphere. In addition to a prevailing eastward and westward wind flow called *zonal flow,* there are many smaller-scale weather patterns, as evidenced by such features as the Great Red Spot.

The Great Red Spot

The Great Red Spot was first reported by the British scientist Robert Hooke (1635–1703). It is a storm, a swirling hurricane or whirlpool, of gigantic dimensions (twice the size of the earth), at least 300 years old. It rotates once every six days and is accompanied by other smaller storms. Neptune has a similar storm called the Great Dark Spot.

Star Words

Zonal flow is the prevailing east-west wind pattern that is found on Jupiter.

How could a storm last for three centuries or more? We know from our experience on the earth that hurricanes form over the ocean and may remain active there for days. Once they move over land, however, they are soon spent (albeit often destructively); the land mass disrupts the flow pattern and removes the source of energy. On Jupiter, however, there is no land. Once a storm starts, it continues indefinitely, until a larger storm disrupts it. The Great Red Spot is the biggest storm on the planet.

Bands of Atmosphere

The atmospheric bands that are Jupiter's most striking feature are the result of *convective motion* and zonal wind patterns. Warm gases rise, while cooler gases sink. The location of particular bands appear to be associated with the wind speed on Jupiter at various latitudes.

Anyone who watches an earthly television weather forecast is familiar with high-pressure and low-pressure areas. Air masses move from high pressure regions to low pressure regions. But we never see these regions on the earth as regular zones or bands that circle the planet. That's because the earth doesn't rotate nearly as fast as Jupiter. The rapid rotation of the giant planet spreads the regions of high and low pressure out over the entire planet.

Layers of Gas

On July 13, 1995, *Galileo* released an atmospheric probe, which plunged into Jupiter's atmosphere and transmitted data for almost an hour before it was destroyed by intense atmospheric heat and pressure. After analysis of this data (and earlier data from *Voyager*), astronomers concluded that Jupiter's atmosphere is arranged in distinct layers. Since there is no solid surface to call sea level, the *troposphere* (the region containing the clouds we see) is considered zero altitude, and the atmosphere is mapped in positive and negative distances from this.

Just above the troposphere is a haze layer, and just below it are white clouds of ammonia ice. Temperatures in this region are 125–150 K. Starting at about –40 miles (60 km) below the ammonia ice level is a cloud layer of ammonium hydrosulfide ice,

Astro Byte

Astronomers don't measure the jovian rotation rates visually, but by monitoring the periodic fluctuations in the strength of radio emissions from the planets' powerful magnetospheres. Since the magnetic fields are anchored deeper in the core of the planet, this measurement is more fundamental.

Star Words

Convective motion is any flow pattern created by the rising movement of warm gases (or liquids) and the sinking movement of cooler gases (or liquids). Convective motion is known to occur in planetary atmospheres, the sun's photosphere, and a tea kettle on the stove.

Astro Byte

The total thickness of Saturn's 3 cloud layers—ammonia ice, ammonium hydrosulfide ice, and water ice—is about 155 miles (250 km), compared to about 50 miles (80 km) on Jupiter.

in which temperatures climb to 200 K. Below this level are clouds of water ice and water vapor, down to about –60 miles (100 km). Further down are the substances that make up the interior of the planet: hydrogen, helium, methane, ammonia, and water, with temperatures steadily rising the deeper we go.

Saturnine Atmosphere

Saturn's atmosphere is similar to Jupiter's—mostly hydrogen (92.4 percent) and helium (7.4 percent) with traces of methane and ammonia; however, its weaker gravity results in thicker cloud layers that give the planet a more uniform appearance, with much subtler banding, than Jupiter. Temperature rises much more slowly as a function of depth in the atmosphere on Saturn.

The Atmospheres of Uranus and Neptune

The atmospheres of Uranus and Neptune have not been probed by unmanned space vehicles, but they have been studied spectroscopically from the earth, revealing that,

Astro Byte

The core temperature of Jupiter must be very high, perhaps 40,000 K. Astronomers speculate that the core diameter is about 12,500 miles (20,000 km), a bit larger than the earth. As the jovian planets collapsed, part of their gravitational energy was released as heat. Some of this heat continues to be released, meaning that Jupiter, Saturn, and Neptune have internal heat sources.

like Jupiter and Saturn, they are mostly hydrogen (about 84 percent) and helium (about 14 percent). Methane makes up about 3 percent of Neptune's atmosphere, and 2 percent of Uranus's, but ammonia is far less in abundance on either planet than on Jupiter and Saturn. Because Uranus and Neptune are colder and have much lower atmospheric pressure than the larger planets, any ammonia present is frozen. The lack of ammonia in the atmosphere and the significant presence of methane give both Uranus and Neptune a bluish appearance, since methane absorbs red light and reflects blue. Uranus, with slightly less methane than Neptune, is blue-green, while Neptune is quite blue.

Uranus reveals almost no atmospheric features. Those that are there are submerged under layers of haze. Neptune, as seen by *Voyager 2,* reveals more atmospheric features and even some storm systems, including a Great Dark Spot, an area of storm comparable in size to the earth. Discovered by *Voyager 2* in 1989, the Great Dark Spot had vanished by the time the *Hubble Space Telescope* observed the planet in 1994.

Inside the Jovians

How do you gather information about the interior of planets that lack a solid surface and that are so different from the earth? You combine the best observational data you have with testable, constrained speculation known as theoretical modeling. Doing just this, astronomers have concluded that the interiors of all four jovians consist largely of the elements found in their atmospheres: hydrogen and helium. As we go deeper into the planet, the gases, at increasing pressure and temperature, become liquid.

In the case of Jupiter, it is believed that the hot liquid hydrogen is transformed from molecular hydrogen to metallic hydrogen and behaves much like a molten metal, in which electrons are not bound to a single nucleus, but move freely, conducting electrical charge. As we shall see in just a moment, this state of hydrogen is likely related to the creation of Jupiter's magnetosphere—the result of its powerful magnetic field. Astronomers are less confident about the nature of the very core of Jupiter, though most believe that it is a rocky core the diameter of the earth. Of course, the incredible temperatures and pressures at this depth in Jupiter mean that the material in the core might behave very differently from materials that we have studied on Earth.

Saturn's internal composition is doubtless similar to Jupiter's, though its layer of metallic hydrogen is probably proportionately thinner, while its core is slightly larger. Temperature and pressure at the Saturnine core are certainly less extreme than on Jupiter.

Uranus and Neptune are believed to have rocky cores of similar size to those of Jupiter and Saturn surrounded by a slushy layer consisting of water clouds and, perhaps, the ammonia that is largely absent from the outer atmosphere of these planets. Because Uranus and Neptune have significant magneto-spheres, some scientists speculate that the ammonia might create an electrically conducting layer, needed to generate the detected magnetic field. Above the slushy layer is molecular hydrogen. Without the enormous internal pressures present in Jupiter and Saturn, the hydrogen does not assume a metallic form.

The Jovian Magnetospheres

Jupiter's magnetosphere is the most powerful in the solar system. Its extent reaches some 18,600,000 miles (30 million km) north to south. Saturn has a magnetosphere that extends about 600,000 miles (1 million km) toward the sun. The magnetospheres of Uranus and Neptune are smaller, weaker, and (strangely) offset from the gravitational center of the planets.

Astro Byte

Io (as we'll see in the next chapter) is volcanically active. Due to its small mass (Io is just slightly smaller in diameter than the earth's moon), its eruptions have enough velocity to send some charged particles into orbit around Jupiter. These particles are trapped by Jupiter's enormous magnetic field.

The rapid rate of rotation and the theorized presence of electrically conductive metallic hydrogen inside Jupiter and Saturn account for the generation of these planets' strong magnetic fields. While Uranus and Neptune also rotate rapidly, it is less clear what internal material generates the magnetic fields surrounding these planets, since they are not thought to have metallic hydrogen in their cores. With charged particles trapped by their magnetospheres, the jovian planets experience *Aurora Borealis,* or "Northern Lights," just as we do here on Earth. These "lights" occur when charged particles escape the magnetosphere and spiral along the field lines onto the planet's poles. The *Hubble Space Telescope* has imaged such auroras at the poles of Jupiter and Saturn.

The Least You Need to Know

➤ The jovian planets are Jupiter, Saturn, Uranus, and Neptune.

➤ While the jovians are the largest, most massive planets in the solar system, they are on average less dense than the terrestrial planets. Their outer layers of hydrogen and helium gas cover a dense core.

➤ Uranus was discovered in 1781, partly because of a bit of numerology called the Bode (or Titius–Bode) Law.

➤ Neptune was discovered in 1846, partly because astronomers were searching for an explanation of Uranus's slightly irregular orbit. Newton's theory of gravity had an explanation—the mass of another planet.

➤ The jovian planets all have in common thick atmospheres, ring systems, and strong magnetic fields.

➤ The missions to the outer planets, *Voyager 1, Voyager 2,* and *Galileo* have had a huge impact on our understanding of the jovian planets. The *Cassini* mission to Saturn and its moons, which will arrive in 2004, is likely to have a similar effect.

Chapter 15

The Far End of the Block

> ### In This Chapter
>
> ➤ Anatomy of Saturn's rings
>
> ➤ The other jovian ring systems
>
> ➤ The large moons of Jupiter and the other jovians
>
> ➤ Europa and Titan: Is life possible there?
>
> ➤ Atmospheres of the jovian moons
>
> ➤ Tiny and distant: Pluto and Charon

Among the great showpieces of our solar system are the fabulous rings of Saturn. Visible through a pair of binoculars, Saturn's ring system appears as solid as the planet. The truth is that the rings are not solid at all, but are composed of small bits of ice and rock. Perhaps more surprisingly, the *Voyager* spacecraft and ground-based adaptive optics images show that all of the jovian planets have ring systems—not just Saturn.

If you were somehow disappointed to discover, in the preceding chapter, that the jovian planets are just big balls of gas, you can take satisfaction in knowing that some of their moons at least are solid, large, and much more like the terrestrial planets. Jupiter's Ganymede and Saturn's Titan, for example, are larger than the planet Mercury. And two of the jovian moons, Jupiter's Europa and Saturn's Titan, present some of the most promising locations for life, outside of Earth, in the solar system.

And then there is Pluto. Unknown before 1930 because of its great distance and diminutive size (it is smaller than the earth's moon and similar in size to many of the jovian moons), Pluto is unlike either the terrestrial or jovian worlds. But we first have unfinished business from the preceding chapter. We begin with the rings of the outer planets.

Lord of the Rings

The spectacular rings of Saturn, a thrill to view in remarkable images from *Voyager 2* and the *Hubble Space Telescope,* are, if anything, even more exciting when seen through your own telescope.

Looking from Earth

In Chapter 4, "Astronomy Reborn: 1543–1687," we observed that Galileo's telescope, a wondrous device in 1610, would be no match even for a decent amateur instrument today. When he first observed the planet, all Galileo could tell about Saturn was that it seemed to have "ears." He speculated that this feature might be topographical, great mountain ranges of some sort. Or perhaps that Saturn was a triple planet system. It wasn't until a half-century later, in 1656, that Christian Huygens, of the Netherlands, was able to make out this feature for what it was: a thin ring encircling the planet. A few years later, in the 1670s, the Italian-born French astronomer Gian Domenico Cassini (1625–1712) discovered the dark gap between what are now called rings A and B. This feature is now called the *Cassini division*.

Six major rings, all lying in the equatorial plane of Saturn, have been identified, of which three, in addition to the Cassini division and a subtler demarcation called the Encke division, can be seen from the earth with a good telescope. With a typical amateur instrument you should be able to see ring A (the outermost ring), the Cassini division, and inside the Cassini division, ring B.

If you become a serious Saturn observer, you will notice that the rings of Saturn are seen at different angles at different times. Sometimes we look down on the top of the ring system, and at other times we see it "edge-on." When the angle is right, it is possible to see the dramatic image of Saturn's shadow cast onto its rings. Consult any of the guides in Appendix E for information on where to look for Saturn and when to view it.

The rings readily visible from the earth are vast, the outer radius of the A ring stretching more than 84,800 miles.

Star Words

The **Cassini division** is a dark gap between rings A and B of Saturn. It is named for its discoverer, Gian Domenico Cassini (1625–1712). A NASA mission named in his honor will arrive at Saturn in 2004.

Close Encounter

Colored filters can help to enhance surface detail on a planet or in the rings of Saturn. For example, a deep yellow filter can enhance the polar caps of Mars, and also aid in identifying the Cassini division of Saturn's rings. Filters, often attached to the telescope eyepiece, pass a smaller band in the optical spectrum and are available relatively cheaply from camera, telescope, and binocular suppliers.

Big as the rings are, they are also very thin—in places only about 65 feet (20 m) thick. If you wanted to make an accurate scale model of the rings and fashioned them to the thickness of this sheet of paper, they would have to be a mile wide to remain in proper scale.

Speculation as to the composition of the rings began with their discovery in the mid-seventeenth century. In 1857, James Clerk Maxwell, the British physicist who had been critical of the nebular hypothesis of the formation of the solar system, concluded that the rings must consist of many small particles in orbit around Saturn. By the end of the century, the instrumentation existed to measure reflectivity, the differences in the way sunlight was reflected from the rings. These observations showed that the rings behaved as was to be expected if they were made up of particles; that is, orbital speeds closer to the planet were faster than those farther out—they were in differential rotation, not rotating as a solid disk might.

Where do the rings come from? There are two ways to think about the question, and both involve the gravitational field of the host planet. First, the rings may be the result of a shattered moon. According to this theory, a satellite could have been orbiting too close to the planet and have been torn apart by tidal forces (the same sort of forces that pulled comet Shoemaker-Levy 9 into pieces), or it might have been shattered by a collision. In either case,

Astronomer's Notebook

The point at which a moon orbiting a planet might be pulled apart is not random. Edouard Roche, a nineteenth-century mathematician, calculated the distance as 2.4 times the radius of the planet, assuming the moon is of density similar to that of the parent planet. Saturn's "Roche limit" is 89,280 miles (144,000 km) from the planet's center, 2.4 times its radius. Any moon that crosses this boundary will be pulled to pieces.

the fragments of the former moon continued to orbit the planet, but now as fragmentary material. The other possibility is that the rings are material left over from the formation of the planet itself, material that was never able to coalesce into planets due to the strong gravitational field of the host planet.

Astronomer's Notebook

The biggest gap in Saturn's rings, the Cassini division, was, of course, discovered long before the *Voyager* missions. It is the result not of moonlets, but of the gravitational influence of Mimas, which deflected some of the ring's particles into different orbits, creating a gap large enough to be visible from the earth.

A Voyager 2 *computer-enhanced view of Saturn's rings from 1981.*

(Image from NASA)

Looking with Voyager

The *Voyager* probes told us much more about the rings than we could have discovered from our earthly perspective. First, data from *Voyager* confirmed that the rings are indeed made up of particles, primarily of water ice. *Voyager* also revealed additional rings, invisible from an earthly perspective. The F ring is more than twice the size of the A ring, stretching out to 186,000 miles.

The D ring is the innermost ring—closer to the planet than the innermost ring visible from the earth, the C ring. F and E are outside of the A ring.

But these additional rings are only part of what *Voyager* told us. *Voyager 2* revealed that the six major rings are composed of many thousands of individual ringlets, which astronomers liken to ripples or waves in the rings.

Voyager 2 also revealed many gaps within the rings, which are believed to be caused by small moonlets, which may be considered very large ring particles—a few miles in diameter—in orbit around Saturn. The gaps are, in effect, the wake of these bodies.

Perhaps the most remarkable *Voyager* discovery concerning Saturn's rings concerns the outermost F ring. Its structure is highly complex, sometimes appearing braided. Apparently, the structure of the F ring is influenced by two small outlying moons that bracket the ring called shepherd satellites, which seem to keep the F ring particles from moving in or out.

More Rings on the Far Planets

During a 1977 Earth-based observation of Uranus in the course of a *stellar occultation* (the passage of Uranus in front of the star), the star's light dimmed several times before disappearing behind the planet. That dimming of the star's light revealed the presence of nine thin, faint rings around the planet. *Voyager 2* revealed another pair. Uranus's rings are very narrow—most of them less than 6 miles (10 km) wide—and are kept together by the kind of shepherd satellites that are found outside of Saturn's F ring. Neptune has rings similar to those of Uranus.

Star Words

A **stellar occultation** occurs when a planet or moon passes in front of a star, dimming the star's light (as seen from the earth). The exact way in which the light dims can reveal details, for example, in the occulting planet's or moon's atmosphere.

On the Shoulders of Giants

One of the key differences between the terrestrial and many jovian planets is that, while the terrestrials have few if any moons, the jovians each have several: 16 (at least) for Jupiter, over 25 for Saturn, 15 for Uranus, and 8 for Neptune. Of these known moons, only 6 are classified as *large bodies,* comparable in size to the earth's moon. Our own moon is all the more remarkable when compared to the moons of the much larger jovian planets. It is larger than all of the known moons except for Ganymede, Titan, Callisto, and Io. The largest Jovian moons (in order of decreasing radius) are …

➤ Ganymede orbits Jupiter; approximate radius: 1,630 miles (2,630 km)

➤ Titan orbits Saturn; approximate radius: 1,600 miles (2,580 km)

➤ Callisto orbits Jupiter; approximate radius: 1,488 miles (2,400 km)

➤ Io orbits Jupiter; approximate radius: 1,130 miles (1,820 km)

➤ Europa orbits Jupiter; approximate radius: 973 miles (1,570 km)

➤ Triton orbits Neptune; approximate radius: 856 miles (1,380 km)

It is interesting to compare these to the earth's moon, with a radius of about 1,079 miles (1,740 km), and the planet Pluto, smaller than them all, with a radius of 713 miles (1,150 km).

The rest of the moons are either *medium-sized bodies*—with radii from 124 miles (200 km) to 465 miles (750 km)—or *small bodies,* with radii of less than 93 miles (150 km). Many of the moons are either entirely or mostly composed of water ice, and some of the smallest bodies are no more than irregularly shaped rock and ice chunks.

Faraway Moons

Thanks to the *Voyager* and *Galileo* space probes, we have some remarkable images and data about the moons at the far end of our solar system. Those that have received the most attention, since they are the largest, are the so-called Galilean moons of Jupiter; Saturn's Titan; and Neptune's Triton. They were first observed in 1610 by Galileo Galilei.

Jupiter's Four Galilean Moons

The four large moons of Jupiter are very large, ranging in size from Europa, only a bit smaller than the earth's moon, to Ganymede, which is larger than the planet Mercury. Certainly, they are large enough to have been discovered even through the crude telescope of Galileo Galilei, after whom they have been given their group name. In his notebooks, Galileo called the moons simply I, II, III, and IV. Fortunately, they were eventually given more poetic names, Io, Europa, Ganymede, and Callisto, drawn from Roman mythology. These four are, appropriately, the attendants serving the god Jupiter.

A montage of Jupiter and its four planet-sized Galilean moons, photographed in 1979 by Voyager 1.

(Image from NASA)

Io is closest to Jupiter, orbiting at an average distance of 261,640 miles (421,240 km); Europa comes next (416,020 miles or 669,792 km); then Ganymede (663,400 miles or 1,068,074 km); and finally Callisto (1,165,600 miles or 1,876,616 km). Intriguingly, data from *Galileo* suggests that the core of Io is metallic, and its outer layers rocky—much like the planets closest to the sun. Europa has a rocky core, with a covering of ice and water. The two outer large moons, Ganymede and Callisto, also have more icy surfaces surrounding rocky cores.

This pattern of decreasing density with distance from the central body mimics that of the solar system at large, in which the densest planets, those with metallic cores, orbit nearest the sun, while those composed of less dense materials orbit farthest away. This similarity is no mere coincidence and can be used to discover more about how the Jupiter "system" formed and evolved.

Let's look briefly at each of Jupiter's large moons.

Because of our own moon, we are accustomed to thinking of moons generally as geologically dead places. Nothing could be further from the truth in the case of Io, which has the distinction of being the most geologically active object in the solar system.

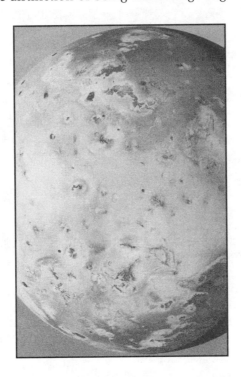

A spectacular portrait of Jupiter's volcanic moon Io by the Galileo *spacecraft. The image was created on March 29, 1998, at a range of 183,000 miles (294,000 kilometers) above Io.*

(Image from JPL/NASA)

Io's spectacularly active volcanoes continually spew lava, which keeps the surface of Io relatively smooth—any craters are quickly filled in—but also angry-looking, vivid orange and yellow, sulfurous. In truth, Io is much too small to generate the kind of heat energy that produces vulcanism (volcanic activity); however, orbiting as close as it does to Jupiter, it is subjected to the giant planet's tremendous gravitational field, which produces tidal forces (see Chapter 10, "The Moon: Our Closest Neighbor"). These forces stretch the planet from its spherical shape and create the geologically unsettled conditions on Io. Think about what happens when you rapidly squeeze a small rubber ball. The action soon makes the ball quite warm. The forces exerted on Io by Jupiter are analogous to this, but on a titanic scale. Don't invest in an Io globe for your desk. Its surface features change even faster than political boundaries on the earth!

In contrast to Io, Europa is a cold world—but probably not an entirely frozen world, and perhaps, therefore, not a dead world. Images from *Galileo* suggest that Europa is covered by a crust of water ice, which is networked with cracks and ridges. It is possible that beneath this frozen crust is an ocean of *liquid* water (not frozen water or water vapor). Liquid water is certainly a requisite of life on Earth, though the presence of water does not dictate the existence of life. Still, the prospects are most exciting. Europa may be a literal lifeboat in the outer solar system, although before we get our hopes up, we need to realize just how cold Europa is at 130 K and how thin its atmosphere is—at a pressure approximately one billionth that on Earth.

Ice on the surface of Europa. The smooth dark regions may be areas where water has welled up from underneath the "ice shelf" that covers the moon.

(Image from NASA/JPL)

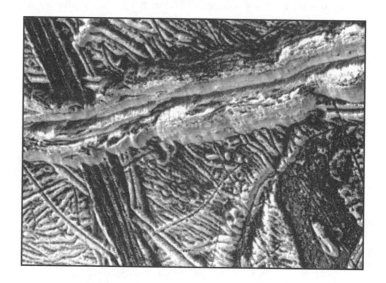

Ganymede is the largest moon in the solar system (bigger than the planet Mercury). Its surface shows evidence of subsurface ice that was liquefied by the impact of asteroids and then refrozen. Callisto is smaller but similar in composition. Both are ancient worlds of water ice, impacted by craters. There is little evidence of the current presence of liquid water on these moons.

Titan: Saturn's Highly Atmospheric Moon

If Io is the most geologically active moon in the solar system and Ganymede the largest, Saturn's Titan enjoys the distinction of having the most substantial atmosphere of any moon. No wispy, trace covering, Titan's atmosphere is mostly nitrogen (90 percent) and argon (nearly 10 percent) with traces of methane and other gases in an atmosphere thicker than the earth's. The earth's atmosphere consists of 78 percent nitrogen, 21 percent oxygen, and 1 percent argon. Surface pressure on Titan is about 1.5 times that of the earth. But its surface is very cold, about 90 K. Remember 90 K is –183 C!

Titan's atmosphere prevents any visible-light view of the surface, though astronomers speculate that the interior of Titan is probably a rocky core surrounded by ice, much like Ganymede and Callisto. Because Titan's temperature is lower than that of Jupiter's large moons, it has retained its atmosphere. The presence of an atmosphere thick with organic molecules (carbon monoxide, nitrogen compounds, and various hydrocarbons have been detected in the upper atmosphere) has led to speculation that Titan might support some form of life. The *Cassini* spacecraft will release a probe into the moon's atmosphere when it arrives at Saturn in 2004.

This artist's conception shows the Titan Probe of the Cassini-Huygens Mission to the planet Saturn arriving on the surface of Titan.

(Image from NASA)

Triton, Neptune's Large Moon

Triton's distinction among the jovian moons is a retrograde (backward) orbit—in the reverse direction of the other moons. Moreover, Triton is inclined on its axis about 20 degrees and is the only large jovian moon that does not orbit in the equatorial plane of its planet. Many astronomers believe that these peculiarities are the result of some violent event, perhaps a collision. Others suggest that Triton did not form as part of the Neptunian system of moons, but was captured later by the planet's gravitational field.

Triton's atmosphere is so thin that *Voyager 2* had no trouble imaging the moon's surface, finding vast lakes of water ice or water-ammonia mixtures there. Nitrogen frost, found at the polar caps, appears to retreat and reforms seasonally.

A Dozen More Moons in the Outer Solar System

Thanks to *Voyager*, the six medium-sized moons of Saturn have also been explored. All of these bodies are *tidally locked* (see Chapter 10) with Saturn, their orbits synchronous, so that they show but one face to their parent planet.

They are frozen worlds, mostly rock and water ice. The most distant from Saturn, Iapetus, orbits some 2,207,200 miles (3,560,000 km) from its parent. Because these moons orbit synchronously, astronomers speak of their *leading faces* and *trailing faces*. That one face always looks in the direction of the orbit and the other in the opposite direction has created asymmetrical surface features on some of these moons. The leading face of Iapetus, for example, is very dark, while the trailing face is quite light. While some astronomers suggest that the dark material covering this moon's leading face is generated internally, others believe that Iapetus sweeps up the dust it encounters.

The innermost moon of Saturn, Mimas is 115,320 miles (186,000 km) out. It is also the smallest of Saturn's moons, with a radius of just 124 miles (200 km). Mimas is very close to Saturn's rings and seems to have been battered by material associated with them. Heavily cratered overall, this small moon has one enormous crater named for the astronomer William Herschel, which makes it resemble the "Death Star" commanded by Darth Vader in *Star Wars*. Whatever caused this impact probably came close to shattering Mimas. Indeed, some astronomers believe that similar impacts may have created some of the debris that formed Saturn's great rings. The Cassini mission will add greatly to our knowledge of these moons and rings.

Star Words

Moons that are **tidally locked** to their parent planet have a **leading face** and a **trailing face;** the leading face always looks in the direction of its orbital motion, the trailing face away from it.

The Cassini-Huygens Mission to Saturn and its mysterious moon Titan was launched on October 15, 1997. The spacecraft will separate into two parts as it approaches Saturn, sending the Huygens Titan probe on a mission to the surface of the atmosphere-enshrouded moon. The mission will study the magnetosphere of Saturn, the planet Saturn itself and its atmosphere and rings, the moon Titan, and finally the other icy moons that orbit the planet. If there were any worries about the performance of the spacecraft, or what it will do when it arrives at Saturn on July 1, 2004, they were substantially allayed in early 2001 when the Cassini-Huygens Mission sped past Jupiter, snapping pictures of the gas giant. You can check on the progress of the mission and view its photos of Jupiter at www.jpl.nasa.gov/cassini/.

Mimas, (or the "Death Star") with its prominent Herschel Crater clearly visible, as seen by Voyager 1 *on November 12, 1980.*

(Image from NASA)

The medium-sized moons of Uranus are Miranda, orbiting 80,600 miles (130,000 km) above the planet; Ariel, 118,400 miles (191,000 km) out; Umbriel, 164,900 miles (266,000 km) out; Titania, 270,300 miles (436,000 km) out; and Oberon, 361,500 miles (583,000 km) out. Of these, the most remarkable is Miranda, which, in contrast to the other moons, is extremely varied geographically, with ridges, valleys, and oval-shaped faults. To the camera of *Voyager 2*, it presented a chaotic, violently fractured, cobbled-together surface unlike that of any other moon in the solar system. Clearly, this moon had a violent past, though it is unclear whether the disruptions it suffered came from within, without, or both. Some astronomers believe that Miranda was virtually shattered, its pieces coming back together in a near-jumble.

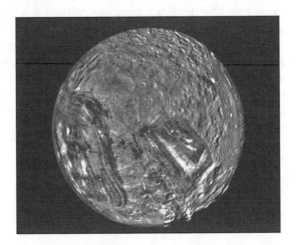

The Frankensteinian face of the moon Miranda.

(Image from NASA)

231

Pluto Found

We are accustomed to thinking of the planets as our ancient companions, so it may come as a surprise to many that one planet, Pluto, was not found until 1930. In Chapter 14, "Great Balls of Gas! The Outer Planets," we saw that irregularities in the orbit of Uranus implied the existence of Neptune, which was not discovered until the middle of the nineteenth century. Yet the discovery of Neptune never did fully account for the idiosyncrasies of the Uranian orbit. Neptune also seemed to be influenced by some as-yet unknown body.

The keen, if eccentric, astronomer Percival Lowell (1855–1916) crunched the numbers for Uranus's orbit and searched for a new planet, in vain, for some ten years. Clyde Tombaugh (1906–1997) was too young to have known Percival Lowell personally, but he took a job as assistant astronomer at the observatory Lowell had built in Flagstaff, Arizona. It was Tombaugh who eventually found the planet by studying photographic images in 1930.

That Pluto was found a mere six degrees from where Lowell had said it would be is more a testament to serendipity than to astronomy. The supposedly persistent irregularities in the orbits of Uranus and Neptune, on which he based his calculations, simply don't exist.

But there was Pluto nevertheless: 3.7 billion miles (5.9 billion km) from the sun, on average, but with an orbit so eccentric that it actually comes closer to the sun than Neptune does about every 248 years. It has yet to be visited by unmanned probes (although the *Hubble Space Telescope* has taken pictures) and, with amateur telescope equipment, is extremely difficult to detect. It is very faint and, therefore, virtually indistinguishable from surrounding stars, except for its movement. Few amateurs have seen the planet.

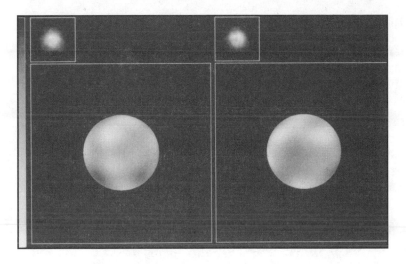

A Hubble Space Telescope view of Pluto, using the space observatory's Faint Object Camera (FOC). Hubble photographed the planet in 1994. The two smaller inset pictures at the top are the actual Hubble images; the larger images (bottom) are from a global map constructed through computer image processing performed on the Hubble data. Opposite hemispheres of Pluto are seen in these two views.

(Image from NASA)

A "New" Moon

If, having been discovered in 1930, Pluto was a late addition to our known solar system, its moon, Charon, is almost brand new, having been found in 1978. Named, fittingly, for the mythological ferryman who rowed the dead across the River Styx to the underworld ruled by Pluto, Charon is a little more than half the size of its parent: 806 miles (1,300 km) in diameter versus Pluto's 1,426 miles (2,300 km). Orbiting 12,214 miles (19,700 km) from Pluto, it takes 6.4 Earth days to make one circuit. Pluto and Charon are tidally locked—forever facing one another; the orbital period and rotation period for both are synchronized at 6.4 days. Like Venus and Uranus, Pluto's rotation is retrograde—spinning on its axis in the opposite direction of most of the planets.

Where Did Pluto Come From?

Except that it doesn't orbit another planet—and, indeed, has a moon of its own—Pluto looks more like a jovian moon than a planet. It fits into neither the terrestrial nor jovian mold. Some

Astro Byte

Astronomical objects (however minor) are typically named by their discoverers. In 1919, the International Astronomical Union (IAU) was given the task of bringing order to the then rather unorganized naming of surface features on Mars and the moon. Since then, this organization has been the only internationally recognized authority on the naming of features on solar system objects and objects outside the solar system. For more on this topic, check out www. iau.org/IAU/Activities/nomenclature/.

astronomers believe that Pluto is really a renegade moon, escaped from Neptune's gravitational influence due to a collision or interaction involving Triton, Pluto, Charon, and Nereid. Others regard it as a kind of spare part, something left over from the creation of the solar system, and perhaps only one of a number of such objects in the outer reaches of the solar system, the Kuiper Belt.

The Least You Need to Know

➤ The jovian realm is rich in moons (over 70 are currently known) and planetary rings. Some jovian moons are volcanically active, others have measurable atmospheres.

➤ Rings are characteristic of all the jovian planets, but only those of Saturn are readily seen from the earth without special optics.

➤ Planetary rings consist of small orbiting particles, including ice crystals, perhaps fragments from pulverized moons that strayed too close to the parent planet and were pulled apart.

➤ Io has frequent volcanic eruptions that loft material into orbit around Jupiter. Europa, also orbiting Jupiter, may have liquid water beneath its frozen, cracked surface.

➤ Pluto, discovered in 1930, does not fit neatly into either the terrestrial or jovian category and is in many ways more like a jovian moon than a planet. It may be an escaped moon of Neptune.

Part 4

To the Stars

By far the closest and most familiar star is the Sun, and before we leave the confines of our solar system to explore the nature of stars in general, we take a close look at our own star, at its structure and at the mechanism that allows it to churn out, each and every second, the energy equivalent of four trillion trillion 100-watt bulbs.

From the Sun, we move to a chapter on how stars are studied and classified. We then explore how stars develop and what happens to stars when they die, including very massive stars that are many times the mass of our Sun. Here, we'll explore the truly strange realm of neutron stars and black holes.

This part comes full circle with a discussion of the stellar birth process.

Chapter 16

Our Star

In This Chapter

➤ The sun: an average star

➤ Sunspots, prominences, and solar flares

➤ The layered structure of the sun and its atmosphere

➤ The sun as a nuclear fusion reactor

➤ Dimensions and energy output of the sun

An evening spent looking up under dark skies will convince you that stars can be breathtaking in their loveliness. We can appreciate why, for thousands of years, human beings thought that the stars were embedded in a perfect sphere, spinning and changeless. Yet, because of their great distance, theirs is a remote beauty. Many amateur astronomers are disappointed to discover that stars (other than the sun) look pretty much the same through even the best telescope. Our common sense sees little similarity between the distant, featureless points of light against a sable sky and the great yellow disk of daytime, whose brilliance overwhelms our vision and warms our world. Yet, of course, our sun *is* a star—and, as stars go, not a particularly remarkable one. We now turn our attention to the very center of our solar system, the parent of the terrestrial and jovian planets and their rings and moons.

We have spent the last three chapters discussing the planets and their moons. But taken together, these objects represent only 0.1 percent of the mass of the solar system. The other 99.9 percent of the mass is found in the sun. Peoples of many times and cultures have worshipped the sun as the source of all life, and in some sense, they were right. The sun is our furnace and our light bulb: the ultimate source of most energy and light here on the earth. And because it contains almost all of the mass, it is the gravitational anchor of the solar system. Indeed, its very matter is ours. The early sun was the hot center of a swirling disk of gas and dust from which the solar system formed some 4.6 billion years ago. If the sun were a cake, the earth and the rest of the planets would be some flour left on the counter.

But the sun is only one star in a galaxy containing hundreds of billions of stars. Astronomers feel fortunate that the sun is so nondescript a citizen of the galaxy. It is, of course, the star closest to us and its very averageness lets us generalize about the many stars that lie far beyond our reach. In this chapter we examine our own star, and begin to explore how the sun (and stars in general) generate the enormous energies that they do.

The Solar Furnace

Recall from Chapter 2, "Ancient Evenings: The First Watchers," that the Greek philosopher Anaximenes of Miletus believed the sun, like other stars, was a great ball of fire. His was an important insight, but not entirely accurate. The sun is not so simple.

In terms of human experience, the sun is an unfailing source of energy. Where does all of that energy come from? In the nineteenth century, scientists knew of two possible sources: thermal heat (like a candle burning) and gravitational energy.

The problem with thermal energy is that even the sun doesn't have enough mass to produce energy the way a candle does—at least, not for billions of years. Calculations showed that the sun "burning" chemically, would last only a few thousand years.

Astro Byte

The description of how mass is converted into energy is perhaps the best known equation of all time: $E=mc^2$. E stands for energy, m for mass (in kg), and c, the speed of light, 3×10^8 m/s. A tiny bit of mass can produce an enormous amount of energy.

While a sun that was a few thousand years old might have pleased some theologians at the time, there was a variety of evidence showing that the earth was much older.

So scientists turned their attention to gravitational energy, that is, the conversion of gravitational energy into heat. The theory went this way: As the sun condensed out of the solar nebula, its atoms fell inward and collided more frequently as they got more crowded. These higher velocities and collisions converted gravitational energy into heat. Gravitational energy could power the sun's output at its current rate for about 100 million years.

But when it started to become clear that the earth was much older (geological evidence showed that it was at least 3.5 billion years old), scientists went back to the drawing board. The nineteenth century ended without an understanding of the source of energy in the sun.

A Very Special Theory

A brilliant physicist at the beginning of the twentieth century eventually came up with the answer. It turns out that the source of energy in the sun is something that had never been considered before. Something that would have been called alchemy centuries earlier. It turns out that the sun converts a tiny bit of its mass into pure energy through collisions between the cores of atoms, their nuclei.

What's It Made Of?

The sun is mostly hydrogen (about 73 percent of the total mass) and helium (25 percent). Other elements are found in much smaller amounts, adding up to just under two percent of the sun's mass. These include carbon, nitrogen, oxygen, neon, magnesium, silicon, sulfur, and iron. Over 50 other elements are found in trace amounts. There is nothing unique about the presence of these particular elements; they are the same ones that are distributed throughout the solar system and the universe.

In particular, hydrogen atoms of the sun's core plow into one another to create helium atoms. In the process, a little mass is converted into energy. That little bit of energy for each collision means enormous amounts of energy when we count all of the collisions that occur in the core of the sun. The fact that c is a very large number means that a tiny amount of mass results in a very large amount of energy. With this energy source, the sun is expected to last not a thousand years, or even 100 million years, but about 8 to 10 billion years, typical for a star with the sun's mass.

Astronomer's Notebook

Some important measurements used to express the energy output of the sun include: watt (a measure of power; the rate at which energy is emitted by an object); solar constant (the energy each square meter of the earth's surface receives from the sun per second: 1,400 watts); and luminosity (the total energy radiated by the sun per second 4×10^{26} watts).

A Spectacular, Mediocre Star

In terms of its size, mass and energy released, the sun is by far the most spectacular body in the solar system. With a radius of 22.8×10^8 feet (6.96×10^8 m), it is 100 times larger than the earth. Imagine yourself standing in a room with a golf ball. If the golf ball is the earth, the sun would touch the eight-foot ceiling. With a mass of 1.99×10^{30} kg, the sun is 300,000 times more massive than the earth. And with a surface temperature of 5,780 K (compared to the earth's average 290 K surface temperature), the sun would melt or vaporize any matter we know.

Four Trillion Trillion Light Bulbs

Next time you are screwing in a light bulb, notice its wattage. A watt is a measure of power, or how much energy is produced or consumed each second. A 100 watt bulb uses 100 joules of energy every second. For comparison, the sun produces 4×10^{26} watts of power. That's a lot of light bulbs—four trillion trillion of them, to be exact. This rate of energy production is called the sun's *luminosity*. Many stars have luminosities much higher than that of the sun.

The source of the sun's power—and that of all stars, during most of their lifetimes—is the fusing together of nuclei. Stars first convert hydrogen into helium, and heavier elements come later (we discuss this process in Chapter 19, "Black Holes: One-Way Tickets to Eternity"). The only fusion reactions that we have been able to produce on the earth are uncontrolled reactions known as hydrogen bombs. The destructive force of these explosions gives insight into the enormous energies released in the core of the sun. Nuclear fusion could be used as a nearly limitless supply of energy on the earth; however, we are not yet able to create the necessary conditions on Earth for controlled fusion reactions.

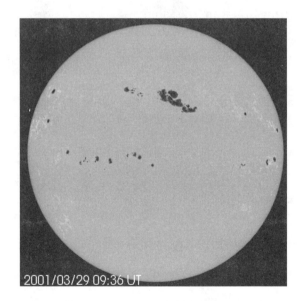

The solar surface as seen with the Solar and Heliospheric Observatory (SOHO), an international solar observatory jointly operated by the European Space Agency and NASA. This image shows an impressive array of sunspots visible on March 29, 2001.

(Image from NASA/ESA)

The Solar Atmosphere

The sun does not have a surface as such. What we call its surface is just the layer that emits the most light. Let's begin our journey at the outer layers of the sun (the layers that we can actually see), and work our way in. When you look up at the sun during the day, what you are really looking at is the sun's photosphere. The layer from which the visible photons that we see arise, the photosphere has a temperature of about 6,000 K. Lower layers are hidden behind the photosphere, and higher layers are so diffuse and faint (though very hot) that we only see them during total solar eclipses or with special satellites. Above the photosphere in the solar atmosphere are the chromosphere, the transition zone, and the corona. As we move higher in the sun's atmosphere, the temperatures rise dramatically.

Not That Kind of Chrome

The sun's lower atmosphere is called the chromosphere, normally invisible because the photosphere is far brighter. However, during a total solar eclipse, which blots out the photosphere, the chromosphere is visible as a pinkish aura around the solar disk. The strongest emission line in the hydrogen spectrum is red, and the predominance of hydrogen in the chromosphere imparts the pink hue.

The chromosphere is a storm-racked region, into which *spicules,* jets of expelled matter thousands of miles high, intrude.

Above the chromosphere is the transition zone. As mentioned earlier, the temperature at the surface of the photosphere is 5,780 K, much cooler than the temperatures in the solar interior, which get hotter the closer one approaches the core. Yet, in the chromosphere, transition zone, and into the corona, the temperature rises sharply the *farther* one goes from the surface of the sun! At about 6,000 miles (10,000 km) above the photosphere, where the transition zone becomes the corona, temperatures exceed 1,000,000 K. (For detailed real-time views of the solar photosphere, chromosphere, and corona, see http://sohowww.estec.esa.nl.)

How do we explain this apparent paradox? It is believed that the interaction between the sun's strong magnetic field and the charged particles in the corona heat it to these high temperatures.

Star Words

Spicules are jets of matter expelled from the sun's photosphere region into the chromosphere above it.

A Luminous Crown

Corona is Latin for "crown," and it describes the region beyond the transition zone consisting of elements that have been highly ionized (stripped of their electrons) by the tremendous heat in the coronal region. Like the chromosphere, the corona is normally invisible, blotted out by the intense light of the photosphere. It is only during total solar eclipses that the corona becomes visible, at times when the disk of the moon covers the photosphere and the chromosphere. During such eclipse conditions, the significance of the Latin name becomes readily apparent: The corona appears as a luminous crown surrounding the darkened disk of the sun.

When the sun is active—a cycle that peaks every 11 years—its surface becomes mottled with sunspots, and great solar flares and prominences send material far above its surface.

Close Encounter

A solar eclipse occurs when the moon moves across the disk of the sun so that the moon's shadow falls across the face of the earth. In the heart of that shadow, called the *umbra*, the sun's disk will appear completely covered by that of the moon: a total solar eclipse. The umbra, however, only falls on a small region of the earth. Thus a total eclipse can be observed only within the zone of totality, a very narrow area of the earth (where this shadow falls as the earth rotates). For this reason, total eclipses are rare events in any given geographical area. Much more common are partial eclipses, in which the moon obscures only part of the sun. Observers located in the much broader outer shadow of the moon (the penumbra) see such an eclipse.

Certainly, partial eclipses are interesting, but a total eclipse can be spectacular, not only dramatically darkening the world, but allowing sight of such solar features as feathery prominences, the chromosphere, and, most thrilling of all, the corona. These features are fleeting, since totality lasts only a few minutes at any one observing location.

As mentioned elsewhere in this book, observing the sun directly is very dangerous. ***Looking at the sun through an unfiltered telescope or binoculars will cause irreversible damage to your eyesight.*** The sun is no more or less dangerous during an eclipse than at any other time; but the point is that looking directly at the sun is *always dangerous and harmful.*

The sun, during an eclipse or at any time, is most safely observed by projecting its image onto a piece of paper or cardboard. You can project a telescope or binocular image onto a white card held at the correct distance from the eyepiece. But you don't need a telescope or binoculars to project an image. Just make a pinhole in a stiff piece of cardboard and project the pinhole image onto a white card or paper. (By the way: *Do* not *look through the pinhole directly at the sun!*)

If you want to look at the sun through your telescope during an eclipse or at another time, purchase a solar filter (glass or Mylar) from any of the major telescope manufacturers. This type of filter attaches to the *front* of your telescope tube, it *does not* screw onto the eyepiece.

See Appendix B for the dates of some upcoming eclipses.

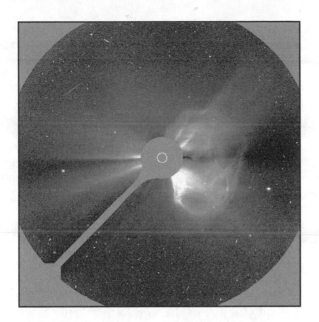

This image was generated by the Large Angle and Spectrometric Corona-graph of SOHO. The Solar corona (or crown) can be viewed only if the light from the photosphere is naturally or artificially blocked. The small white circle indicates the size of the solar disk.

(Image from NASA/ESA)

Solar Wind

The sun does not keep its energy to itself. Its energy flows away in the form of electromagnetic radiation and particles. The particles (mostly electrons and protons) do not move nearly as fast as the radiation, which escapes the sun at the speed of light, but they move fast nevertheless—at more than 300 miles per second (500 km/s). It is this swiftly moving particle stream that is called the solar wind.

The solar wind is driven by the incredible temperatures in the solar corona. As a result, the gases are sufficiently hot to escape the tremendous gravitational pull of the sun. The surface of the earth is protected from this wind by its magnetosphere, the magnetic "cocoon" generated by the rotation of the earth's molten core. As with many other planets, the motion of charged molten material in the earth's core generates a magnetic field around the planet. This magnetic field either deflects or captures charged particles from the solar wind. Some of these particles are trapped in the Van Allen Belts, doughnut-shaped regions around the earth named after their discoverer. Some of the charged particles rain down on the earth's poles and collide with its atmosphere, giving rise to displays of color and light called *aurora* (in the Northern Hemisphere,

Astro Byte

Will the sun eventually evaporate into solar wind? Every second, a million tons of solar substance is emitted as solar wind. Yet less than a tenth of one percent of the sun has been thus spent since the beginning of the solar system 4.6 billion years ago.

the Aurora Borealis, or Northern Lights, and in the Southern Hemisphere, the Aurora Australis, or Southern Lights). The Auroras are especially prominent when the sun reaches its peak of activity every 11 years.

Fun in the Sun

We have described the layers in the sun's outer atmosphere, but have ignored some of their more interesting aspects, the storms in the atmosphere. The sun's atmosphere is regularly disturbed by solar weather in the form of *sunspots,* prominences, and solar flares. With the proper equipment—or an Internet connection (http://sohowww.estec.esa.nl)—you can observe some of the signs of activity on the sun's surface.

A Granulated Surface

If we look at the sun, its surface usually appears featureless, except, perhaps, for sunspots, which we'll discuss in a moment. However, viewed at high-resolution, the surface of the sun actually appears highly granulated. Now, *granule* is a relative concept when we are talking about a body the size of the sun. Each granule is about the size of an earthly continent, appearing and disappearing as a hot gas bubble rises to the surface of the sun.

Galileo Sees Spots Before His Eyes

People must have seen sunspots before 1611, when Galileo (and, independently, other astronomers) first reported them. (As recently as March 2001, sunspots easily visible to the unaided eye have appeared.) The largest spots are visible to the naked eye (at least when the sun is seen through clouds). Yet, at the time, the world was reluctant to accept imperfections on the face of the sun. Sunspots were not (as far as we know) studied before Galileo.

Galileo drew a profound conclusion from the existence and behavior of sunspots. In 1613, he published three letters on sunspots, explaining that their movement across the face of the sun showed that the sun rotated.

Sunspots: What They Are

Sunspots are irregularly shaped dark areas on the face of the sun. They look dark because they are cooler than the surrounding material. The strong local magnetic fields push away some of the hot ionized material rising from lower in the photosphere. A sunspot is not uniformly dark. Its center, called the umbra, is

Star Words

Sunspots are irregularly shaped dark areas on the face of the sun. They appear dark because they are cooler than the surrounding material. They are tied to the presence of distorted magnetic fields at the sun's surface.

darkest and is surrounded by a lighter penumbra. If you think of them as blemishes on the face of the sun, just remember that one such blemish may easily be the size of the earth or larger.

Sunspots may persist for months, and they may appear singly, although, usually, they are found in pairs or groups. Such typical groupings are related to the magnetic nature of the sunspots. Every pair of spots has a leader and a follower (with respect to the direction of the sun's rotation), and the leader's magnetic polarity is always the opposite of the follower. That is, if the leader is a north magnetic pole, the follower will be a south magnetic pole.

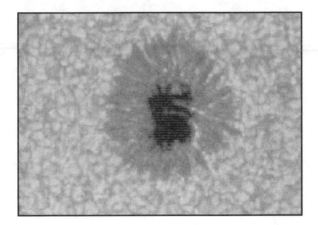

Sunspot photographed by the National Solar Observatory at Sacramento Peak, Sunspot, New Mexico. Note the small "granules" and the larger sunspot umbra, surrounded by the penumbra.

(Image from NSO)

Sunspots are never seen exactly at the equator or near the solar poles, and leaders and followers in one hemisphere of the sun are almost always opposite in polarity from those across the equator. That is, if all the leaders in the northern hemisphere are south magnetic poles, all the leaders in the southern hemisphere will be north magnetic poles.

We have said that sunspots are thought to be associated with strong local magnetic fields. But why are the fields strong in certain regions of the photosphere?

A meteorologist from Norway, Vilhelm Bjerknes (1862–1951) concluded in 1926 that sunspots are the erupting ends of magnetic field lines, which are distorted by the sun's differential rotation. That is, like the gas giant jovian worlds, the sun does not rotate as a single, solid unit, but differentially, at different speeds for different latitudes. The sun spins fastest at its equator—the result being that the solar magnetic field becomes distorted. The field lines are most distorted at the equator, so that the north-south magnetic field is turned to an east-west orientation. In places where the field is sufficiently distorted, twisted like a knot, the field becomes locally very strong, powerful enough to escape the sun's gravitational pull. Where this happens, field lines "pop" out of the photosphere, looping through the lower solar atmosphere and forming a sunspot pair at the two places where the field lines pass into the solar interior.

The Sun as seen in the extreme ultraviolet part of the electromagnetic spectrum.

(Image from NASA/ESA)

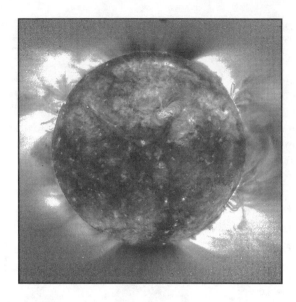

Sunspot Cycles

Long before the magnetic nature of sunspots was perceived, astronomer Heinrich Schwabe, in 1843, announced his discovery of a solar cycle, in which the number of spots seen on the sun reaches a maximum about every 11 years (on average). In 1922, the British astronomer Annie Russel Maunder charted the latitude drift of sunspots during each solar cycle. She found that each cycle begins with the appearance of small spots in the middle latitudes of the sun, followed by spots appearing progressively closer to the solar equator until the cycle reaches its maximum level of activity. After this point, the number of spots begins to decline. The most recent maximum occured in early 2001.

Actually, the 11-year period is only half of a 22-year cycle that is more fundamental. Recall that the leading spots on one hemisphere exhibit the same polarity; that is, they are all either north magnetic poles or south (and the followers are the opposite of the leaders). At the end of the first 11 years of the cycle, polarities reverse. That is, if the leaders had north poles in the southern hemisphere, they become, as the second half of the cycle begins, south poles.

The cyclical nature of sunspot activity is very real, but not exact and inevitable. Studying historical data, Maunder discovered that the cycle had been apparently dormant from 1645 to 1715. At present, there is no explanation for this dormancy and other variations in the solar cycle.

Close Encounter

The safest and most convenient way to observe sunspots is to project a telescope image of the sun onto a white card. Accessory screens and rods to mount them on can be purchased for most refracting and reflecting telescopes. Adjust the distance of the screen from the eyepiece so that the sun's projected disk is about six inches in diameter. ***Never look at the sun through the telescope eyepiece! You* will *permanently damage your eyesight almost instantly.***

You may draw and count the projected sunspots. Try recording sunspot activity each day over a period of time. Do this by counting individual sunspots and by looking for changes in position. Daily images of the solar surface are archived at http://sohowww.estec.esa.nl.

Coronal Fireworks

Most frequently at the peak of the sunspot cycle, violent eruptions of gas are ejected from the sun's surface. The *prominences* and *flares* may rise to some 60,000 miles (100,000 km) and may be visible for weeks.

Solar flares are more sudden and violent events than prominences. While they are thought to also be the result of magnetic kinks, they do not show the arcing or looping pattern characteristic of prominences. Flares are explosions of incredible power, bringing local temperatures to 100,000,000 K. Whereas prominences release their energy over days or weeks, flares explode in a flash of energy release that lasts a matter of minutes or, perhaps, hours.

Astro Byte

To generate the amount of energy that the sun does, 600 million tons of hydrogen must be fused into helium each and every second. Yet the sun has so much hydrogen available that it will not exhaust its fusion fuel for at least five billion years.

At the Core

The sun is a nuclear fusion reactor. Impressive as its periodic outbursts are, its *steady* production of energy should generate even more wonder—and thought. The sun has been churning out energy every second of every day for the last 4 to 5 billion years.

Gone Fission

On December 2, 1942, Enrico Fermi, an Italian physicist who had fled his fascist-oppressed native land for the United States, withdrew a control rod from an "atomic pile" that had been set up in a squash court beneath the stands of the University of Chicago's Stagg Field. This action initiated the world's first self-sustaining atomic chain reaction. Fermi and his team had invented the nuclear reactor, and the world hasn't been the same since.

Nuclear fission is a nuclear reaction in which an atomic nucleus splits into fragments, thereby releasing energy. In a fission reactor, such as the one Fermi was instrumental in creating, the process of fission is controlled and self-sustaining, so that the splitting of one atom leads to the splitting of others, each fission liberating more energy.

Nuclear fission is capable of liberating a great deal of energy, whether in the form of a controlled sustained chain reaction or in a single great explosion, like an atomic bomb. Yet even the powerful fission process cannot account for the tremendous amount of energy the sun generates so consistently. We must look to another process: *nuclear fusion.*

Whereas nuclear fission liberates energy by splitting atomic nuclei, nuclear fusion produces energy by joining them, combining light atomic nuclei into heavier ones. In the process, the combined mass of two nuclei in a third nucleus is *less* than the total mass of the original two nuclei. The mass is not simply lost, but converted into energy. A *lot* of energy.

One of the by-products of nuclear fusion reactions is a tiny neutral particle called the *neutrino*. The fusion reactions themselves produce high energy gamma ray radiation, but those photons are converted into mostly visible light by the time their energy reaches the surface of the sun. Neutrinos, with no charge to slow them down, come streaming straight out of the sun's core. The numbers that we detect give us great insight into a region of the sun that is otherwise inaccessible.

Star Words

Nuclear fission is a nuclear reaction in which an atomic unit is split into fragments, thereby releasing energy that previously held the nucleus together. In a fission reactor, the split-off fragments collide with other nuclei, causing them to fragment, until a chain reaction is underway.
Nuclear fusion is a nuclear reaction that produces energy by joining atomic nuclei. The mass of a nucleus produced by joining two nuclei is slightly less than that of the sum of the original two nuclei. The mass is not lost, but instead converted into pure energy.

Chain Reactions

The sun generates energy by the converting the hydrogen in its core to helium. The details are complex, but we may content ourselves with an overview. When temperatures and pressures are sufficiently high (temperatures of about 10 million K are required) 4 hydrogen nuclei (which are protons, positively charged particles) can combine to create the nucleus of a helium atom (2 protons and 2 neutrons). Now the mass of the helium nucleus created is slightly less than that of the four protons that were needed to create it. That small difference in mass is converted into energy in the fusion process. One of the simplest fusion reactions involves the production of deuterium (a hydrogen isotope) from a proton and a neutron. When these two particles collide with enough velocity, they create a deuterium nucleus (consisting of a proton and a neutron) and the excess energy is given off as a gamma ray photon. In the sun, this process proceeds on a massive scale, liberating the energy that lights up our daytime skies. That's a 4×10^{26} watt lightbulb up there, remember.

Your Standard Solar Model

By combining theoretical modeling of the sun's (unobservable) interior with observations of the energy that the sun produces, astronomers have come to an agreement on what is called a *standard solar model,* a mathematically-based picture of the structure of the sun. The model seeks to explain the observable properties of the sun and also describe properties of its unobservable interior.

With the standard solar model, we can begin to describe some of the interior regions—regions hidden, beneath the photosphere, from direct observation. Below the photosphere is the convection zone, some 124,000 miles (200,000 km) thick. Below this is the radiation zone, 186,000 miles (300,000 km) thick, which surrounds a core with a radius of 124,000 miles (200,000 km).

The sun's core is tremendously dense (150,000 kg/m³) and tremendously hot: some 15,000,000 K. We can't stick a thermometer in the sun's core, so how do we know it's that hot? If we look at the energy emerging from the sun's surface, we can work backward to the conditions that must prevail at the sun's core. At this density and temperature, nuclear fusion is continuous, with particles always in violent motion. The sun's core is a giant nuclear fusion reactor.

At the very high temperatures of the core, all matter is completely ionized—stripped of its negatively charged electrons. As a result, photons (packets of electromagnetic energy) move slowly out of the core into the next layer of the sun's interior, the radiation zone.

Here the temperature is lower, and photons emitted from the core of the sun interact continuously with the charged particles located there, being absorbed and re-emitted. While the photons remain in the radiation zone, heating it and losing energy, some of their energy escapes into the convection zone, which in effect, boils like water on a stove so that hot gases rise to the photosphere and cool gases sink back into the convection zone. Convective *cells* become smaller and smaller, eventually becoming visible as granules at the solar surface. Thus, by convection, huge amounts of energy reach the surface of the sun. At the sun's surface, a variety of processes give rise to the electromagnetic radiation that we detect from the earth. Atoms and molecules in the sun's photosphere absorb some of the photons at particular wavelengths, giving rise to the sun's absorption-line spectrum. Most of the radiation from a star that has the surface temperature of the sun is emitted in the visible part of the spectrum.

Close Encounter

Astronomers use a technique called *helioseismology* to study the sun's interior. By measuring the frequencies of various regions on the surface of the sun, conclusions can be drawn about the sun's internal structure. This process is analogous to the way in which seismologists monitor and measure the waves generated by earthquakes in order to draw conclusions about the unseen internal structures of our own planet.

The Least You Need to Know

➤ Staggering as the sun's dimensions and energy output are, the sun is no more nor less than a very average star.

➤ The sun is a complex, layered object with a natural nuclear fusion reactor at its core.

➤ The gamma rays generated by the fusion reactions in the core of the sun are converted to optical and infrared radiation by the time the energy emerges from the sun's photosphere.

➤ Never look directly at the sun, especially not through binoculars or a telescope. The safest way to observe the sun is by *projecting* its image, either through binoculars or a telescope or simply through a pinhole punched in cardboard.

➤ Although the sun has been a dependable source of energy for the last 4 billion years, its atmosphere is frequently rocked by such disturbances as sunspots, prominences, and solar flares, peaking every 11 years.

Chapter 17

Of Giants and Dwarfs: Stepping Out into the Stars

Our own star, the sun, is relatively accessible. We can make out features on its surface and can track the periodic flares and prominences that jut from its surface. But our sun is only one star of over 100 billion in our Galaxy alone, and all the other stars that exist are much farther away, parts of other distant galaxies.

Earlier, we said that if the earth were a golf ball, the planet Pluto would be a chickpea eight miles away. But even distant Pluto is very close in comparison to our stellar neighbors. On the same scale, the nearest star (in the Alpha Centauri system) would be 50,000 miles away. That distance wraps around the earth's equator twice. The other stars are even farther. Much farther.

In this chapter, we reach out to the myriad cousins of our sun.

Sizing Them Up

The planets of the solar system appear to us as disks. Once we know a planet's distance from us, it is quite simple to measure the disk and translate that figure into a real measurement of the planet's size. But train your telescope on any star, and all you will see is a point of light. Pop in a higher-power eyepiece and guess what? It's still a point of light!

Only very recently (with the advent of the *Hubble Space Telescope*) have we been able to resolve the disks of *any* stars. In 1996, the HST took a picture of the star Betelgeuse. This image was the *first* resolved image of a star other than the sun. The Very Large Array has also been used to image Betelgeuse. Although far away (500 light-years), Betelgeuse is a red giant, and as we'll soon see, that means it has a very large radius.

The Very Large Array in Socorro, New Mexico, recently made an image of the nearby supergiant star Betelgeuse, the bright red star in the constellation Orion. Like the sun, Betelgeuse appears to send plumes of gas far above its optically visible atmosphere, as seen in this radio-frequency image.

(Image from NRAO)

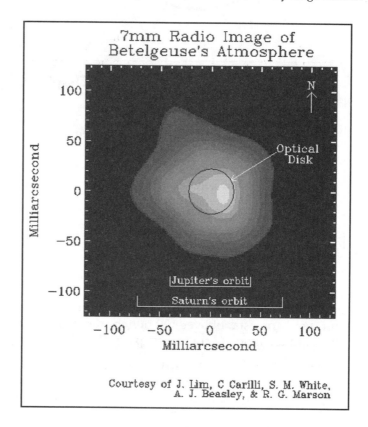

Radius, Luminosity, Temperature: A Key Relationship

We don't have to give up on measuring the sizes of stars, however. We just have to be more clever. What astronomers do is determine the temperature and mass of a star, which can be done using a star's color and spectrum. Then, using numerical models of how stars hold together, they derive the quantity that they are interested in (radius, for example). It is akin to looking out over a parking lot and seeing a Cadillac. Now, you may not know its size, but you know (consulting a chart) that this model of Cadillac is 18.5 feet long. You can see clearly that it is indeed this particular model of Cadillac, so you know its length, even though you didn't actually measure it with a ruler.

Recall from Chapter 7, "Over the Rainbow," that Stefan's law states that a star's luminosity (its wattage, or the rate at which it emits energy into space) is proportional to the fourth power to the star's surface temperature. This relationship can be extended further. A star's luminosity is not only related to its temperature, but to its surface area. Heat the head of a pin to 400 degrees F and a large metal plate to the same temperature. Which will radiate more heat? Obviously, the object with the larger surface area. Given the same surface temperature, a larger body will always radiate more energy than a smaller one.

This relationship can be expressed in this way: A star's luminosity is proportional to the square of its radius (that's the surface area term) times its surface temperature to the fourth power (luminosity × radius² × temperature⁴). Thus, if we know a star's luminosity and temperature (which can be measured by available astronomical instruments), we can calculate its radius. How do we measure a star's luminosity and temperature? Let's see.

The Parallax Principle

First, how do we know that the nearest stars are so far away? For that matter, how do we know how far away *any* stars are? We've come a long way in this book, and, on our journey, we have spoken a good deal about distances—by earthly standards, often extraordinary distances. Indeed, the distances astronomers measure are so vast that they use a set of units unique to astronomy. When measuring distances on the earth, meters and kilometers are convenient units. But in the vast spaces between stars and galaxies, such units are inadequate. As we'll see in this chapter and those that follow, the way astronomers measure distances, and the units they use depend upon how far away the objects are.

Distances between a given point on the earth and many objects in the solar system can be measured by radar ranging. *Radar,* a technology developed shortly before and during World War II, is now quite familiar. Radar can be used to detect and track distant objects by transmitting radio waves, then receiving the echo of the waves the object bounces back (*sonar* is a similar technique using sound waves). If we multiply the round-trip travel

Star Words

In astronomy, **radar** (short for "radio detection and ranging") is the use of radio signals to measure the distance of planets and other objects in the solar system.

Astronomer's Notebook

If you have a stopwatch, you could also clap your hands and see how long it takes for the echo to return. Using the technique described for radar ranging, but multiplying the round-trip time by the speed of sound in air—about 340 m/s, you get twice the width of the canyon.

time of the outgoing signal and its incoming echo by the speed of light (which, you'll recall, is the speed of all electromagnetic radiation, including radio waves), we obtain a figure that is twice the distance to the target object.

Radar ranging works well with objects that return (bounce back) radio signals. But stars, including our sun, tend to absorb rather than return electromagnetic radiation transmitted to them. Moreover, even if we could bounce a signal off a star, most are so distant that we would have to wait thousands of years for the signal to make its round trip—even at the speed of light! Even the nearby Alpha Centauri system would take about eight years to detect with radar ranging, were it even possible.

Another method is used to determine the distance of the stars, a method that was available long before World War II. In fact, it is at least as old as the Greek geometer Euclid, who lived in the third century B.C.E.

The technique is called *triangulation*—an indirect method of measuring distance derived by geometry using a known baseline and two angles from the baseline to the object. Triangulation does not require a right triangle, but the establishment of one 90-degree angle does make the calculation of distance a bit easier. It works like this. Suppose you are on one rim of the Grand Canyon and want to measure the distance from where you are standing to a campsite located on the other rim. You can't throw a tape measure across the yawning chasm, so you must measure the distance indirectly. You position yourself directly across from the campsite, mark your position, then turn 90 degrees from the canyon and carefully pace off another point a certain distance from your original position. This distance is called your baseline. From this second position, you sight on the campsite. Whereas the angle formed by the baseline and the line of sight at your original position is 90 degrees (you arranged it to be so), the angle formed by the baseline and the line of sight at the second position will be somewhat less than 90 degrees. If you connect the campsite with Point A (your original viewpoint) and the campsite with Point B (the second viewpoint), both of which are joined by the baseline, you will have a right triangle.

Now, you can take this right triangle and, with a little work, calculate the distance across the canyon. If you simply make a drawing of your setup, making sure to draw the angles and lengths that you know to scale, you can measure the distance across the canyon from your drawing. Or if you are good at trigonometry, you can readily use the difference between the angles at Points A and B and the length of the baseline to arrive at the distance to the remote campsite.

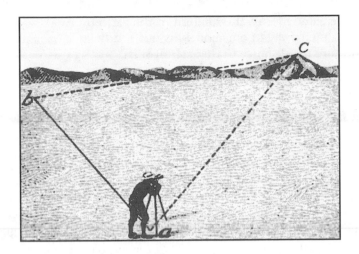

Old print of a surveyor using triangulation. He knows the distance from his viewpoint "a" to point "b" (his baseline) and, sighting on distant "c," he can use the angles ∠CBA and ∠CAB to work out the distance from "a" to "c."

(Image from arttoday.com*)*

How Far Away Are the Stars?

Like the campsite separated from you by the Grand Canyon, the stars are not directly accessible to measurement. However, if you can establish two view points along a baseline, you can use triangulation to measure the distance to a given star.

There is just one problem.

Take a piece of paper. Draw a line one inch long. This line is the baseline of your triangle. Measure up from that line, say, one inch, and make a point. Now connect the ends of your baseline to that point. You have a nice, normal looking triangle. But if you place your point several feet from the baseline, then connect the ends of the baseline to it, you will have an extremely long and skinny triangle, with angles that are very difficult to measure accurately, because they will both be close to 90 degrees. If you move your point several *miles* away, and keep a 1-inch baseline, the difference in the angles at Points A and B of your baseline will be just about impossible to measure. They will both seem like right angles. For practical purposes, a 1-inch baseline is just not long enough to measure distances of a few miles away. Now recall that if our Earth is a golf ball (about 1 inch in diameter), that the nearest star, to scale, would be 50,000

Astronomer's Notebook

Stellar distances are so great that miles, kilometers, and even astronomical units (A.U.s) quickly become unwieldy. Astronomers use a unit called the *parsec*, which stands for "parallax in arc seconds." One *parsec* is the distance a star must be from the sun in order to show a parallax of one arc second (1"). That distance is 206,265 A.U. or 10.7×10^{16} feet (3.1×10^{16} meters) or 3.26 light-years. Another way to think of it is this: A parsec is the distance you would have to be from the solar system for the angular separation between the earth and the sun to be 1 arcsecond.

miles away. So the baseline created by, say, the rotation of the earth on its axis—which would give 2 points 1 inch away in our model—is not nearly large enough to use triangulation to measure the distance to the nearest stars.

The diameter of the earth is only so wide. How can we extend the baseline to a useful distance?

The solution is to use the fact that our planet not only rotates on its axis, but also orbits the sun. Observation of the target star is made, say, on February 1, then is made again on August 1, when the earth has orbited 180 degrees from its position six months earlier. In effect, this motion creates a baseline that is 2 A.U. long—that is, twice the distance from the earth to the sun. Observations made at these two times (and these two places) will show the target star apparently shifted relative to the even more distant stars in the background. This shift is called *stellar parallax,* and by measuring it, we can determine the angle relative to the baseline and thereby use triangulation to calculate the star's distance.

To get a handle on parallax, hold your index finger in front of you, with your arm extended. Using one eye, line up your finger with some vertical feature, say the edge of the window. Now, keeping you finger where it is, look through the other eye. The change in viewpoint makes your finger appear to move with respect to a background object. In astronomy, your eyes are the position of the earth separated by 6 months, your finger is a nearby star, and the window edge is a distant background star. This method works as long as the star (your finger) is relatively close. If the star is too far away, parallax is no longer effective.

Nearest and Farthest

Other than the sun, the star closest to us is Alpha Centauri, which has the largest known stellar parallax of 0.76 arc seconds. In general, the distance to a star in parsecs (abbreviated pc) is equal to 1 divided by the stellar parallax in arcseconds—or conversely, its parallax will be equal to 1 divided by the distance in parsecs. The measured parallax, in any case, will be a very small angle (less than an arcsecond). Recall that the moon takes up about 1,800" on the sky when full, so the parallax measured for Alpha Centauri is about $\frac{1}{2000}$ the diameter of the full moon! Using the rule above to convert parallax into distance, we find that Alpha Centauri is about 1.3 pc or 4.2 light-years away. On average, stars in our Galaxy are separated by 7 light-years. So Alpha Centauri is even closer than "normal." If a star were 10 pc away, it would have a parallax of $\frac{1}{10}$ or 0.1".

The farthest stellar distances that can be measured using parallax are about 100 parsecs (333 light-years). Stars at this distance have a parallax of $\frac{1}{100}$" or 0.01". That apparent motion is the smallest that we can measure with our best telescopes. Within our own Galaxy, most stars are even farther away than this. As telescope resolutions improve with the addition of adaptive optics, this outer limit will be pushed farther out.

Do Stars Move?

As we saw in Chapter 1, "Naked Sky, Naked Eye: Finding Your Way in the Dark," the ancients believed that the stars were embedded in a distant spherical bowl and moved in unison, never changing their relative positions. We know now, of course, that the daily motion of the stars is due to the earth's rotation. Yet the stars move, too; however, their great distance from us makes that movement difficult to perceive, except over long periods of time. A jet high in the sky, for example, can appear to be moving rather slowly, yet we know that it has to be moving fast just to stay aloft and its apparent slowness is a result of its distance.

Astronomers think of stellar movement in three dimensions:

➤ The *transverse component* of motion is perpendicular to our line of sight—that is, movement across the sky. This motion can be measured directly.

➤ The *radial component* is stellar movement toward or away from us. This motion must be measured from a star's spectrum.

➤ The actual motion of a star is calculated by combining the transverse and radial components.

The transverse component can be measured by carefully comparing photographs of a given piece of the sky taken at different times and measuring the angle of displacement of one star relative to background stars (in arcseconds). This stellar movement is called *proper motion*. A star's distance can be used to translate the angular proper motion thus measured into a transverse *velocity* in km/s. In our analogy: If you knew how far away that airplane in the sky was, you could turn its apparently slow movement into a true velocity.

Determining the radial component of a star's motion involves an entirely different process. By studying the spectrum of the target star (which

Star Words

The **proper motion** of a star is determined by measuring the angular displacement of a target star relative to more distant background stars. Measurements are taken over long periods of time, and the result is an **angular velocity** (measured, for example, in arcseconds/year). If the distance to the star is known, this **angular displacement** can be converted into a **transverse velocity.**

Astro Byte

Does the sun move? Yes, the sun and its planets are in orbit around the center of the Milky Way. The solar system orbits the Galaxy about once every 250 million years. Thus, since the sun formed, the solar system has gone around the merry-go-round of our Galaxy only 15 to 20 times.

shows the light emitted and absorbed by a star at particular frequencies), astronomers can calculate the star's approaching or receding velocity. Certain elements and molecules show up in a star's spectrum as *absorption lines* (see Chapter 7). The frequencies of particular absorption lines are known if the source is at rest, but if the star is moving toward or away from us, the lines will get shifted. A fast-moving star will have its lines shifted more than a slow-moving one. This phenomenon, more familiar with sound waves, is known as the Doppler effect.

How fast do stars move? And what is the fixed background against which the movement can be measured? For a car, it's easy enough to say that it's moving at 45 miles per hour relative to the road. But there are no freeways in space. Stellar speeds can be given relative to the earth, relative to the sun, or relative to the center of the Milky Way. Astronomers always have to specify which reference frame they are using when they give a velocity. Stars in our neighborhood typically move at tens of kilometers per second relative to the sun.

Close Encounter

We have all heard the Doppler effect. It's that change in pitch of a locomotive horn when a fast-moving train passes by. The horn doesn't actually change pitch, but the sound waves of the approaching train are made shorter by the approach of the sound source, whereas the waves of the departing train are made longer by the receding of the sound source.

Electromagnetic radiation behaves in exactly the same way. An approaching source of radiation emits shorter waves relative to the observer than a receding source. Thus the electromagnetic radiation of a source moving toward us will be blueshifted; that is, the wavelength received will be shorter than what is actually emitted. From a source moving away from us, the radiation will be redshifted; we will receive wavelengths longer than those emitted. Color is not material in these terms, since they apply to any electromagnetic radiation, whether visible or not. It is just that blue light has a shorter wavelength than red, so the terms are convenient.

By measuring the degree of a blueshift or redshift, astronomers can calculate the oncoming or receding velocity (the radial velocity) of a star. As we'll see in the next chapter, the same method is used to measure the motions of clouds of gas in regions of star formation.

How Bright Is Bright?

In ordinary English, *luminosity* and *brightness* would be pretty nearly synonymous. Not so in astronomy. You are standing beside a quiet road. Your companion, a couple feet away from you, shines a flashlight in your eyes. Just then a car rounds a curve a quarter-mile away. Which is more luminous, the flashlight or the headlights? Which is brighter?

Luminosity Versus Apparent Brightness

Ask an astronomer this question, and she will respond that the flashlight, a few feet from your eyes, is apparently brighter than the distant headlights, but that the headlights are more luminous. Luminosity is the total energy radiated by a star each second. Luminosity is a quality intrinsic to the star; brightness may or may not be intrinsic. *Absolute brightness* is another name for luminosity, but *apparent brightness* is the fraction of energy emitted by a star that eventually strikes some surface or detection device (including our eyes). Apparent brightness varies with distance. The farther away an object is, the lower its apparent brightness.

Simply put, a very luminous star that is very far away from the earth can appear much fainter than a less luminous star that is much closer to the earth. Thus, although the Sun is the brightest star in the sky, it is not by any means the most luminous.

Creating a Scale of Magnitude

So astronomers have learned to be very careful when classifying stars according to *apparent brightness.* Classifying stars according to their magnitude seemed a good idea to Hipparchus (in the second century B.C.E.) when he came up with a 6-degree scale, ranging from 1, the brightest stars, to 6, those just barely visible. Unfortunately, this somewhat cumbersome and awkward system (higher magnitudes are fainter, and the brightest objects have negative magnitudes) has persisted to this day.

Hipparchus' scale has been expanded and refined over the years. The intervals between magnitudes have been regularized, so that a difference of 1 in

259

magnitude corresponds to a difference of about 2.5 in brightness. Thus, a magnitude 1 star is $2.5 \times 2.5 \times 2.5 \times 2.5 \times 2.5 = 100$ times brighter than a magnitude 6 star. Because we are no longer limited to viewing the sky with our eyes, and larger apertures collect more light, magnitudes greater than (that is, fainter than) 6 appear on the scale. Objects brighter than the brightest stars may also be included, their magnitudes expressed as negative numbers. Thus the full moon has a magnitude of –12.5 and the sun, –26.8.

In order to make more useful comparisons between stars at varying distances, astronomers differentiate between *apparent magnitude* and *absolute magnitude*, defining the latter, by convention, as an object's apparent magnitude when it is at a distance of 10 parsecs from the observer. This convention cancels out distance as a factor in brightness and is therefore an intrinsic property of the star.

Some key comparative magnitudes are:

➤ Sun: –26.2

➤ Full moon: –13

➤ Venus (at its brightest): –4.4

➤ Vega: 0

➤ Deneb: 1.6

➤ Faintest stars visible to the naked eye: 6

Astronomer's Notebook

Some of the most beautiful visual binary stars are those that have distinct colors. Alberio (also called Beta Cygni in the constellation Cygnus, the Swan) is a lovely double, separated by about 35 arcseconds. The blue and yellow colors of the pair are quite distinct and indicate the different surface temperatures of the stars. If you have a telescope and a sky chart, see if you can find it.

How Hot Is Hot?

Stars are too distant to stick a thermometer under their tongue. We can't even do that with our own star, the sun. But you can get a pretty good feel for a star's temperature simply by looking at its color.

As we discussed in Chapter 7, the temperature of a distant object is generally measured by evaluating its apparent brightness at several frequencies in terms of a blackbody curve. The wavelength of the peak intensity of the radiation emitted by the object can be used to measure the object's temperature. For example, a hot star (with a surface temperature of about 20,000 K) will peak near the ultraviolet end of the spectrum and will produce a blue visible light. At about 7,000 K, a star will look yellowish-white. A star with a surface temperature of about 6,000 K—such as our sun—appears yellow. At temperatures as low as 4,000 K, orange predominates, and at 3,000 K, red.

So simply looking at a star's color can tell you about its relative temperature. A star that looks blue or white has a much higher surface temperature than a star that looks red or yellow.

Close Encounter

Career opportunities for women were limited at the turn of the century. However, women with specialized training in astronomy were able to find employment at the nation's observatories. The Harvard College Observatory first hired women in 1875 to undertake the daunting task of the classification of stellar types. Under the direction of Edward Pickering, the observatory employed 45 women over the next 42 years. Photography and the telescopes of the HCO were beginning to generate vast amounts of astronomical data, in particular photographic plates filled with individual stellar spectra. It was more economical to pay college-educated women to pore over these data sets than to hire the same number of male astronomers.

The observatory employed a series of women whose experience, intelligence, and familiarity with the data allowed them to make valuable contributions to astronomy. Williamina Fleming (appointed Curator of Astronomical Photographs in 1898) made the spectral classifications of the bulk of the stars in the first Henry Draper catalog. (Henry Draper was a wealthy physician whose widow gave money to the observatory in his memory.) Annie Jump Cannon rearranged the spectral classification of Fleming and, in the end, classified over 500,000 stars. She was so expert at classifying stellar spectra from photographic plates that she could classify them at a rate of three per minute, calling out the star name and its spectral type to an assistant. These catalogs of stellar spectra were invaluable to later astronomers, and the HD prefix is still used for stars originally classified by these women.

Finally, Henrietta Swan Leavitt made one of the most fundamental astronomical discoveries of the early part of the century. She was involved in the effort to catalog variable stars, and identified a number of so-called "Cepheid" variable stars, with variation in brightness that occurs over a period of 2 to 40 days. Much like massive bells that ring with a lower tone than a tiny bell, massive (intrinsically brighter) stars have a longer period than low-mass (intrinsically fainter) stars. This result meant that astronomers could make a leap from the timing of the period of a star's fluctuations in brightness directly to its intrinsic brightness. And knowledge of the intrinsic brightness of a star (and its observed brightness) let us know how far away the star might be. This discovery earned Leavitt a posthumous nomination for the Nobel Prize in 1925.

Stellar Pigeonholes

The color of stars can be used to separate them into rough classes, but the careful classification of stellar types did not get under way until photographic studies of many spectra were made.

Using the Spectrum

Precise analysis of the absorption lines in a star's spectrum gives us information not only about the star's temperature, but also about its chemical make-up (see Chapter 7). Using spectral analysis to gauge surface temperatures with precision, astronomers have developed a system of *spectral classification,* based on the system worked out at the Harvard College Observatory. The presence or absence of certain spectral lines is tied to the temperatures at which we would expect those lines to exist. The stellar spectral classes and the rough temperature associated with the class are given in the following table.

Spectral Class	Surface Temperature
O (violet)	>28,000 K
B (blue)	10,000–28,000 K
A (blue)	~10,000 K
F (blue/white)	~7,000 K
G (yellow/white)	~6,000 K
K (orange)	~4,000 K
M (red)	<3,500 K

Star Words

Spectral classification is a system for classifying stars according to their surface temperature. The presence or absence of certain spectral lines is used to place stars in a spectral class.

The most massive stars are the hottest, so astronomers refer to the most massive stars they study as "0 and B" stars. The least massive stars are the coolest. The letter classifications have been further refined by ten subdivisions, with 0 (zero) the hottest in the range and 9 the coolest. Thus a B5 star is hotter than a B8, and both are hotter than any variety of A star. The sun is a spectral type G2 star. Our Galaxy and others are chock full of type G2 stars.

From Giants to Dwarfs: Sorting the Stars by Size

The radius of a star can be determined from the luminosity of the star (which can be determined if the distance is known) and its surface temperature (from its spectral type). It turns out that stars fall into several distinct classes. In sorting the stars by size, astronomers use a vocabulary that sounds as if it came from a fairytale:

➤ A giant is a star whose radius is between 10 and 100 times that of the sun.

➤ A supergiant is a star whose radius is more than 100 times that of the sun. Stars of up to 1,000 solar radii are known.

➤ A dwarf star has a radius similar to or smaller than the sun.

Making the Main Sequence

Working independently, two astronomers, Ejnar Hertsprung (1873–1967) of Denmark and Henry Norris Russell (1877–1957) of the United States studied the relationship between the luminosity of stars and their surface temperatures. Their work (Hertsprung began about 1911) was built on the classification scheme of another woman from the Harvard College Observatory, Antonia Maury. She first classified stars both by the lines observed and the *width* or *shape* of the lines. Her scheme was an important step toward realizing that stars of the same temperature could have different luminosity. Plotting the relationship between temperature and luminosity graphically (in what is now known as a *Hertzsprung-Russell diagram* or *H-R diagram*), these two men discovered that most stars fall into a well-defined region of the graph. That is, the hotter stars tend to be the most luminous, while the cooler stars are the least luminous.

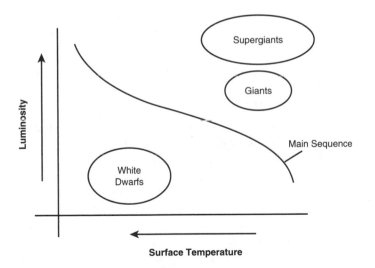

A Hertzsprung-Russell diagram or H-R diagram. This plot of a star's temperature, and intrinsic brightness, shows where in its lifespan a particular star is. The solid line (main sequence) is where stars spend the majority of their lifetimes.

(Image from authors' collection)

H-R Diagram Schematic

263

The region of the temperature luminosity plot where most stars reside is called the *main sequence*. Most stars are there, because as we will discover, that is where they spend the majority of their lives. Stars that are not on the main sequence are called giants or dwarfs, and we will see in coming chapters how stars leave the main sequence and end up in the far corners of the temperature-luminosity plot.

Off the Beaten Track

While some 90 percent of stars fall into the region plotted as the main sequence, about 10 percent lie outside this range. These include *white dwarfs,* which are far less luminous than we might expect from their high surface temperatures, and *red giants,* which are far more luminous than we would expect from their relatively low surface temperatures. In Chapter 18, "Stellar Careers," we'll describe how stars leave the main sequence and end up here. Briefly, these are stars that have used up their fuel, and are in the process of dying.

Astro Byte

Red dwarfs are the most common stars, probably accounting for about 80 percent of the star population in the universe.

Star Words

Binaries or **binary stars** are two-star systems, in which the stars orbit one another. The way the companion stars move can tell astronomers much about the individual stars, including their masses.

Stellar Mass

The overall orderliness of the main sequence suggests that the properties of stars are not random. In fact, a star's exact position on the main sequence and its evolution are functions of only two properties: composition and mass.

Composition can be evaluated if we have a spectrum of the star, its fingerprint. But how can we determine the mass of a star?

Fortunately, most stars don't travel solo, but in pairs known as *binaries*. (Our sun is an exception to this rule.) *Binary stars* orbit one another.

Some binaries are clearly visible from the earth and are called *visual binaries,* while others are so distant that, even with powerful telescopes, they cannot be resolved into two distinct visual objects. Nevertheless, these can be observed by noting the Doppler shifts in their spectral lines as they orbit one another. These binary systems are called *spectroscopic binaries*. Rarely, we are positioned so that the orbit of one star in the binary system periodically brings it in front of its partner. From these *eclipsing binaries* we can monitor the variations of light emitted from the system, thereby gathering information about orbital motion, mass, and even stellar radii.

However we observe the orbital behavior of binaries, the key pieces of information sought are orbital period (how long it takes one star to orbit the other) and the size of the orbit. Once these are known, Kepler's third law (see Chapter 4, "Astronomy Reborn: 1542–1687,") can be used to calculate the combined mass of the binary system.

Why is mass so important?

Mass *determines* the fate of the star. It sets the star's place along the main sequence and it also dictates its life span.

The Life Expectancy of a Star

A star dies when it consumes its nuclear fuel, its mass. We might be tempted to conclude that the greater the supply of fuel (the more massive the star), the longer it will live; however, a star's life span is also determined by how rapidly it burns its fuel. The more luminous a star, the more rapid the rate of consumption. Thus stellar lifetime is directly proportional to stellar mass and inversely proportional to stellar luminosity (how fast it burns). An analogy: A car with a large fuel tank (say a new Ford Excursion that gets 4–8 mpg) may have a much smaller range than a car with a small fuel tank (a Saturn which might get 30–40 mpg). The key? The Saturn gets much better mileage, and thus can go farther with the limited fuel it has.

Thus, while O- and B-type giants are 10 to 20 times more massive than the our G-type sun, their luminosity is *thousands* of times greater. Therefore, these most massive stars live much briefer lives (a few million years) than those with less fuel but more modest appetites for it.

Star Words

Visual binaries are binary stars that can be resolved from the earth. **Spectroscopic binaries** are too distant to be seen as distinct points of light, but they can be observed with a spectroscope. In this case, the presence of a binary system is detected by noting Doppler shifting spectral lines as the stars orbit one another. If the orbit of one star in a binary system periodically eclipses its partner, it's possible to monitor the variations of light emitted from the system and thereby gather information about orbital motion, mass, and radii. These binaries are called **eclipsing binaries.**

A B-type star such as Rigel, 10 times more massive than the sun and 44,000 times more luminous, will live 20×10^6 years, or 20 million years. For comparison, 65 million years ago, dinosaurs roamed the earth! The G-type sun may be expected to burn for $10,000 \times 10^6$ years (ten *billion* years). Our red dwarf neighbor, Proxima Centauri, an M-type star that is $\frac{1}{10}$ the mass of the sun (and $\frac{1}{100}$ that of Rigel), is only 0.00006 times as luminous as the sun, so will consume its modest mass at a much slower rate and may be expected to live more than the current age of the universe.

In the next two chapters we will see how stars go through their lives, and how they grow old and die.

The Least You Need to Know

➤ The distance to nearby stars cannot be measured directly (such as by radar ranging), but can be determined using stellar parallax. Distances to farther stars can be determined by measuring the period of variable stars.

➤ Stellar motion, velocity, size, mass, temperature, and luminosity can all be measured. A rough measurement of a star's temperature can be derived from its color. Hot stars are blue. Cool stars are red.

➤ By plotting the relationship between the luminosity and temperature of a large numbers of stars, astronomers have noticed that most stars fall along a band in the plot called the main sequence. Stars spend most of their lives on the main sequence.

➤ The lifetime of a star is determined primarily by its mass. High-mass stars have short lives, low-mass stars have long lives, and all stars die.

Chapter 18

Stellar Careers

In This Chapter

➤ Red giants and supergiants

➤ From supergiant to planetary nebula

➤ The death of a low-mass star

➤ Novae and supernovae

➤ The death of a high-mass star

➤ Supernovae as creators of elements

The stern Protestant theologian John Calvin would have liked stellar evolution. Each and every star appears "predestined" to follow a certain path in its life, and that path is set only by the mass of the star at its birth. There are some complications. If stars have nearby companions, they can be "revived" late in life, and the details of evolution for some stellar types are not entirely figured out. But the mass of a star at its birth does unalterably determine where it will reside on the main sequence, whether it will be a relatively cool M-type dwarf star, a hot and massive O or B star, or something in between (like our sun). Once the forces within a star—gravity pulling inward and the pressure of the heat of fusion pushing outward—reach equilibrium, stars enter the main sequence and the bulk of their lives—middle age, if you will—is relatively dull. But when the forces get out of balance and stars leave the main sequence to enter their death throes, the fireworks begin. The final years of a star can be spectacular, as this chapter shows.

Astronomer's Notebook

When a main sequence star becomes a red giant, it swells up to many times its original radius, and becomes far more luminous. When the sun swells up, its luminosity will be about 2,000 times greater than at present, and its radius will be about 150 times greater. That means that the sun's outer layers will reach out to about 0.7 A.U. (or to about the orbit of Venus). The earth is at 1 A.U. When this happens, the earth's oceans and atmosphere will be boiled away, leaving only the planet's rocky core. But don't lose any sleep—those days are probably 5 billion years away.

Star Words

Core hydrogen burning is the principal fusion reaction process of a star. The hydrogen at the star's core is fused into helium, producing enormous amounts of energy in the process.

A Star Evolves

You probably know people with a Type A personality—they are always keyed up, hyper, superoverachievers, and you may even call them workaholics. These people often make it big, only to burn out quickly. Then there's the Type B personality: the kind of person who moves through life calmly, doing what's necessary, but no more. These people might not be very spectacular, but as with that tortoise in the fable, slow and steady sometimes wins the race.

Stellar careers are similar, though the letters of the alphabet used to label them are different. The massive, hot O and B stars are born, mature, and enter the main sequence, only to burn out, their hydrogen fuel exhausted, after a few million years. They are the gas-guzzlers of the galaxy. In contrast, the red dwarfs, stars of type M, low in mass and low in energy output, may not have as much fuel, but they will burn for hundreds of billions of years. Low-mass stars are econo-boxes. They are not as flashy as their more massive cousins, but you don't see their steaming hulks by the side of the road, either.

The Main Sequence—Again

Once it has matured and assumed its rightful place on the main sequence (see Chapter 17, "Of Giants and Dwarfs: Stepping Out into the Stars"), a star (regardless of its mass) enjoys a steady job. Indeed, it does only one thing in its incredibly hot core: fuses hydrogen into helium, producing great amounts of energy exactly as we described for the sun in Chapter 16, "Our Star." This *core hydrogen burning* keeps a star alive—or, more precisely, maintains its equilibrium, the balance between the radiation pressure sustained by fusion pushing out and the force of gravity pulling in.

From Here to Eternity

From our human perspective, the main sequence life span of an average G-type star such as the sun seems like an eternity. A good 10 billion years goes by before such a star enters the late stages of its life, having fused a substantial amount of hydrogen into helium.

At this point, the delicate equilibrium between gravity and radiation pressure shifts, and the structure of the star begins to change. At the core of the star is a growing amount of helium "ash." It is referred to as ash because it is the end product of the burning or fusion of hydrogen. While hydrogen is fusing in the core, the more massive helium nucleus cannot reach sufficient temperature to fuse into a heavier element. As the supply of hydrogen in the core is diluted by the helium ash, the force of gravity starts to win out over the pressure of the slowing fusion reactions. As a result, the core begins to shrink.

Swelling and *Shrinking*

Yet a star does not simply shrink to its core or implode. Changes beget other changes. The process of shrinking releases gravitational energy throughout the star's interior. This energy release increases the core temperature, allowing hydrogen that was located outside the star's core to begin burning while the helium core continues to contract and heat up. At this point, the star's situation is rather paradoxical. Its outer layers are swelling up and burning brightly (called *shell hydrogen burning*), while its core, filled with helium ash, continues to collapse. It is a relatively short-lived state.

Stellar Nursing Homes

The transition from G-type star (like the sun) on the main sequence to the next stage of its career (a red giant) consumes perhaps 100 million years. That's a lot of time if your calculating a return on your investment, but it's a blink of the eye relative to the 10-billion-year lifetime of a G-type star—about $\frac{1}{100}$ of its total main sequence lifetime.

Red Giant

The core of the aging star shrinks. As it does, the gravitational energy released raises the temperature in the hydrogen-burning shell, increasing the pace of fusion. With this increase, more and more energy is dumped into the outer layers of the star, increasing the outward pressure. So while the core shrinks, the outer layers of the star expand dramatically, cooling as they do. As this happens, the star becomes more luminous and cooler, moving to the upper right-hand side of the H-R diagram (see Chapter 17). The resulting star is a called a *red giant*. For a solar mass main sequence star, a red giant will be about 100 times larger than the sun, with a core that is a mere $\frac{1}{1000}$ the size of the star.

Star Words

A **red giant** is a late stage in the career of stars about as massive as the sun. More massive stars in their giant phase are referred to as **supergiants.** Their relatively low surface temperature produces their red color.

Close Encounter

The night sky offers many examples of red giants. Two of the most impressive are Aldebaran, the brightest star in the constellation Taurus, and Arcturus, in Boötes. Look to the constellation Orion for an example of a red supergiant, Betelgeuse (pronounced *Beetlejuice*). You'll find very large array image of Betelgeuse in Chapter 17.

A Flash in the Pan

A red giant may continue on its unstable career for a few hundred million years, outwardly expanding and inwardly shrinking. At some point, however, the shrinking of the core raises its temperature high enough to finally ignite the helium that has been waiting there patiently. It's like the piston in your car engine on the compression stroke. Eventually, the combustion reaction is triggered. Only, in a star, it's a fusion reaction, not chemical combustion. Now, along a timeline that has been measured in tens of millions of years, something very sudden occurs. In a process that consumes only a few hours, not millions or billions of years, helium starts to burn in an explosion of activity called the helium flash—the explosive onset of helium burning in the core of a red giant star. The helium is fused into carbon, and the star settles into a (short-lived) equilibrium.

After the helium flash, the helium in the core is rapidly fused into carbon and (occasionally) oxygen. The star's core begins to heat up again, and with helium being fused in the core comes another equilibrium. The star's outer layers shrink, but become hotter, so the star becomes bluer and less luminous. Once helium is exhausted in the star's core, it's nearly the end of the road for a low-mass star. It won't be able to again raise the core temperature sufficiently to fuse carbon into heavier elements.

Astronomer's Notebook

The Messier Catalog (see Appendix D) lists several planetary nebula visible with a good amateur telescope. Among the best: The Dumbbell Nebula (M27) is associated with the constellation Vulpecula, and the Ring Nebula (M57) is in Lyra. To view these is to see a low-mass star like the sun in its death throes. It is the fate, some billions of years off, that awaits our own star.

Red Giant Revisited

By the prevailing standards of our stellar timeline, the equilibrium that results from helium core burning doesn't last long, only some tens of millions of years, because fusion proceeds rapidly, as the star fiercely goes about fusing helium into carbon and some oxygen.

Now something very much like the process that created the first red giant phase is replayed, only with helium and carbon instead of hydrogen and helium. As helium is exhausted in the core of the star, the carbon ash settles and the core shrinks, releasing heat. This release of heat triggers helium-burning in one shell, which, in turn, is enveloped by a hydrogen-burning shell. The heat generated in these burning shells causes the star's outer layers to swell, and the star becomes a red giant once more. When stars revisit the red giant phase, they are generally bigger and more luminous than the first time.

Core and Nebula

A star may last in this second red giant phase for a mere 100,000 years before its carbon core shrinks to an incredibly dense inert mass only about $\frac{1}{1000}$ the current radius of the sun—or about the size of the earth. The surrounding shells continue to fuse helium and carbon, and the outer layers of the star continue to expand and cool.

The outer layers of the star are now so far from the core, that they are able to lift off and move out into interstellar space, often in several distinct shells. These outer layers are puffed off from the star like spherical smoke rings, leaving behind the bare, hot core of the star. The cast-off outer layer of the star (which can contain 10 to 20 percent of the mass of the star) is misleadingly called a planetary nebula. Planetary nebulae are the ejected gaseous envelopes of red giant stars. These shells of gas are lit up by the ultraviolet photons that escape from the hot, white dwarf star that remains. In truth, a planetary nebula has nothing to do with planets. The source of the misnomer is that, in early observations, these objects were thought to resemble planetary disks.

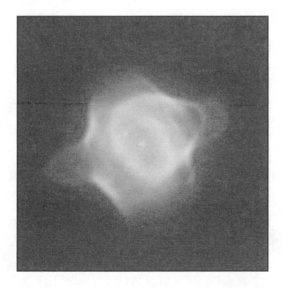

A Hubble Space Telescope *portrait of a young planetary nebula, the Stingray Nebula, located in the direction of the southern constellation Ara. The bright central star is visible within the great envelope of ionized gas.*

(Image from NASA)

White Dwarf

The expansion of a planetary nebula continues until the nebula fades and becomes too faint to be observable. The carbon core of the parent star, at the center of the planetary nebula, continues to glow white hot with surface temperatures of 100,000 K. Nuclear fusion in the star has now ceased; so what keeps it from collapsing further? What is balancing gravity?

It turns out that the white dwarf is so dense that the electrons themselves (which usually have plenty of room to move around freely) are close to occupying the same position with the same velocity at the same time. Electrons are forbidden to do so, and the gas, therefore, can contract no farther. Such a gas is called a degenerate electron gas, and the pressure of degenerate electrons is all that holds the star up at this point. The great astrophysicist S. Chandrasekhar first showed that stars could support themselves from further collapse in this way. This white-hot core, about the size of the earth but much more massive (about 50 percent as massive as the sun), is called a *white dwarf*. One teaspoon of the carbon core would weigh a ton on the earth.

As the white dwarf continues to cool, radiating its stored energy, it changes color from white to yellow to red. Ultimately, when it has no more heat to radiate, it will become a black dwarf, a dead, inert ember, saved from gravitational collapse by the resistance of electrons to being compressed beyond a certain point. Some astonomers have proposed that the carbon atoms will eventually assume a lattice structure. The stellar corpse may become, in effect, a diamond with half the mass of the sun.

Star Words

A **white dwarf** is the remnant core of a red giant after it has lost its outer layers as a planetary nebula. Since fusion has now halted, the carbon–oxygen core is supported against further collapse only by the pressure supplied by densely packed electrons. Their small size makes them relatively faint objects, despite their high surface temperatures.

This Hubble Space Telescope *image (right) reveals a population of faint white dwarfs (the circled stars) in globular cluster M4 (Messier object number 4), in the direction of Scorpius. The panel on the left is a view of this region of the sky from a ground-based telescope.*

(Image by Harvey Richer of the University of British Columbia and NASA)

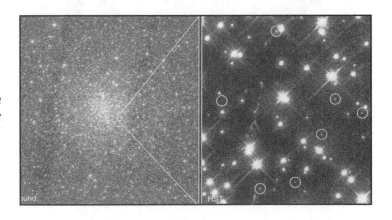

Going Nova

Nova is the Latin word for "new," and it seemed to early astronomers an appropriate term for stars that suddenly appeared in the sky. They were obviously new stars. Well, they were new in that they hadn't been seen before. But, in fact, a nova is a phenomenon associated with an *old* star.

When a binary pair forms, it is very unlikely that they will have the exact same mass. And since the mass of a star determines its lifetime, the stars in a binary pair will (more than likely) be at different points in their evolution. Sometimes, when a white dwarf is part of binary system in which the companion star is still active (on the main sequence or a red giant), the dwarf's gravitational field will pull hydrogen and helium from the outer layers of its companion. This gas accumulates on the white dwarf, becoming hotter and denser until it reaches a temperature sufficient to ignite the fusion of hydrogen into helium. The flare up is brief, a matter of days, weeks, or months, but, to earthly observers, spectacularly brilliant.

Given the right circumstances, a white dwarf in a binary pair can go nova repeatedly, reigniting every few decades as enough material is borrowed from its companion star.

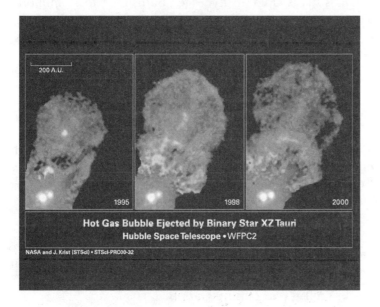

The binary system XZ Tauri consists of two stars separated by approximately the distance between the Sun and Pluto. Images taken over the course of five years reveal that the binary system is associated with an enormous expanding bubble of hot gas.

(Image from NASA and J. Krist, STScI)

The Life and Death of a High-Mass Star

Up to this point, we've concentrated on the life of a star like the sun. In truth, everything that we have discussed (up to the planetary nebula phase) applies equally well to a high-mass star. It has already been mentioned that stars of significantly lower mass than our sun enter the main sequence and stay there not for millions or billions of years, but much longer. Stars of significantly greater mass than the sun (somewhere between 5 and 10 solar masses) have a very different destiny.

273

Fusion Beyond Carbon

Astronomers think of 5 to 10 solar masses as the dividing line between low- and high-mass stars. That is, stars about this much more massive than the sun die in a way very different from stars of lesser mass. The major difference in their evolution is that high-mass stars are able not only to fuse hydrogen, helium, carbon, and oxygen, but heavier elements as well. As core burning of one element ends, it begins burning in a shell above the core.

Hydrogen fusion produces helium ash that settles to the star's core. Then helium fusion produces carbon ash, which again settles to the star's core. As the fusion of each heavier element proceeds, the core layers progressively contract, producing higher and higher temperatures. Unlike low-mass stars, in a high mass star core temperatures are high enough to fuse carbon into oxygen, oxygen into neon, neon into magnesium, and magnesium into silicon. The end of the road is iron. When a massive star has iron building up in its core, the final curtain is near. The reason is that for every element up until iron, energy was released when nuclei were fused. But to fuse iron into heavier elements *absorbs* energy. In terms of fusion, iron is a dead end.

The evolution of the high-mass star is rapid. Its hydrogen burns for 1 to 10 million years, its helium for less than 1 million years, its core of carbon for a mere 1,000 years, oxygen for no more than a year, and fusion of silicon consumes only a week. An iron core grows as a result, but for less than a day. Just before its spectacular death, a massive star consists of nested shells of heavier elements within lighter elements, all the way down to its iron core.

Over the Edge

At this point, the core of a high-mass star is, in effect, a white dwarf. It is supported by its degenerate electrons. But there is a problem. The mass of a high-mass stellar remnant is so large that gravity overwhelms even the resistance of electrons to having the same position and velocity. A mass of 1.4 solar masses is sufficient to overwhelm those electrons, which are combined with protons to create neutrons. The temperatures in the core of the star become so high that all of the work of fusion is rapidly undone. The iron nuclei are split into their component protons and neutrons in a process called photodisintegration.

As the core of the star collapses under its own gravity, the electrons combine with protons to become neutrons and neutrinos. The neutrinos escape into space, heralds of disaster. The core of the star only stops its collapse when the entire core has the density of an atomic nucleus. This sudden halt in the collapse causes a shock wave to move through the outer layers of the star and violently blow off its outer layers.

Star Words

Core-collapse supernova is the extraordinarily energetic explosion that results when the core of a high-mass star collapses under its own gravity.

Supernova: So Long, See You in the Next Star

"Violently" is an understatement. The process of evolution from a hydrogen-burning star to a collapsing core consumed 1 to 10 million years. The final collapse of a high-mass core takes less than a second and ends in a *core-collapse supernova*.

Types of Supernovae

Two types of supernovae are recognized. Type I supernovae contain little hydrogen, whereas Type II are rich in hydrogen. Only Type II supernovae are associated with the core collapse of high-mass stars. Type I supernovae are associated with our friends the white dwarfs.

Close Encounter

No supernova has appeared in our Galaxy, the Milky Way, since 1604. Since supernovae are among the most energetic processes known, it is not surprising that the light of a supernova can outshine the combined light of the entire Galaxy. Theory predicts a supernova occurrence in our Galaxy every 100 years or so. We are, therefore, more than a bit overdue, and we may be in for a spectacular display any day now. The cosmic rays and electromagnetic radiation that would rain down on the earth if a nearby supernova were to go off (say within 30–50 light-years) would have catastrophic results. Fortunately, there aren't any stars that close to us massive enough to generate a Type II supernova.

But you don't have to wait for a supernova to occur in the Milky Way. You can search for them (as well as novae) in other galaxies. The best areas for searching are in the regions of the constellations Leo and Virgo, which contain many galaxies. It is sometimes possible to see novae and supernovae, which appear as bright stars, with binoculars, if you don't have a telescope. Be aware that, if you witness a supernova, you are viewing the most powerful explosion you will ever see. Amateur astronomers have been credited with many supernova discoveries. For an impressive example of supernova discoveries by Puckett Observatory (33 as of March 2001), see www.astronomyatlanta.com/nova.html.

The Supernova as Creator

As you might expect, an explosion as tremendous as that of a supernova creates a great deal of debris. The Crab Nebula, in the constellation Taurus, is the remnant of a supernova that appeared in C.E. 1054. Chinese astronomers left records of that event, reporting a star so brilliant that it was visible for a month in broad daylight. The bright radio source Cassiopeia A is also a supernova remnant.

But supernovae create far more than glowing remnants.

Hydrogen and helium, the two most basic elements in the universe, are also the most primitive. They existed before the creation of the stars. A few other elements (carbon, oxygen, neon, silicon and sulfur) are created by nuclear fusion in low and high-mass stars. But all of the other elements in the universe are created only in supernova explosions. Only in these explosions is there enough energy to bring nuclei together with sufficient force to create elements heavier than iron.

The only elements that existed at the beginning of the universe were hydrogen and a little bit of helium, beryllium, and lithium. The rest of the periodic table was generated by stars. Each one of us contains the debris of a supernova explosion.

Discovery image of Type II Supernova 2001. Bars indicate postition of the supernova.

(Image from T. Puckett and M. Peoples)

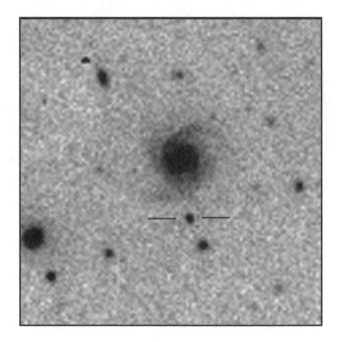

Neutron Stars

What could possibly survive a supernova? A supernova explosion pushes things out, away from the star's core. But, depending on the original mass, there will be something left behind. Either a bizarre object known as a neutron star or, even more strange, a black hole.

In Theory

While the explosive shock wave originates in the core, it does not start at the core's very center. If the core's mass was between 1.4 and 3 solar masses, the remnant at the center will be a ball of neutrons known as a neutron star.

In size, this so-called *neutron star* is small by astronomical standards—and tiny, compared to the high-mass star (many times the radius of our sun) of which it was a part. The neutron star's diameter would be something over twelve miles (20 km), and its density a staggering 10^{17} kg/m^3. All of humanity (if compressed to the density of a neutron star) would be the size of a pea.

What the Pulsars Tell Us

Neutron stars don't stand still. Just like the stars of which they are a remnant, they rotate. Since they have collapsed from a much larger size, they spin rapidly. Recall from Chapter 12, "Solar System Family Snapshot," the discussion of conservation of angular momentum: A rotating body (like a whirling skater drawing in her arms) spins faster as it shrinks. A neutron star has shrunk from a body hundreds of times larger than the sun to one that is smaller than the Earth. The earth takes 24 hours to make one revolution. A very massive but very small neutron star does it in a fraction of a second.

The rate of rotation is not the only property that intensifies in the neutron star. Its magnetic field is many times stronger than that of the parent star because the lines of the magnetic field are compressed along with the matter of the core itself. The combination of rapid rotation and a powerful magnetic field serves to announce the presence of some neutron stars in the universe.

In the late 1960s, S. Jocelyn Bell Burnell was a graduate student at Cambridge University working with Anthony Hewish, looking for interesting sources of radio emission. One very strange signal was detected: an an instantaneous burst of emission followed by a brief pause, and then another pulse. The pulses and pauses alternated with great precision—as it turns out, with a precision greater than that of the most advanced and accurate timepieces in the world.

In 1974, Hewish received the Nobel Prize in Physics for the discovery of the radio signals now called *pulsars*.

Star Words

A **neutron star** is the super-dense, compact remnant of a massive star, one possible survivor of a supernova explosion. It is supported by degenerate neutron pressure, not fusion. It is a star with the density of an atomic nucleus.

Star Words

A **pulsar** is a rapidly rotating neutron star whose magnetic field is oriented such that it sweeps across Earth with a regular period.

A Stellar Lighthouse

Imagine the pulsar as a stellar lighthouse. At the magnetic poles of the neutron star (though not necessarily aligned with the star's rotational axis) are regions in which charged particles are accelerated by the star's magnetic field. These regions, which rotate with the star, radiate intense energy. As the neutron star rotates, a beam of electromagnetic radiation (especially intense in the radio regime) sweeps a path through space. If the earth lies in that path, we see the pulsar.

Thus, all pulsars are neutron stars, but not all neutron stars will necessarily be pulsars. If the beam of a particular neutron star does not sweep past the earth, we will not detect its radio pulsations. Pulsar periods range anywhere from milliseconds to a few seconds.

Astro Byte

When pulsars were first detected in the late 1960s, their signals were so regular that some astronomers thought they might be a sign of extraterrestrial intelligence. However, the large number of pulsars detected, and the neutron star theory of pulsars provided an adequate and more mundane explanation.

I Can't Stop!

In the next chapter, we explore the nature of an even more strange supernova leftover: a black hole. If the core of the star is more massive than 3 solar masses, not even neutron degeneracy pressure can support it. In this case, the gravitational collapse continues with nothing to halt it, and a black hole is born.

The Least You Need to Know

➤ A star's mass is the primary determinant of the course its life will take. High mass stars are short-lived, while low-mass stars are long-lived.

➤ When a star has fused most of the hydrogen in its core, its days are numbered.

➤ Low-mass stars evolve into red giants and, ultimately, into white dwarfs and planetary nebulae. While some white dwarfs in binary systems periodically reignite as novae, others gradually cool, finally becoming burned-out embers of carbon and oxygen.

➤ High-mass stars (stars greater than 5-10 solar masses) die spectacularly when their cores collapse, creating a supernova. The remnant is either a neutron star, or a black hole.

➤ Most of the elements in the periodic table are produced only in supernova explosions.

Chapter 19

Black Holes: One-Way Tickets to Eternity

> ## In This Chapter
>
> ➤ Black holes
>
> ➤ The architecture of the universe
>
> ➤ Relativity theory

As the last couple of chapters have demonstrated, contemplating stars and stellar processes can require a vigorous stretching of the mind. It is not just that the time spans involved are so long, the objects so enormous, the distances so great, and the temperatures so high, but that reality itself seems very unfamiliar to us under such extreme conditions. Can we really fathom our sun swelling up to the size of the orbit of Venus, or the power of a supernova explosion, briefly bright enough to outshine its host galaxy?

As unfamiliar as things may have become in the last two chapters, we have saved perhaps the strangest object in the universe for this chapter.

We now explore another possible end state of a massive star, a *black hole*. When the iron core of a massive star is collapsing, it may stop when the entire core of the star has the density of an atomic nucleus. If the core is massive enough (more than 3 solar masses) the collapse is unstoppable, overwhelming neutron degeneracy pressure, and we call the resulting object a black hole.

Astronomer's Notebook

The parent star of a collapsing neutron star core that is more than 3 solar masses would have to be 20 to 30 times more massive than the sun. In other words, only main sequence stars of at least 20 to 30 solar masses will ever collapse into a black hole. Which stars are they? These are a small subset of the massive, hot O and B stars that we have mentioned. We'll study the nurseries of these stars in the next chapter.

Astronomer's Notebook

If the earth were compressed to the size of a garbanzo bean—yet retained its current mass—a velocity greater than the speed of light would be required to escape its pea-sized surface. Such an object would be, effectively, a black hole.

Is There No End to This Pressure?

Is the neutron star a dead end?

In the case of ordinary stars, remember, equilibrium is reached when the outward-directed forces of heat (from fusion reactions) are in balance with the inward-directed forces of gravity. Neutron stars, however, produce no new energy, simply radiating away the heat that is stored in them. They resist the crush of gravity not with the countervailing radiation pressure from fusion, but with neutrons so densely packed that they simply cannot be squeezed any more. Astronomers call these degenerate neutrons.

If this is the case, the neutron star is not so much in equilibrium as it is in stasis. A stalemate exists between the irresistable force of gravity and immovable objects in the form of a supremely dense ball of neutrons.

But if the force of gravity is sufficient, the collapse is apparently unstoppable.

Black Holes: The Ultimate End

Incredible though it seems, if a star is massive enough it will continue to collapse on itself. Forever.

Recall from Chapter 18, "Stellar Careers," that a white dwarf evolves from a low-mass parent star (a star less than 5 to 10 solar masses) and that the resulting white dwarf can be no more massive than 1.4 solar masses. If it has a higher mass, gravity will overwhelm the tightly packed degenerate electron pressure, and the core will continue to collapse. When a star's mass is greater than 1.4 solar masses, its core collapse continues, and it will blow off its outer layers as a supernova. If the mass of the core is less than 3 solar masses, the remnant will be a neutron star. However, the "specs" for a neutron star also have an *upper* mass limit. It is believed that a neutron star can be no more massive than about three times the mass of the sun. Beyond this point, even its apparently uncrushable core of neutrons will yield to gravity's pull.

So what? Do we just get an even smaller and denser neutron star?

Not exactly. We get an object from which there is literally no escape.

When an extremely massive star is ripped apart in a supernova explosion, it may produce a supernova remnant so massive that the subsequent core collapse cannot be stopped. When this happens, *nothing* escapes the attractive forces near the core— not even electromagnetic radiation, including visible light.

This fallen star, the end-result of the collapse of an extremely massive stellar core is an object from which no light can escape. It is called a *black hole.*

What's That on the Event Horizon?

Star Words

A stellar mass **black hole** is the end-result of the core collapse of a high mass star. It is an object from which no light can escape. Although space behaves strangely very close to a black hole, at astronomical distances the black hole's only effect is gravitational.

Although light cannot escape from a black hole, the black hole has certainly not ceased to exist. It is still a physical object with mass. That is the reason it creates a gravitational field, just as the earth or any other object with mass does. But how do we talk about the size of a black hole when we have just described it collapsing without end? Is it infinitely small?

It turns out that there is a dimension to ascribe to a black hole. We can talk about the mass of a black hole, and its radius, but it is a very different kind of radius, as we'll see in a moment. First, a quick detour.

In Chapter 9, "Space Race: From *Sputnik* to the International Space Station," we described the fundamental difficulty of getting rockets into space. Building a rocket capable of escaping the earth's gravitational pull requires an engine capable of delivering sufficient thrust to achieve a velocity of about 7 miles per second (11 km/s). This *escape velocity* (as Newton determined in the eighteenth century) depends upon two factors: the mass of the planet, and its radius. For a fixed-mass object, the smaller the radius, the greater the escape velocity required to get free of its gravitational field.

So as the core of a star collapses (with its mass remaining constant), the escape velocity from its surface increases rapidly. But what about when it has no more "surface" that we can talk about? What is the surface of a black hole? And what is the limit to the increase in the velocity?

Star Words

Escape velocity is the velocity necessary for an object to escape the gravitational pull of another object.

Nature has one very strict speed limit: the speed of light. Nothing in the universe, not even photons carrying information from distant reaches of the universe, can move faster than the speed of light, or 984,000,000 feet per second (300,000,000 m/s). So there is an upper limit to escape velocity as well. When a body of a given mass reaches a certain—very small—size, objects would have to be moving faster than the speed of light to escape. It makes Alcatraz look like a piece of cake.

Star Words

The **Schwarzschild radius** of a black hole is the radius of an object with a given mass at which the escape velocity equals the speed of light. As a rule of thumb: The Schwarzschild radius of a black hole (in km) is approximately 3 times its mass in solar masses. So a 5-solar mass black hole has a Schwarzschild radius of about 5 × 3 = 15 km. The **event horizon** coincides with the Schwarzschild radius and is an imaginary boundary surrounding a collapsing star or black hole. Within the event horizon, no information of the events occurring there can be communicated to the outside.

Where's the Surface?

But how can we talk about the "surface" of a black hole?

The German astronomer Karl Schwarzschild (1873–1916) first calculated what is now called the *Schwarzschild radius* of a black hole. This value is the radius at which escape velocity would equal the speed of light (and, therefore, the radius within which escape is impossible) for a star of a given mass. For the earth, the Schwarzschild radius is the size of a garbanzo bean, about 0.4 inches, or 1 centimeter. For a neutron star at 3 solar masses the Schwarzschild radius is 5.58 miles (9 km). So this radius doesn't define a literal "surface" so much as a characteristic property of a black hole.

Remember that the collapse of a black hole is in some sense infinite. Our three-or-more solar-mass stellar core will not stop shrinking just because it has reached the Schwarzschild radius. It keeps collapsing. Once it is smaller than the Schwarzschild radius, however, it will effectively disappear. Its electromagnetic radiation (and the information that it carries) is thereafter unable to escape. We spoke earlier about electromagnetic radiation carrying energy and information. Since we cannot get radiation from within the Schwarzschild radius, we cannot get any information from there, either. Events that occur within that radius are hidden from our view. For this reason, the Schwarzschild radius is also called the *event horizon*. We cannot see past this ultimate horizon.

Relativity

A full understanding of black holes and the phenomena associated with them requires knowledge of Einstein's theory of general relativity. Einstein's most famous works were his two relativity theories, special relativity and general relativity. Special relativity dealt with the ultimate speed limit, the speed of light. And general relativity is a theory of gravity. General relativity gives a more complete description of gravity's effects, and can explain some anomalies that Newtonian mechanics cannot. Newtonian mechanics is really a special case of the more encompassing general relativity.

What Is Curved Space?

Whereas Newton introduced the concept of gravitational force as a property of all matter possessing mass, Einstein proposed that matter does not merely attract matter, but that all matter warps the space around it. For Newton, the trajectory of an orbiting planet is curved because it is subject to the gravitational influence of, for example, the sun. For Einstein, the planet's trajectory is curved because space itself has been curved by the presence of the massive sun. This change represents a fundamental shift in the way we think about the universe. One major difference between Newton's and Einstein's theory of gravitation is that if mass distorts space, then massless photons of light (in addition to matter) should also feel the effect of gravity.

And like any good theory, Einstein's ideas explained questions that had gone unanswered, and made testable predictions. In particular, his theory explained some tiny peculiarities in the orbit of Mercury and successfully predicted that the sun's mass was enough to bend light rays passing very close to it.

Astronomer's Notebook

In *isolation*, the mass of an ever-collapsing stellar core does not change. However, a black hole will draw in matter that wanders too near to the event horizon. With such accretion, the star's mass will increase—even as its collapse continues. This means that its Schwarzschild radius (and event horizon) will increase. The famous astrophysicist Steven Hawking has also proposed that black holes can actually evaporate when pairs of particles are created close to the event horizon.

No Escape

The popular (well, popular among beleaguered physics teachers) image of space in Einstein's view is a vast rubber sheet with a bowling ball sitting in a local depression. The mass of the bowling ball, representing a very massive object, distorts the sheet, which represents space. In this way, a massive object distorts space itself.

Now, in this example, the sheet is two-dimensional. But space is three-dimensional! We can't picture that distortion applied to our three-dimensional universe, but we call its effect gravity.

Say that instead of making a dimple in the sheet (as the bowling ball does), an object were to make an infinitely deep sinkhole. That object would be a black hole.

The Black-Hole Neighborhood

We have said that no radiation can escape from within the Schwarzschild radius. But what of the region just outside it? It turns out that material near the black hole does produce observable radiation. Matter that strays close enough to the event horizon to be drawn into it does not remain intact, but is stretched and torn apart by enormous tidal forces. In the process, energy can be released in the form of x-rays. Bright x-ray sources may point the way to black holes in the neighborhood.

Astronomer's Notebook

If the sun were suddenly whisked away and replaced with a 1–solar mass black hole, would all of the planets be sucked in? In a word, no. The planets would continue in their normal orbits around the new "black" sun. However, we'd probably miss the sun's glow. (Actually, without that glow, we wouldn't be around to miss anything.)

Star Words

A **thought experiment** is a systematic hypothetical or imaginary simulation of reality; used as an alternative to actual experimentation when such experimentation is impractical or impossible.

Thought Experiments

No one could ever visit a black hole and live to tell about it. A spaceship, let alone a human being, would be torn to pieces by tidal forces as it approached the event horizon.

Faced with situations impractical or impossible to observe directly or to test physically, scientists typically construct *thought experiments,* methodical exercises of the imagination based on whatever data are available.

Postcards from the Edge

So here's a thought experiment: Suppose it were possible to send an indestructible probe to the event horizon of a black hole.

Next, suppose we equipped the probe with a transmitter broadcasting electromagnetic radiation of a known frequency. As the probe neared the event horizon, we would begin to detect longer and longer wavelengths. This shifting in wavelength is known as a gravitational redshift. This shift occurs because the photons emitted by our transmitter lose some energy in their escape from the strong gravitational field near the event horizon. The reduced energy would result in a frequency reduction (and, therefore, a wavelength increase) of the broadcast signals—that is, a redshift.

The closer our probe came to the event horizon, the greater the redshift. At the event horizon itself, the broadcast wavelength would lengthen to infinity, each photon having used all of its energy in a vain attempt to climb over the event horizon. Suppose we also equipped the probe with a large digital clock that ticked away the seconds and that was somehow visible to us. (Remember, this is a *thought* experiment.)

Through a phenomenon first described in Einstein's special relativity called *time dilation,* the clock would appear (from our perspective) to slow until it actually reached the event horizon, whereupon it—and time itself—would appear to slow to a crawl and stop. Eternity would seem to exist at the event horizon—the process of falling into the black hole would appear to take forever.

As the wavelength of the broadcast is stretched to infinite lengths, so the time between passing wavecrests becomes infinitely long. Realize, however, that if you could somehow survive aboard the space probe and were observing from inside rather than from a distance, you would perceive no changes in the wavelength of electromagnetic radiation or in the passage of time. Relative to you, nothing strange would be happening. Moreover, as long as the physical survival of your craft in the enormous tidal forces of the black hole were not an issue, you would have no trouble passing beyond the event horizon. But to remote observers, you would have stalled out at the edge of eternity. It's all a matter of point of view.

Into the Abyss

What's inside a black hole, you ask? We have no idea. Not because we're not curious, mind you, but because we can literally get no information from beyond the event horizon.

Theoreticians refer to the infinitely dense result of limitless collapse as a *singularity.*

We do not yet have a description of what happens to matter under the extreme conditions inside a black hole. Newton and Einstein moved our knowledge of the universe ahead in two great leaps. An understanding of these strange objects may be the next great leap forward. Whose name will be added to the list of great explicators?

Black-Hole Evidence

Lost in a black hole? Or just lost in thought? It is all too easy to start thinking of black holes as purely theoretical constructs, fantasies, mind games. Can we actually *see* a black hole?

Star Words

Time dilation is the apparent slowing of time (as perceived by an outside observer) as an object approaches the event horizon of a black hole, or moves at very high relative velocities, approaching the speed of light.

Star Words

A **singularity** is the infinitely dense remnant of a massive core collapse.

Not directly, but we can certainly see its effects. Black holes are like the monsters in old (and new) movies. Before we ever see the monster itself, we see its footprints and its claw marks on the trees, the muddy trail that leads back to the swamp. What are the muddy trails of black holes?

In our own galaxy, there is a bright source of x-rays in the constellation Cygnus (the Swan) known as Cygnus X-1. It is the binary companion of a B star, and the x-rays from it flare up and fade quickly, indicating that it is very small in radius. Remember that we mentioned the x-rays are often emitted in the neighborhood of a stellar-remnant black hole. In addition, the x-ray source has no visible radiation and a mass (inferred from the orbit of its companion) of about 10 solar masses. In this case circumstantial evidence may be enough to convict.

The black hole in Cygnus is what is called a stellar remnant black hole. In theory, black holes can have much higher masses—masses far greater than a stellar core. These are called supermassive black holes and are found in the cores of galaxies. The *Hubble Space Telescope* image reproduced here is about as close as we have gotten to a supermassive black hole. Released on May 25, 1994, the image shows a whirlpool of hot (10,000 K) gas swirling at the center of an elliptical galaxy (see Chapter 22, "A Galaxy of Galaxies") known as M87, 50 million light-years from us in the direction of the constellation Virgo.

A Hubble Space Telescope image of a giant gas disk apparently swirling about a black hole at the center of galaxy M87, 50 million light-years from the earth.

(Image from NASA)

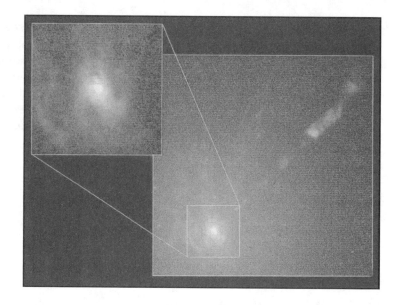

Using the *Hubble's* Faint Object Spectrograph, astronomers Holland Ford and Richard Harms were able to measure how light from the gas is redshifted and blueshifted as one side of the 60-light-year-radius disk of gas spins toward us and the other away from us. The radius at which gas is spinning, and the velocity of its rotation tell us how much matter must be within that radius. With the high resolution of the *Hubble Space Telescope* and radio telescopes like the Very Long Baseline Array (VLBA), astronomers can trace the rotation of gas to smaller and smaller distances from the center. Doing so, they find that even at very small radii, the gas is rotating at velocities that indicate something very massive still lies within that radius.

At the heart of M87 and other galaxies, indications are strong that supermassive black holes, perhaps with the mass of many billions of suns, lie in residence.

The Least You Need to Know

➤ A neutron star is one possible remnant of a massive star that has exploded as a supernova.

➤ Black holes are the other possible remnant of collapsed massive star cores. If the core has a mass greater than three solar masses, the collapse cannot stop and a black hole is born.

➤ If you don't get too close to a black hole, its effects (in terms of gravity) are no different than a star of the same mass. They are *not* giant galactic "vacuum cleaners."

➤ According to general relativity, black holes, stars and anything with mass distorts the space around it. This distortion can influence the path of both particles *and* light.

➤ Because light cannot escape from them, we can't see black holes, but we can indirectly observe evidence that their mass is present.

Stellar Nurseries

When you look at the night sky far from city lights, it's hard to miss that there is a lot of blackness up there. There is "space" between the stars. A whole lot of it, apparently.

But, as we've just seen, just because something is black doesn't mean nothing is there. We may just have to look harder.

Like ripples in a pond, planetary nebulae and supernova remnants move away from their parent stars, gradually fading and becoming part of the space between the stars. All of the heavy elements produced in a supernova explosion are sent back out into space. Some astronomers exclusively study these dark places between the stars. These regions are called interstellar space, and the material that we find there is called the *interstellar medium*.

To be sure, most of space is a more perfect vacuum than is found in any earthly laboratory. But what we call the interstellar medium is filled with radiation, magnetic fields, and (in some places) vast clouds of gas and dust. These gas clouds are waiting in the wings of the galaxy. They are the raw materials from which the next generation of stars will form.

Nor is interstellar space all dark. In fact, some of those points of light that you see in the night sky aren't stars at all. A few of them (like the Great Nebula in Orion) are regions of ionized gas around hot O- and B-type stars. While much of the Milky Way is hidden from optical view by the interstellar medium, radio-frequency and other observations have opened new windows. With high-resolution observations available at many wavelengths, we are now able to witness the birth, evolution, and death of stars. Of course, we have to view a number of different stars at each of these stages, but that is always a limitation in astronomy.

"Nothing?" Shakespeare's King Lear observed to his daughter Cordelia. "Nothing will come of nothing." But, in space, many things come from what is apparently nothing. In this chapter we look at these dark (and not so dark) places between the stars.

An Interstellar Atlas

The space between the stars is filled with two basic components, gas and dust. The vast majority (99 percent) of *interstellar matter* is what we call gas, and the majority of the gas consists of the most abundant element in the universe, hydrogen. The remainder of the material is what is referred to as "dust," though smoke might be a more accurate description. In fact, we'll see that interstellar dust puts up quite a smoke screen, keeping us from viewing many regions in our own Galaxy optically.

A multifrequency view of our own Milky Way galaxy, viewed from our perspective within it, makes it clear that there is more than meets the eye. The gas and dust, apparent in the radio and infrared wavelength views of the galaxy, all but disappear at optical wavelengths. The individual wavelengths are labeled, with low frequencies (long wavelengths) at the top and high frequencies (short wavelengths) at the bottom.

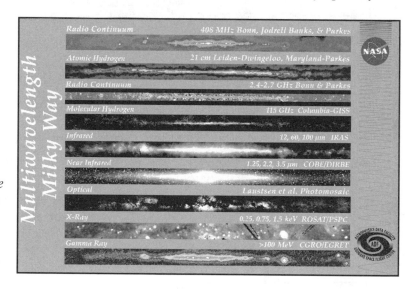

(Image from NASA)

The material between the stars is not evenly distributed. Since gas and dust have mass, the material pulls together into clouds and clumps via its own gravity. The patchy distribution of the interstellar medium means that in some regions of the sky, astronomers can observe objects that are very distant. In other regions, where the interstellar matter is more concentrated, our range of vision is more limited. Think of the last time you looked out an airplane window. If the clouds were patchy, there were some directions in which you could see the ground and others in which you could not.

The lowest-density interstellar clouds consist mostly of atomic hydrogen (called HI—no, not "Hi," but H-one, an H followed by a roman numeral I). Until the advent of radio astronomy, this atomic material was impossible to detect. Cooler, higher-density clouds contain molecular hydrogen (H_2). These regions are the fuel tanks of the universe, ready to produce new stars. The regions closest to young stars are ionized—their electrons stripped away. The 10,000 K gas in these regions (called HII—"H-two"—regions) emits strongly in the optical portion of the spectrum.

Let's look more closely at these different types of matter between the stars.

Astronomer's Notebook

Gas and dust is thinly distributed throughout space, and is the matter from which the stars are formed. About 5 percent of our Galaxy's mass is contained in its gas and dust. The remaining 95 percent is in stars.

Blocking Light

Why is it that dust blocks our optical view of the Milky Way? It's due to the size of the dust grains. Let's think about this for a moment. A satellite dish can be made out of a wire mesh, perforated by small holes. Why doesn't this structure let the radio waves slip through, like water through a sieve? Because it's catching radio waves, and radio waves are *big*. So big, in fact, that as long as the holes are small enough, the radio waves don't even know the holes are there. The waves reflect off the surface of the dish as if it were solid. In fact, all electromagnetic radiation (light included) works this way. The waves interact only with things that are about the same size as their wavelength. As luck would have it, optical wavelengths are about the same size as the diameter of a typical dust grain. As a result, optical photons are absorbed or scattered by dust, while long-wavelength radio waves pass right through.

The combined effect of the scattering and absorption caused by the dust results in *extinction*. The dust *absorbs* blue light more than red and also *scatters* blue light more than red. As a result, the visible light that makes it through is reddened. *Interstellar reddening* is increased when objects are farther away, and astronomers must take this into account when determining the true color of a star.

You may picture interstellar dust as a kind of fog. Certainly, fog, which consists of water molecules and often particulate matter (a.k.a. dust), interferes with the transmission of light, as anyone who has driven in terror along a foggy mountain road can attest. We often speak of a thick fog and say "You couldn't see your hand in front of your face."

Close Encounter

If a cloud of dust is between us and a distant star, the light from the star must pass through the dust before it can get to us. Dust allows the longer (redder) wavelengths to pass, but the shorter (bluer) wavelengths are scattered and absorbed. As a result, the light that makes it through is reddened. The setting sun looks redder for the very same reason. As the sunlight passes through a thicker slab of the atmosphere near the horizon, its blue frequencies are absorbed and scattered by the atmosphere and the sun looks red.

But the fact is, you usually have no trouble seeing such nearby objects as your hand. Indeed, fog seems to play with us. You never really walk *into* the fog. It always seems to be just ahead of you. This is because the obscuring effect of all those water molecules and dust particles is cumulative. The matter stacks up with depth. A cubic yard of thick fog (about the distance of a hand held at arm's length from your face) may not be especially impressive, but ten or twenty or fifty cubic yards, let alone a hundred or more, create the well-known pea-soup effect: a thick fog, and photons scatter before they can get to you.

Now, interstellar dust is so diffuse that many thousands of miles of it may not obscure anything. But stack billions upon billions of miles of material between you and a star, and it will certainly begin to have an effect on what you see.

Dusty Ingredients

Astronomers have been able to analyze the content of interstellar gas quite accurately by studying spectral absorption lines, the fingerprint elements create by allowing some wavelengths to pass while absorbing others (see Chapter 7, "Over the Rainbow"). The precise composition of the interstellar dust is less well understood; however, astronomers have some clues.

The 1 percent of interstellar gas that isn't hydrogen or helium contains far less carbon, oxygen, silicon, magnesium, and iron than would be expected based on the amounts of these elements found in our solar system or in the stars themselves. It is believed that the interstellar dust forms out of the interstellar gas, in the process drawing off some of the heavier elements from the gas. So the dust probably contains silicon, carbon, and iron, as well as ice consisting mainly of water, with traces of ammonia and methane as well as other compounds. The dust is rich in these substances, while the gas is poor in them.

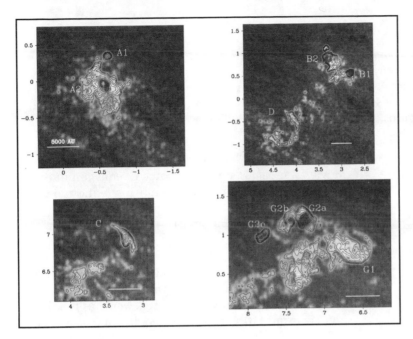

Details of the W49 star-forming region. These radio-frequency images were made with the Very Large Array and have a resolution of .04 arcsecond (40 milliarcseconds). The emission comes from the hot gas that surrounds a number of young O and B stars. The bar in each corner has a length of 5,000 A.U. The letters help astronomers refer to individual regions of hot gas.

(Image from authors' collection)

Flipping Out

Most of the gas in the interstellar medium, some 90 percent, consists of the simplest element, hydrogen. Helium, the second simplest element, accounts for another 9 percent. The remaining 1 percent consists of other elements. And the gas is mostly cold. Recall that hydrogen consists of a proton and an electron. If there is enough ambient energy, the electron gets bumped up the energy ladder into an *excited state*. But most of these clouds of hydrogen are far from energy sources—stars—and emit no detectable visible light.

But they do emit at radio wavelengths.

Dutch astronomers were the first to appreciate, in the 1940s, that radio waves would travel unimpeded through clouds of dust, and that hydrogen would produce a radio frequency spectral line. The way it happens is this: Both the proton and the electron can be pictured to be spinning like tops. They are either spinning in the same direction or in opposite directions. If they are spinning in the same direction, the hydrogen atom has a little more energy than if they were spinning in opposite directions. Every so often (it takes a few million years), an electron will spontaneously go from the high energy state to the low energy state (flip its spin), and the atom will give off a photon with a wavelength of 21 cm. This photon travels unimpeded through the Galaxy to our radio telescopes.

Since the 21 cm line (or HI—pronounced *H-one*—line) has a particular frequency, it tells us not only *where* the gas is, but *how* its moving. As we'll see in the next chapter, the HI line has been invaluable in mapping our own Milky Way Galaxy.

Star Light, Star Bright

By definition, there is *nothing* for the amateur astronomer to see with an optical telescope in the regions of the interstellar medium that are cold. The best she can hope for is to see these relatively cold regions a dark patches in the foreground of bright optical emission. But, closer to stars, the interstellar medium can be lit up to spectacular effect. These regions of gas that are illuminated by a nearby massive star or stars are called *emission nebulae*.

Star Words

Emission nebulae (singular, **nebula**) are glowing clouds of hot, ionized interstellar gas, located near a young, massive star.

As we learned in Chapter 11, "Solar System Home Movie," a *nebula* is any wispy patch of emission in the sky. Sky watchers didn't pay too much attention to these objects—which, after all, lacked the sharp brilliance of stars and the solidity of planets—until a Frenchman, Charles Messier (1730–1817), decided to compile a systematic catalog of nebulae as well as star clusters. His purpose in this undertaking was to enable himself and others to discriminate between these fuzzy patches and comets for which many astronomers were busily searching. There are 110 star clusters, nebulae, and galaxies in what is now called the Messier Catalog. Amateur astronomers still take pride in attempting to locate and study all 110 Messier objects (as they are called). You will find the Messier Catalog in Appendix D.

Messier objects M8 and M16 (the Lagoon Nebula and the Eagle Nebula) are two of the more famous hot clouds of interstellar matter, or emission nebulae.

Emission nebulae (also sometimes called HII, pronounced *H-two*, regions) form around young type O and B stars. Recall that these stars are tremendously hot, emitting most of their energy in the ultraviolet. This ultraviolet radiation ionizes the gas surrounding the star, causing it to glow when the liberated electrons reunite (or recombine) with atomic nuclei. As the electrons cascade down the rungs of the energy ladder, they give off electromagnetic radiation with particular wavelengths. The spectral lines that result can be detected from radio frequencies through the infrared and into the optical and ultraviolet.

But there's a problem: If there are large amounts of gas and dust between us and an emission nebula or HII region, we won't see it optically. The dust grains will absorb or scatter the optical photons from the ionized hydrogen. The solution? As long as you have a high-resolution infrared or radio telescope, even the most obscured emission nebula can be seen. In fact, toward heavily obscured regions of our Milky Way, such as the Galactic center, radio and infrared observations have given us almost all the information that we have.

If you view any optical emission nebula with a good telescope on a dark, clear night, you will notice dark patches superimposed on the nebula. These dark regions are clouds of interstellar gas and dust in the foreground of the cloud that obscure the background light emitted from the ionized gas.

Optically observed emission nebulae are often huge objects, with diameters typically measured in parsecs. As you may recall from Chapter 17, "Of Giants and Dwarfs: Stepping Out into the Stars," a parsec is equal to 206,265 A.U.; that is, one parsec is equivalent to 206,265 times the distance between the sun and the earth.

However, regions of ionized gas near young stars (HII regions) have a wide variety of sizes. Diameters range from several tens of pc for the so-called giant HII regions, all the way down to tiny HII regions (called compact or ultracompact HII regions) with diameters as small as 1/100 of a pc. The smallest HII are deeply embedded in clouds of gas and dust, and thus are only observable at radio and infrared wave lengths.

A Matter of Perspective

When we are able to see part of the interstellar medium visually, it is often because a nearby star has lit it up. But there are other ways to see. In some directions, if we are fortunate, a cold dark cloud of gas will fall between the earth and an emission nebula. When that happens, we can see the dark cloud as a black patch against the emission from the ionized gas. In other directions, we see small patches of the sky where there are few or no stars. It is unlikely that there are truly few stars in that direction. Almost certainly, a dense cloud of gas in our line of sight is absorbing the starlight from the stars that are indeed behind it.

These gas pillars in the Eagle Nebula, an emission nebula 7,000 light-years from the earth, were im-aged by the Hubble Space Telescope. *They consist primarily of dense molecu-lar hydrogen gas and dust. This image shows both the emission from ionized hydrogen at the surface of the pillars, as well as the absorption caused by the gas and dust contained in the pillars.*

(Image by Jeff Hester and Paul Scowen of Arizona State University and NASA)

But we don't have to depend on luck. Even if an interstellar cloud of gas does not fall between the earth and a background source, we have ways to detect it. Even at the relatively cold temperatures of these clouds (100 K as compared to the nearly 10,000 K in most emission nebulae), atoms and molecules are in motion. As molecules collide with one another, they are occasionally set spinning about.

It turns out (according to quantum mechanics) that molecules can spin only at very particular rates, like the different speeds of an electric fan. And when a molecule goes from spinning at one rate to spinning at another rate, it gives off (or absorbs) a small amount of energy. We can detect that energy in the form of electromagnetic radiation. For many molecules, these photons have wavelengths that are about a millimeter long. Special telescopes (called millimeter telescopes and millimeter interferometers) can detect these photons.

With the addition of millimeter-wave telescopes, there is no portion of the interstellar medium that can escape our notice. With radio telescopes we can image neutral (cold) hydrogen atoms. With optical, infrared, and radio telescopes, we get pictures of the hot gas near young stars. And with millimeter telescopes, we can even seek out the cold clouds of gas that contain molecular hydrogen and other molecules.

The Berkeley Illinois Maryland Association (BIMA) array has done extensive studies of the molecules in the Orion nebula. These images show where particular molecules are located in this star-forming region. From the molecules listed, it is apparent that the interstellar medium is teeming with complex organic molecules.

(Image from NCSA)

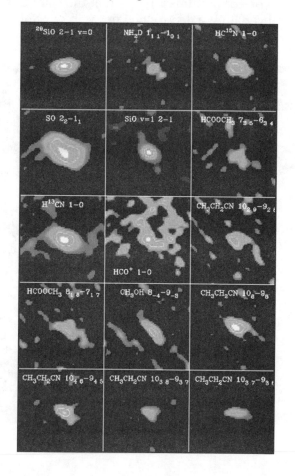

The Interstellar Medium: One Big Fuel Tank

Why so much fuss over gas and dust? Because, as far as we can tell, this is the raw material from which stars are born. Now, how does a cloud of gas become a star?

Tripping the Switch

A *giant molecular cloud* is subject to a pair of opposing forces. Gravity (as always) tends to pull the matter of the cloud inward, causing it to collapse and coalesce, yet as the constituent atoms of the cloud come together, they heat up, and heat tends to cause expansion, movement outward. Unless some event occurs to upset the balance, the cloud will remain in equilibrium.

Whatever the cause, it is clear (since we have stars!) that some molecular clouds become gravitationally unstable and begin to collapse. As a cloud collapses, its density and temperature increase, allowing smaller pieces of the cloud (with less mass) to collapse. The result is that once a cloud begins to collapse, it breaks into many fragments, which will form scores, even hundreds (depending on the original cloud mass) of stars of various masses. The size of each fragment will determine the mass of the star that forms.

Letting It All Out

We now join a single fragment in the collapsing cloud that will become a 1-solar mass star. Over a period of perhaps a million years, the cloud fragment contracts. In this process, most of the gravitational energy released by the contraction escapes into space, because the contracting cloud is insufficiently dense to reabsorb the radiation. At the *center* of the coalescing cloud (where densities are the highest), more of the radiated energy is trapped and the temperature increases. As the cloud fragment continues to contract, photons have a harder

Star Words

Giant molecular clouds are huge collections of cold (10 K to 100 K) gas that contain mostly molecular hydrogen. These clouds also contain other molecules that can be imaged with radio telescopes. The cores of these clouds are often the sites of the most recent star formation.

Astro Byte

What causes a molecular cloud to collapse and form stars? We're not exactly sure, but there are several likely possibilities. The expanding shock wave of a nearby supernova explosion might be sufficient to cause a cloud to collapse. Or, as we'll see in the next chapter, a ripple in a galaxy called a density wave could also be the trigger. Some astronomers even think that a fast-moving massive star, punching through a molecular cloud, could cause parts of it to collapse.

and harder time getting out of the increasingly dense material, thereby causing the temperature at the core to rise even higher.

If the original fragment had any slight rotation (as undoubtedly it did), it will be spinning faster now—the spinning cloud is contracting, like the skater pulling in her arms. If the original giant molecular cloud was 10 to 100 parsecs across, then the first cloud fragments would still have been much larger than our solar system. Depending on the eventual stellar type, a cloud fragment on the verge of becoming a protostar might be somewhat smaller than our solar system. This stage of star formation is typically accompanied by dramatic jets of outflowing material. These objects have been dubbed Herbig-Haro objects.

Herbig-Haro objects are jetlike structures that appear to arise during the process of low-mass star formation. High-mass stars might experience very brief, scaled-up versions of these impressive outflows.

(Image from NASA, Alan Watson (Universidad Nacional Autonoma de Mexico, Mexico et al.)

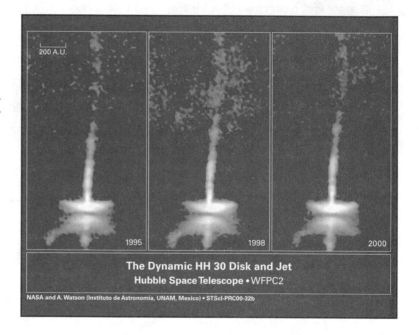

200 A.U.

1995   1998   2000

The Dynamic HH 30 Disk and Jet
Hubble Space Telescope • WFPC2

NASA and A. Watson (Instituto de Astronomía, UNAM, Mexico) • STScI-PRC00-32b

Not Quite a Star

The evolution of a protostar is characterized by a dramatic increase in temperature, especially at the core of the protostar. Still too cool to trigger nuclear fusion reactions, its core reaches a temperature of 1 million Kelvin. The protostar is also still very large, about 100 times larger than the sun. While its surface temperature at this stage is only half that of the sun, its area is so much larger that it is about 1,000 times more luminous than the sun. At this stage, it has the luminosity and radius of a red giant. A solar mass star will not look like this again until it is on its death bed, about 10 billion years later.

Despite the tremendous heat produced by its continuing collapse, which exerts an outward-directed force on the protostar, the inward-directed gravitational force is still greater.

Through the course of some 10 million years, the protostar's core temperature increases 5 fold, from 1 million Kelvin to 5 million, while its density greatly increases and its diameter shrinks, from 100 times to 10 times that of the sun.

Despite the increase in temperature, the protostar becomes less luminous, because its surface area is reduced. Contraction continues, but slows as the protostar approaches equilibrium between the inward-directed force of gravity and the outward-directed force of heat energy. It is approaching that part of its life that we called the *main sequence*.

The "On" Switch

Once the star's core temperature reaches about 10 million Kelvin, the star begins to fuse hydrogen into helium. The force of gravity will be balanced by the pressure of the fusion-produced heat.

A Collapsed Souffle

Not all cloud fragments become stars. If a fragment lacks sufficient mass, it will still contract, but its core temperature will never rise sufficiently to ignite nuclear fusion. Failed stars are known as *brown dwarfs*. Students using the Very Large Array recently discovered a brown dwarf doing a very unexpected thing: flaring in brightness. Astronomers are now trying to figure out what might make brown dwarfs flare.

Star Words

A **brown dwarf** is a failed star; that is, a star in which the forces of heat and gravity reached equilibrium before the core temperature rose sufficiently to trigger nuclear fusion in the core.

Multiple Births

We have traced the birth of a single star, but a collapsing molecular cloud actually produces a star cluster, a collection of a few high-mass stars and many smaller, low-mass stars, and brown dwarfs. Messier noted and catalogued a number of star clusters (see Appendix D, "The Messier Catalog"), which are readily observed.

In the Delivery Room

Analyzing the process of human gestation and birth is relatively easy. Little theorizing is required because all one needs is about nine months of free time to make some direct observations.

The birth of a low-mass star, however, may consume 40 to 50 million years—obviously more time than any observer can spare. High-mass stars, as we have seen, have shorter lives and more spectacular deaths. They also appear to collapse and fall onto the main sequence more rapidly than low-mass stars.

While we don't have time to watch a single star go through all of the stages outlined, there are fortunately many different star formation regions in different parts of our Galaxy. Taken as a whole, these regions (in different stages of evolution) give us a more complete picture of how stars form from the matter between them.

The Least You Need to Know

➤ The spaces between the stars are not empty, but contain gas and dust, from which future generations of stars form.

➤ Dust grains in interstellar space absorb and scatter visible light, so that distant objects are reddened or not optically observable.

➤ Emission nebulae, or HII regions, are regions of ionized gas that occur near hot, young stars. These clouds of hydrogen absorb ultraviolet emission from the young star and re-emit the light at longer wavelengths.

➤ When giant molecular clouds collapse, they fragment into smaller clumps, some of which form stars.

Part 5
Way Out of This World

We turn from individual stars and stellar phenomena to some of the largest structures in the universe. the galaxies. We begin at home, with a chapter on our Galaxy, the Milky Way, then move on to an explanation of the three major galaxy types—spiral, elliptical, and irregular—and how galaxies, themselves vast, are often grouped into even larger galaxy clusters, which, in turn, may be members of superclusters. Here is cosmic structure on the greatest scale.

The first two chapters in this part deal with normal galaxies, but there is another galaxy type, the active galaxy, which emits large amounts of energy. We explore such objects as Seyfert galaxies and radio galaxies, as well as one of the most spectacular cosmic dynamos of all, quasars. We also explore the nature of a phenomenon that appears to be occuring in distant galaxies: gamma ray bursters.

The Milky Way: Much More Than a Candy Bar

Ancient societies were very aware of the Milky Way, a fuzzy band that arcs across the sky. Without the aid of a telescope, it is not possible to see that this band consists of billions of individual stars. For this reason, ancient cultures described the Milky Way variously: as a bridge across the sky, as a river, as spilled corn meal from a sack dragged by a dog, or as the backbone of the heavens. No one described it as what it is—unresolved stars. So imagine Galileo's surprise when he first looked through his telescope at the Milky Way. To his eyes, the Milky Way appeared to be a fuzzy band of light arcing across the sky. But through a telescope, the fuzz was resolved into an enormous number of individual stars. Galileo must have been stunned. Why were those stars all lying in an arc on the sky? It was several centuries before we determined our place in that grand arc.

With many of us living in atmospherically polluted and light-polluted cities and suburbs, the Milky Way can be difficult or impossible to see. Under the best viewing conditions, it is a stunning sight, a majestic band of light extending high above the horizon.

When we look at the band, we see our own galaxy from the inside. The Milky Way is our home, and, in this chapter we take a closer look at our place in the Milky Way, and how it may have come to be.

Where Is the Center and Where Are We?

Galileo and other astronomers soon realized that the stars around us were distributed in a distinct way, confined to a narrow band of the sky. In the late eighteenth century, William Herschel proposed the first model of our Galaxy, suggesting that it was disk shaped, and that our sun lay near the center of the disk. This model held some psychological comfort (the earth having been so recently elbowed from the center of the universe), but it had a problem. Herschel failed to account for what is now known as interstellar extinction. He had assumed that spore was transparent. But the disk of our galaxy is "foggy," and we can see only so far into the "fog."

As a result, we appear to be at the center of the disk— (from star counts in various directions) not because we actually are, but because the dust in our Galaxy absorbs visible light so that we can only see out into the disk for a limited distance.

It was not until early in the twentieth century that we realized that we were not in the center of our own Galaxy, but at its far outer reaches.

Star Words

The **Galactic bulge** or **nuclear bulge** is a swelling at the center of our Galaxy. The **bulge** consists of old stars and extends out a few thousand light-years from the Galactic center.

Home Sweet Galaxy

The universe is not evenly populated with stars and other objects. Just as there are large distances between individual stars, there are large distances between collections of stars called galaxies. Galaxies (of which the Milky Way is one) are enormous collections of stars, gas, and dust.

Typical galaxies contain several hundred *billion* stars. There are about 100 times as many stars in our Galaxy as there are people on the earth. Astronomers sometimes refer to our own Galaxy with a capital "G" to distinguish it from all of the other galaxies in the universe.

A Thumbnail Sketch

Since we live within the Galaxy, we see it in profile. The Milky Way, viewed at this angle, is shaped rather like a flying saucer—that is, it resembles a disk that bulges toward its center and thins out at the periphery. The thickest part we call the *bulge,* and the thin part, the *disk.* (Astronomers aren't *always* in love with highly technical terms!) We will see later how we figured it out, but we know now that the center of our own Galaxy is in the *Galactic bulge,* in the direction of the constellation Sagittarius.

Our solar system is located in part of the thinned-out area, the so-called *Galactic disk.* Our location explains what we see when we look at the Milky Way on a clear summer night in the country, far from city lights and smog. The wispy band of light arcing across the sky is our view *into* this disk. The band of light is the merged glow of the stars close enough to be seen optically—so many that they are not differentiated as separate points of light without a telescope. A dark band of obscuration runs the length of the arc. This light is blocked by the presence of large amounts of dust in the disk.

When we look away from the arc in the sky, we don't see much of the Milky Way, because we are looking out of the disk. Looking into the disk is like looking at the horizon on the earth: There's lots of stuff there to block our view, including houses, trees, and cars. But look straight up into the sky, and you can see much farther—you might even see a plane flying overhead. When we look more or less perpendicular to the disk of the Galaxy, we can see much farther, and this is the direction to look to see other galaxies.

Star Words

The **Galactic disk** is the thinnest part of our Galaxy. The disk surrounds the nuclear bulge, and contains a mixture of old and young stars, gas, and dust. It extends out some 50,000 light-years from the Galactic center, but is only about 1000 light-years thick. It is the dust in the disk (a few hundred light-years thick) that creates the dark ribbon that runs the length of the Milky Way and limits the view of our own Galaxy.

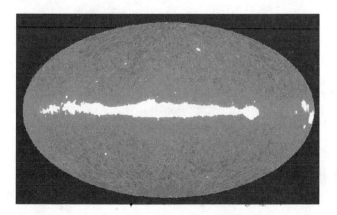

The Milky Way, as seen by the Compton Gamma Ray Observatory. This black-and-white reproduction of a false-color image of our home clearly shows the galactic disk. Transient sources called gamma-ray bursters (GRBs) are seen to flare on average once a day.

(Image from NASA, Compton Gamma Ray Observatory)

305

It was looking in this direction—up in the air, as it were—that astronomers discovered another component of the Galaxy, the *globular clusters*. Globular clusters are collections of several hundred thousand older stars held together by their mutual gravitational attraction, that are generally found well above and below the disk of the Galaxy. Reasoning that globular clusters should cluster around the gravitational center of the galaxy, Harlow Shapley used these collections of stars in the early twentieth century to determine where in the Galaxy we were located.

Along with these collections of stars in globular clusters are single stars.

Star Words

The **Galactic halo** is a large (50,000 light–year radius) sphere of old stars surrounding the Galaxy.

Simple diagram of our view of the Milky Way, showing the principal parts. Studies have shown that the Milky Way is an ordinary spiral galaxy.

(Image from authors' collection)

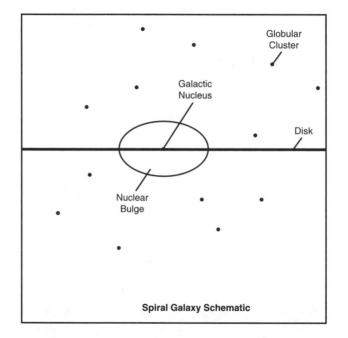

Globular Cluster

Galactic Nucleus

Disk

Nuclear Bulge

Spiral Galaxy Schematic

Keeping up with the Joneses

Not even Dick Clark will live long enough to watch a single star evolve. So, to overcome the limitation of the human life span, remember that we can chart the development of a certain mass star by observing many similar stars in various stages of development. The same is true for galaxies. The Milky Way (at our radius) takes 225 million years to rotate once! We can sometimes look to galaxies similar to our own in order to gain the necessary perspective and make generalizations about the Milky Way.

The Andromeda galaxy (Messier Object 31) is about 2 million light-years distant, and is clearly visible with any decent telescope pointed toward its namesake constellation. Like the Milky Way, Andromeda is what we call a spiral galaxy, with distinct, bright, curved structures called spiral arms (we'll talk about different kinds of galactic shapes in the next chapter). The universe contains many spiral galaxies which astronomers can observe in order to study aspects of the Milky Way that are hidden by our location in it.

Take a Picture, It'll Last Longer

While Andromeda may be seen with a modest amateur telescope as a fuzzy patch of light, photographic equipment or an electronic CCD camera is required to make out the kind of detail in a galaxy that we are used to seeing in astronomy magazine pictures. In order to see sweeping arms and dark bands of dust in other galaxies, we need to collect more light, either by using a larger aperture (a big telescope), or waiting longer (taking a sufficiently exposed conventional or electronic photograph).

The Andromeda galaxy, photographed by the Yerkes Observatory in the 1920s.

(Image from arttoday.com*)*

Measuring the Milky Way

One way to gauge the size of the Milky Way would be to look at other, similar galaxies, such as Andromeda. If we know how far away such a galaxy is, and we can measure its angular size on the sky, we can calculate how big it is.

Star Words

A **variable star** is a star that periodically changes in brightness. A **cataclysmic variable** is a star, such as a nova, that changes in brightness suddenly and dramatically, as a result of interaction with a binary companion star. An **intrinsic variable** changes brightness because of rapid changes in its diameter. **Pulsating variables** are intrinsic variables that vary in brightness in a fixed period or span of time.

But determining the distance of Andromeda and the other galaxies can be a serious challenge. Parallax as a distance indicator is out, since we know that the apparent angular shift is equal to 1 divided by the distance in parsecs. Stars farther than 100 pc are too far to use the parallax method, and those stars are still in our own Galaxy. Half a microarcsecond resolution (0.0000005") would be required to see parallax of even the nearest galaxy, Andromeda.

One very good distance indicator was discovered in 1908 by Henrietta Swan Leavitt, working at the time under the direction of Edward Pickering at the Harvard College Observatory.

Over many centuries of star gazing, astronomers had noted many stars whose luminosity was variable—sometimes brighter, sometimes fainter. These *variable stars* fall into one of two broad types: *cataclysmic variables* and *intrinsic variables*. We have already discussed one type of *cataclysmic variable* star. Novae are stars that periodically change in luminosity (rate of energy output) suddenly and dramatically when they accrete material from a binary companion.

Another type of variable star is an *intrinsic variable,* whose variability is not caused by interaction with a binary companion, but by factors internal to the star. The subset of intrinsic variable stars important in distance calculations are *pulsating variable* stars. Cepheid variable stars and RR Lyrae stars are both pulsating variable stars.

Astronomer's Notebook

The names of the class of variable stars, RR Lyrae and Cepheid, are derived from the names of the first stars of these types to be discovered. RR Lyra is a variable star (labeled RR) in the constellation Lyra; RR Lyrae is a genitive form of this name. The Cepheid class is named after Delta Cepheus, a variable star in the constellation Cepheus. Cepheid variables have longer periods (as a class) and are more luminous than RR Lyrae stars. RR Lyrae stars have periods of less than a day. Cepheid periods range upward of 50 days.

Close Encounter

Henrietta Swan Leavitt was one of a number of talented astronomers on the staff of the Harvard College Observatory. She was the first to propose, in 1908, that the period of a certain type of intrinsic variable star (a Cepheid variable) was directly proportional to its luminosity.

She had observed a large number of variable stars in the Magellanic clouds (companions to our own Galaxy). The advantage of studying these clouds is that whatever the distance to them, the stars within them are all roughly at the same distance from the earth. What Leavitt noticed was that the brightest Cepheid variable stars in the Magellanic clouds always had the longest periods, and the faintest always had the shortest periods.

What her discovery meant was that astronomers could simply measure the period of a Cepheid variable star and its apparent brightness to derive its distance directly. What Leavitt accomplished was to greatly extend the astronomer's ruler from a few hundred light–years (using the parallax method) to tens of millions of light–years.

The pulsating variables vary in luminosity because of regular changes in their diameter. Why does this relationship exist? Stars are a little like ringing bells. When struck, large bells vibrate more slowly, producing lower tones. Tiny bells vibrate rapidly, producing higher tones.

A pulsating variable star is in a late evolutionary stage and has become unstable, its radius first shrinking and its surface heating. Then its radius expands and its surface cools. These changes produce measurable variations in the star's luminosity (because luminosity depends on surface area, and the star is shrinking and expanding).

The two types of pulsating variables are named after the first known star in each group. The RR Lyrae stars all have the similar average luminosity of about 100 times that of the sun. Cepheid luminosities range from 1,000 to 10,000 times the luminosity of the sun. Because both pulsating variables can be recognized by their pulsation pattern and average luminosity, they make convenient markers for determining distance. Because they are intrinsically more luminous, Cepheid variable stars are useful for measuring greater distances.

An observer simply locates such a star and measures its apparent brightness and period of variation. Then, using the star's known luminosity (from its period), she can calculate its distance.

Where Do We Fit In?

We mentioned earlier that it was the globular clusters (which lie far out of the plane of the Galaxy) that pointed the way to the Galactic center. How did they do so?

Armed with Leavitt's period-luminosity relation, an American astronomer named Harlow Shapley (1885–1972) used the period of RR Lyrae variable stars in almost 100 globular clusters in the Galaxy to determine their distances. With this distance information, Shapley could see that that the globular clusters were all centered on a region in the direction of Sagittarius, about 25,000 light-years from the earth. He concluded from his study of globular clusters that the Galaxy was perhaps 30,000 parsecs (100,000 light-years) across—much, much larger than anyone had ever before imagined.

Then he dropped another bombshell.

In an eerie twentieth-century replay of the days of Copernicus and Galileo, Shapley demonstrated that the hub of the Galaxy, the Galactic center, was certainly not us. It lay fully 8,000 parsecs (25,000 light-years) from the sun. Our star and our solar system are far from the center of it all.

Astro Byte

Shapley made his discoveries about globular clusters in 1917, but it was not until Edwin Hubble observed Cepheid variables in the Andromeda galaxy, more than a decade later, that Andromeda and other so-called spiral nebulae were recognized as galaxies in their own right, separate from the Milky Way and very distant. Not only were we not at the center of the galaxy, but our Galaxy was revealed as only one of billions of galaxies. The universe was much more vast than we had ever imagined.

Earlier in this chapter we defined various parts of the Milky Way: the nuclear bulge, the disk, the halo. Each of these regions has a characteristic size. Studies in this century have shown that the different parts of the Milky Way are also characterized by distinct stellar populations.

If you look at a true-color picture of the Milky Way or a similar galaxy, such as Andromeda, the bulge would appear yellow, while the disk would appear blue or white-blue. The color of a region tells us the average color of most of the stars that reside there. Globular clusters, the bulge and the halo appear to contain mostly cooler (yellow) stars. Most of the young (hot) stars are gone, and new ones apparently aren't taking their places in those locations.

The presence of large numbers of the youngest, hottest stars of type O and B makes the Galactic disk appear blue. Since O and B stars have such short lifetimes (1 to 10 million years), their presence in the disk tells us that they must be forming there "now" ("now" to an astronomer has a different meaning than it does to most people). Remember that the Milky Way (at our radius) rotates once every 225 million years. That means that tens of generations of massive O and B stars form and explode every time the disk of the Galaxy rotates. Cooler and smaller G, K, and M stars are also present in the disk, but the giant blue stars, far more luminous, outshine them, imparting to the entire disk region its characteristic blue-white color.

Why are the youngest stars in the disk and the oldest in the halo?

As we have said, the Galactic disk is where the interstellar gas clouds reside—what we called giant molecular clouds. With raw materials plentiful, star production is very active here; therefore, young stars are abundant. In contrast, the halo region has very little non-stellar material, so that no new stars can be created there. In the Galactic bulge, there is an abundance of interstellar matter as well as old *and* new stars.

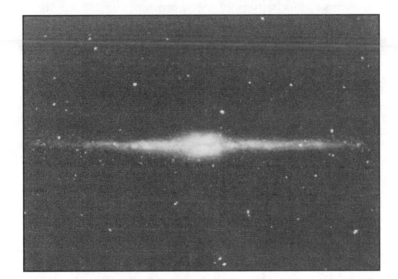

An image of the Milky Way from an edge-on perspective made by the Diffuse Infrared Background Experiment (DIRBE) on the Cosmic Background Explorer (COBE) *probe. The basic architecture of the Milky Way is apparent in this infrared image.*

(Image from NASA)

Milky Way Portrait

The different parts of the Milky Way are not static, but in constant motion. The disk rotates about the Galactic center, and at large radii, the rate of rotation does not trail off, but remains fairly constant. This rotation is in contrast to what we see in a planetary system, where objects (planets) rotate more and more slowly the farther they are from the center. As we'll see in a moment, the constant rotation rate at large distances from the Galactic center betrays the presence of something we cannot see.

The stars in the halo move very differently, plunging through the Galactic disk in elliptical orbits that are randomly oriented. Their orbits are not confined to the Galactic disk, and in some ways seem unaware of it. All of these orbits are centered on what we call the Galactic center region, and give clues as to how the Galaxy formed.

A Monster at the Center?

The central part of our Galaxy (first identified by Harlow Shapley) is literally invisible. It cannot be observed at optical frequencies, but radio and infrared frequency observations have told us much about this hidden realm.

Astro Byte

Our part of the Galactic disk orbits the Galactic center at over 136 miles per second (220 km/s). It takes about 225 million years for our region to complete one orbit around the Galactic center. Our solar system has orbited the Galactic center some 15–20 times since it formed. One-quarter Galactic orbit ago, dinosaurs roamed the earth.

Some radio observations have identified a ring of molecular material orbiting the Galactic center at a distance of perhaps 8 or so parsecs (25 light-years). The rotational velocity of this ring tells us that there are several million solar masses of material located within its radius. Other radio observations (sensitive to hot gas) show that within this ring is a small amount of ionized material, nowhere near a few million solar masses.

Recent infrared observations show that the Galactic center region contains a great many stars, closely packed. But they do not account for all of the required mass.

The last clue: Radio astronomers have picked up strong radio emission from a source in the Galactic center region that appears to be very small—perhaps as small as a solar system. Called Sagittarius A* (pronounced "A-star"), this source may give the location of a black hole of several million solar masses.

Stars near the Galactic center are in orbit around a very small, very high-mass object. Is there a monster at the center? It appears increasingly likely that a black hole lurks at the galactic center. These images show two clusters that are within 100 light years of the center of the galaxy.

(Image from D. Figer, STScI, and NASA)

Star Clusters Near the Center of the Galaxy   HST · NICMOS
PRC99-30 · STScI OPO · D. Figer (STScI) and NASA

The Birth of the Milky Way

The structure, composition, and motion of the Milky Way hold the keys to its origin. Though the theory of galaxy formation is far from complete, we have a fairly good picture of how our Galaxy might have formed.

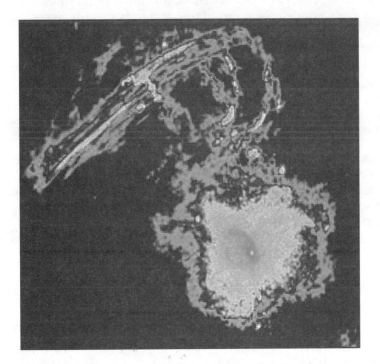

This image of the center of the Milky Way galaxy was generated by the Very Large Array. The loops and filaments apparent are related to the presence of magnetic fields. The Sgr A* source (thought to be the center of our galaxy) is located in the lower right of the image.

(Image from NRAO)

Ten to fifteen billion years ago, an enormous cloud of gas began to collapse under its own gravity. The cloud, like all the universe, would have been mostly hydrogen. It would have had a mass equal to that of all the stars and gas in the Milky Way, or several hundred billion solar masses.

The stars that formed first as the great cloud collapsed assumed randomly oriented elliptical orbits—with no preferred plane. Today, the oldest stars in our Galaxy—which are those in the Galactic halo and Galactic bulge—ring with the echoes of their earliest days. The Galactic halo is a vestige, a souvenir of the galaxy's birth. And the globular clusters in the halo may have even formed prior to the cloud's collapse.

Star Words

Mass-to-light ratio is the ratio of the total mass of a galaxy (detected by its gravitational effect) to the mass in luminous matter. If there were no "dark matter," the mass-to-light ratio would be 1.

In a gravitational collapse process very similar to that which formed the solar system (see Chapter 11, "The Solar System Home Movie,") the great cloud began rotating faster around a growing mass at the center of the Galaxy. The rotation and collapse caused the clouds of gas to flatten into the Galactic disk, leaving only stars in the halo.

Since the Galactic disk is the repository of raw materials, it is the region of new-star formation in the Galaxy. The Galactic halo is out of fuel and consists only of old (cool) redder stars. The orderly rotation of stars and gas in the Galactic disk stands in contrast to the randomly oriented orbits of stars in the halo and the Galactic bulge.

Dark Matters

"What you see is what you get," the popular saying goes. Such is not always the case in Galactic affairs.

Using Kepler's Third Law, we can calculate the mass of the Galaxy. This mass (expressed in solar masses) can be derived by dividing the cube of orbit size (expressed in astronomical units, or A.U.) by the square of orbital period (expressed in years). And Newton told us that at a given radius, all of the mass causing the rotation can be considered to be concentrated at a point at the center of rotation. For a system in "Keplerian rotation" (like a planetary system), we would expect the velocities of rotation to decrease as we looked farther and farther out, much as Jupiter orbits more slowly than, say, Mercury.

Star Words

The **dark halo** is the region surrounding the Milky Way and other galaxies that contains dark matter. The shape of the dark halo can be probed by examining in detail the effects its mass has on the rotation of a galaxy. **Dark matter** is a catch-all phrase used to describe an apparently abundant substance of unknown composition.

Taking this approach, we find that the mass of the Milky Way *within* 15,000 parsecs of the Galactic center—that is, the *radius* of the visible galaxy (diameter ~30,000 parsecs)—is 2×10^{11} solar masses.

We would expect the mass of the Galaxy to drop off precipitously when we run out of matter—at the visible outer edge of the Galaxy. But the puzzling fact is that *more* mass is contained *beyond* this boundary than within it!

Within a radius of 40,000 parsecs from the Galactic center, the mass of the Milky Way is calculated to be about 6×10^{11} solar masses. This means that there is as much, if not more, of the Milky Way unseen as there is seen. Spiral galaxies like the Milky Way have *mass-to-light ratios* ranging up to 10, even in the visible part of the disk.

What is all this other mass?

Whatever its makeup, it apparently emits no radiation of any kind—no visible light, no X-rays, no gamma radiation. But it cannot hide completely. We see it simply because it has mass, and its mass is affecting the way in which the stars and gas of the Milky Way orbit.

Astronomers call the region containing this mass the *dark halo*. And the Milky Way is not alone in possessing such a region. Many, if not all, galaxies have the same signature in the rotation of their stars and gas. The dark halo presumably contains *dark matter*, a catch-all term that is used to describe a variety of candidate objects. The truth is, dark matter is a bit like Spam™. We're not sure what it is, but we know it's there because we can see its effects.

The nature of dark matter remains one of the greatest mysteries of science. Some astronomers have suggested that difficult-to-detect low-mass stars (brown dwarfs or faint red dwarfs) may be responsible for the mass in this region—although recent *Hubble Space Telescope* observations have suggested an insufficient quantity of such objects to account for so much mass. It has recently been established that neutrinos do have non-zero mass, so their presence might contribute. Yet others propose that dark matter consists of massive neutrinos hitherto unknown subatomic particles, which pervade the universe.

In the Arms of the Galaxy

We have referred to our own Galaxy several times in this chapter as a *spiral galaxy*. We will discuss galaxy classification in the next chapter, but what are these *spiral arms?* When observed from afar, they appear to be arcs of bright emission, curving out from the center of the galaxy. There may be only two arms or more, and they may be loosely or tightly wound.

How do we know that the Milky Way also consists of spiral arms? Using the 21 cm hydrogen line discussed in the last chapter, the distribution of neutral hydrogen (by far the most abundant element in the Galactic disk or anywhere) has been plotted. Both position and velocity of the hydrogen clouds are required to make this plot, since it shows a dimension that is not on the sky, namely depth. These radio images confirm the spiral structure of the Milky Way.

These radio studies also confirm that we are far from the Galactic center, located inauspiciously on a cusp between two spiral arms. Spiral arms themselves are not that hard to account for, but their longevity is. In the following chapter, we will discuss how spiral arms might arise, and how they could possibly persist.

Star Words

A **spiral galaxy** is characterized by a distinctive structure consisting of a thin disk surrounding a Galactic bulge. The disk is dominated by bright curved arcs of emission known as **spiral arms.**

NGC 253 is a spiral galaxy experiencing a wave of star formation and is referred to as a starburst galaxy. (The NGC in the designation stands for New General Catalog.) This image shows both the whole visible disk of the galaxy and a detailed view of a region in its bulge.

(Image from J. Gallagher and NASA)

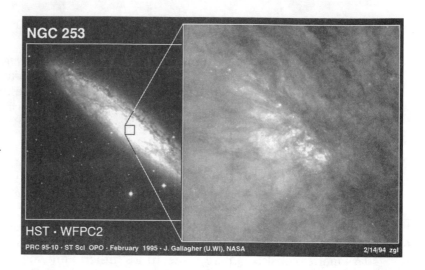

NGC 253

HST · WFPC2

PRC 95-10 · ST ScI OPO · February 1995 · J. Gallagher (U.WI), NASA

2/14/94 zgl

The Least You Need to Know

➤ Our home Galaxy is the Milky Way. We see it from within as an arc of fuzzy emission across the sky.

➤ The main components of the Galaxy are the disk, the stellar halo (containing the globular clusters), the nuclear bulge, and the dark halo.

➤ Our solar system is in the Galactic disk of the Milky Way, 25,000 light-years from the Galactic center.

➤ Because our solar system is in the dusty disk of the Milky Way, it is impossible to gain a bird's-eye view of our Galaxy; astronomers, therefore, look at other, similar galaxies in order to learn about our own.

➤ We can only account for a fraction of our Galaxy's mass with normal matter; most of the Galaxy's mass must be contained in a halo of material, consisting of dark matter.

Chapter 22

A Galaxy of Galaxies

In This Chapter

➤ Hubble's classification of galaxy types: spiral, elliptical, and irregular

➤ Spiral density waves

➤ Determining the distance of galaxies

➤ Galactic clusters and superclusters

➤ Calculating the mass of galaxies and galactic clusters

➤ Receding galaxies, expanding universe

➤ Hubble's Law

Even with the naked eye, it is clear that some "stars" are fuzzier than others. After the invention of the telescope, it became obvious that this was the case. Through a telescope eyepiece some are resolved into the disks of planets, some into regions where stars are forming, and others into collections of old stars. There was great disagreement about one class of objects long called *spiral nebulae*. These clearly were not stars, and yet there was no way to figure out how *big* they were without knowing how *far away* they were.

Before Edwin Hubble extended our conception of the size of the universe in the 1920s, these objects were classified as various nebulae and were thought to lie within the Milky Way, which was thought to be synonymous with the universe. Now, at last, we know that our Galaxy is but one among many, and that the spiral nebulae are other galaxies, some smaller, some bigger than our own, all containing hundreds of billions of stars. And we see galaxies no matter where we look in the sky. When we look deep enough, they are there. Some are so close that they are part of our "Local Group," while others are so distant that their light has barely had time to get to us since the universe began.

In this chapter and the next we talk about galaxies other than our own. Astronomers have two major goals in the exploration of other galaxies. The first is to understand the galaxies themselves and how they have evolved with time. But, it is equally important to use other galaxies to tell us more about the Milky Way, our home.

Sorting Out the Galaxies

A big part of the ongoing excitement of astronomy is the invigorating sensation of awe it can stir. News from other planets in the solar system is interesting, but it has a certain familiarity. However strange, a planet like Venus is about the size of the earth, and has a surface and an atmosphere. We can handle that.

But galaxies are different. Each one of them contains several hundred billion stars. Does that number make you go blank? Remember this: An average galaxy contains about 100 times as many stars as there are people on our planet …

And there are a lot of galaxies out there.

It is easy to get overwhelmed. If astronomers had to explain each galaxy individually, they would be overwhelmed as well. Fortunately, galaxies are not all completely different, but seem to fall into broad groups. They can be classified, just like trees or beetles or clams. But they couldn't be classified until there were photographs of large numbers of them.

The great American astronomer Edwin Hubble (1889–1953) came to the Mount Wilson Observatory in California in 1922 to use its new 100-inch (2.5-meter) telescope to study nebulae. An open, and fundamental, question hung in the air back then. Were these spiral nebulae as big as our Galaxy and incredibly distant (as first suggested by Immanuel Kant in the eighteenth century)? Or were they more mundane, relatively nearby objects? The huge surface area and high resolution of the 100-inch telescope allowed astronomers to see that these nebulae (like our own Milky Way) resolved into individual, faint stars.

Once intrinsic variable stars were identified in some of these "nebulae," the relationship between period and luminosity (see Chapter 21, "The Milky Way: Much More Than a Candy Bar,") could be used to determine their distance. And Hubble's distances, determined from Cepheid variable stars, prompted him to conclude in 1924 that many of the nebulae were not part of the Milky Way at all, but were galaxies in their own right. Andromeda, for example, was determined to be almost a million light-years away, *far* outside the limits of our own Galaxy. Shortly after making his mind-blowing discovery, Hubble—not one to be overwhelmed—set about classifying the galaxies he saw.

Hubble's classifications (like many first impressions) are based on appearance and fall into three broad categories: spiral, elliptical, and irregular.

318

Spirals: Catch a Density Wave

In the preceding chapter, we took a prolonged look at one spiral galaxy, the Milky Way. Hubble labeled all spiral galaxies with the letter "S" and added an *a*, *b*, or *c*, depending on the size of the galactic bulge. Sa galaxies have large bulges, Sb medium-sized bulges, and Sc the smallest bulges. Very clearly defined, tightly wrapped spiral arms are also associated with Sa galaxies, whereas Sb galaxies exhibit more diffuse arms, and Sc galaxies have more loosely wrapped, less defined spiral arms.

Another subtype of the spiral galaxy is the *barred-spiral galaxy*, which exhibits a linear bar of stars running through the galactic disk and bulge out to some radius. In these barred galaxies, the spiral arm structure begins at the ends of the bar. This spiral subgroup, designated SB, also includes the a, b, and c classifications based on the size of the galactic bulge and the winding of the arms. Some astronomers believe that the Milky Way is a barred-spiral galaxy, though our viewpoint (stuck in the disk!) makes it difficult to tell.

Star Words

A **barred-spiral galaxy** is a spiral galaxy that has a linear feature, or bar of stars running through the galaxy's center. The bar lies in the plane of the spiral galaxy's disk, and the spiral arms start at the end of the bar.

WFPC2

NICMOS

Barred Spiral Galaxy NGC 1365
NASA and M. Carollo (Columbia University) • STScI-PRC99-34a

HST • WFPC2 • NICMOS

A region in NGC 1365, a barred-spiral galaxy located in a cluster of galaxies called Fornax. The central image was photographed through a ground-based telescope. The images to the left and right show the central bulge as seen at optical and infrared wavelengths with the Hubble Space Telescope.

(Image from NASA and M. Carollo)

Ellipticals: Stellar Footballs

Elliptical galaxies present a strikingly different appearance from the spirals. When viewed with a telescope, they look a bit dull compared to their flashy spiral cousins.

319

There are no spiral arms, nor any discernable bulge or disk structure. Typically, these galaxies appear as nothing more than round or football-shaped collections of stars, with the most intense light concentrated toward the center and becoming fainter and more wispy toward the edges.

Of course, the orientation of an elliptical galaxy will influence its shape on the sky—that is, its apparent shape may not be its true shape. A football viewed from the side has a sort of oval shape, but, viewed on end, looks like a circle. Regardless, Hubble differentiated within this classification by apparent shape. E0 ("E-zero") galaxies are almost circular, and E7 galaxies are very elongated (or elliptical). The rest—E1 through E6—range between these two extremes.

Typically, elliptical galaxies are made up of only old stars in random orbits—much like the galactic halo of a spiral galaxy (see Chapter 21). Radio astronomers have detected no ionized hydrogen in these galaxies, which means that there is little or/no ongoing star formation.

Hickson Compact Group (HCG) 87 consists of four galaxies. The edge-on spiral at the bottom of the image and the elliptical to the right are both known to have "active nuclei," most likely related to the presence of a black hole.

(Image from AURA/STScl/NASA)

Hickson Compact Group 87

PRC99-31 • Space Telescope Science Institute • Hubble Heritage Team (AURA/STScl/NASA)

Close Encounter

Those of us confined to the earth's northern hemisphere don't get to see the most spectacular of the irregular galaxies, the Small and Large Magellanic Clouds. Named for the sixteenth-century explorer Ferdinand Magellan, whose men brought word of them to Europe at the conclusion of their global voyage, the Magellanic Clouds are rich in hydrogen gas. The clouds, like moons, are believed to orbit the much larger Milky Way. Like any object with mass in the universe, galaxies feel the irresistible tug of gravity and are pulled into groupings by its effects.

There is no such thing as a typical size for an elliptical galaxy. Their diameters range from a thousand parsecs (these dwarf ellipticals are much smaller than the Milky Way) to a few hundred thousand parsecs (giant ellipticals). The giant ellipticals are many times larger than our Galaxy.

Like any classification scheme, Hubble's has its share of duck-billed platypuses, objects that don't quite fit in. Some elliptical-shaped galaxies exhibit more structure than others, showing evidence of a disk and a galactic bulge. They still lack spiral arms and, like the other ellipticals, they do not contain star-forming gas. This type of galaxy (since it does have a disk and a bulge) is designated S0 (pronounced "S-zero"). Some of these galaxies even contain a bar, and are designated SB0.

Are These Reduced? They're All Marked "Irregular"

Finally comes the miscellaneous category of *irregular galaxies,* which lack any regular structure. Galaxies in this class often look as if they are coming apart at the seams or are just plain messy. Unlike the ellipticals, irregulars are rich in interstellar material and sites of active star formation.

Two of the most famous irregular galaxies are very close: the Large and Small Magellanic Clouds. Usually called the LMC and the SMC by acronym-happy astronomers, they are visible from the southern hemisphere. These Milky Way companions

Star Words

Irregular galaxies lack obvious structure, but have lots of raw materials for the creation of new stars.

(10 times closer to us than the closest spiral galaxy) interact with each other and our Galaxy via gravity, and both contain active star formation regions. The Small Magellanic Cloud is stretched into an elongated shape due to the tidal forces exerted on it by our Galaxy.

This radio image of the Small and Large Magellanic Clouds (SMC and LMC) was made with CSIRO's Parkes radio telescope. Think of it as a photographic negative: The dark areas are the brightest in the radio, and the light areas have no radio emission.

(Image from CSIRO)

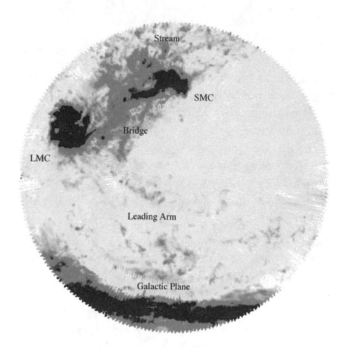

Galactic Embrace

In the constellation Canes Venatici is Messier object 51 (M51), the Whirlpool Galaxy. Its distinct shape resembles an overhead view of a hurricane or, less dramatically, a pinwheel. Curved, bright arms grow out of its hub-like galactic bulge and wrap part way around the galactic disk. It is easy to see the spiral pattern in the Whirlpool Galaxy because it faces us at an angle that gives us a spectacular overhead view rather than a side view.

It is the existence of these spiral arms that give such galaxies as the Whirlpool (M51), Andromeda (M31), and our own Milky Way the name spiral galaxies. But what exactly are spiral arms? We have said a good deal about spiral arms in Chapter 21 without saying how we think they got there to begin with.

Star Words

Spiral density waves are waves of compression that move around the disk of a spiral galaxy. These waves are thought to trigger clouds of gas into collapse, thus forming hot, young stars. It is mostly the ionized gas around these young, massive stars that we observe as spiral arms. Near encounters of galaxaies are thought to trigger these waves.

Catch the Wave

The existence of spiral arms presents a puzzle. We know that the stars and other matter in the galactic disk orbit differentially—faster toward the center, slower toward the periphery. This rotational pattern will stretch any large clump of stars or gas in the disk of a spiral galaxy into a spiral structure relatively quickly.

So quickly, in fact, that if differential rotation were to account for spiral arms, the arms would soon get totally wrapped up around a galaxy's bulge and disappear. Somehow, the Milky Way and other spiral galaxies retain their spiral arm structure for long periods of time, long enough that they are plentiful in the universe.

Most astronomers are convinced that the arms of spiral galaxies are the result of compressions, called *spiral density waves*, moving through a galaxy's disk. These ripples in the pond of a galaxy's disk move around the galaxy center, compressing clouds of gas until they collapse and form stars. It is mostly the ionized gas surrounding hot, young stars (formed by the passage of a density wave) that we see as spiral arms.

The theory of spiral density waves neatly resolves the problem of the effect of differential rotation. As you may recall from our discussion of waves in Chapter 7, "Over the Rainbow," a wave is a disturbance that moves through matter, a ripple in the pond. Thus, spiral density waves move *through* the matter of a galactic disk and are not caught up in differential rotation. The stars that they form may then get stretched out by this differential rotation, but the wave keeps on moving.

How to "Weigh" a Galaxy

In the same way that we cannot directly measure the temperature of stars, we cannot directly measure their masses. We measure the mass of the sun, for instance, by using Kepler's Third Law. If we know the distance to the sun and the period of the earth's orbit, we can calculate the sun's mass. Galaxy masses are calculated in the same way, only the objects orbiting are stars and gas instead of planets.

Astronomer's Notebook

As we study the universe, phenomena that are short-lived are harder to observe than those that are long-lived. It's similar with people: You are more likely to see someone staring than blinking. A blink takes a fraction of a second, whereas people typically go 10 seconds or more between blinks. How about a stellar example? Stars spend the majority of their lives on the main sequence. They spend a short fraction of their lives being born, and an even shorter fraction dying. Thus, for a star of any mass, we are most likely to see it on the main sequence. And massive stars live for a few million years, while low-mass stars live for many billions.

A Big Job

Plotting the *rotation curve* of an individual spiral galaxy—the velocity of disk material versus the distance of that material from the galactic center—yields the mass of the galaxy that lies within that radius. This method gives a good estimate as long as most of a galaxy's mass is contained near its center. In the solar system, for example, the outer planets rotate much more slowly than the inner planets, in accordance with Kepler's Third Law, because the mass of the solar system (99.9 percent of it, anyway) is contained in the sun.

Astronomer's Notebook

Luminous matter is matter that we can detect at any wavelength using any of the telescopes on the earth. This matter, everything we see, seems to make up only about 10 percent of the universe. How do we know? Our old friend gravity gives away the presence of the other 90 percent, what we call dark matter because it is nonluminous.

But astronomers soon noticed a problem with galaxies. There are objects in the outer reaches of spiral galaxies (clouds of gas—in particular, clouds of neutral hydrogen called HI clouds) that orbit in a way that indicates that they "see" more mass out there than we do here at the radius of the sun. They are orbiting faster than they should. And using the 21 cm radio line of hydrogen to see the HI, astronomers have traced the rotation curve of many galaxies far beyond the outermost stars. These curves seem to indicate that there is more matter at large radii in these galaxies. Whatever has this mass, though, is something that we can't see, because it is not "shining" at any wavelength. We call it *dark matter*.

Amazingly, there seems to be about 10 times more matter that we *can't* see than matter that we *can* see in most galaxies. And it gets even worse. In clusters of galaxies, the mass-to-light ratio (see Chapter 21) can be 100 or more. That is, luminous matter on very large scales accounts for only 1 percent of the matter that we "see" via gravity.

"It's Dark Out Here"

Astronomer's Notebook

Most spiral galaxies contain from 10^{11} to 10^{12} solar masses of matter, most of it contained far outside the radius of the galaxy that we see. Large elliptical galaxies contain about the same mass (but can be even more massive), whereas dwarf ellipticals and irregular galaxies typically contain between 10^6 and 10^7 solar masses of material.

As we discussed in Chapter 21, dark matter accounts for a great deal of the mass of a galaxy—up to 10 times more than the mass of the galaxy's visible matter—visible at any wavelength. The shocking conclusion that this observation leads us to is that 90 percent of the universe is dark matter—that is, matter that is invisible in the most profound sense, in that it neither produces nor reflects any electromagnetic radiation of any sort at any wavelength.

You can think of it as an embarrassing truth, or as an exciting unanswered question. The fact remains: We're not quite sure what 90 percent of the matter in the universe is made of.

Let's Get Organized

As early as Chapter 1, "Naked Sky, Naked Eye: Finding Your Way in the Dark," we saw human beings as inveterate pattern makers. Men and women have looked at the sky for centuries and have superimposed upon the stars patterns from their own imaginations. While the mythological heritage of constellations is still with us, scientists long ago realized that the true connections among the planets and among the stars are matters of mass and gravitational force, not likenesses of mythological beings.

Are there gravitationally determined patterns to the distribution of galaxies throughout the universe? For that matter, *are* the galaxies distributed *throughout* the universe?

Measuring Very Great Distances

In Chapter 17, "Of Giants and Dwarfs: Stepping Out into the Stars," we saw that the distance from us to the planets can be measured accurately by radar ranging, but that measuring the distance to farther objects, namely the nearer stars, requires measuring stellar parallax. But beyond about 100 parsecs parallax doesn't work well, because the apparent angular displacement becomes smaller than the angular resolution of our best telescopes.

Gas velocities and a model of the rotation of the Milky Way can be used to measure distances within our own Galaxy (out to about 30,000 light-years). Beyond this, and out to 10 to 20 million parsecs (30 to 60 million light-years), variable stars (the RR Lyrae and Cepheid variables discussed in Chapter 21) can be identified and observed in order to determine distance.

The trouble is, Cepheid variable stars farther than 15 million parsecs are difficult to resolve, and for most telescopes too faint to be detected. And many galaxies are well beyond 15 million parsecs away.

Two methods have been used to estimate intergalactic distances greater than 15 million parsecs. One tool is called the Tully-Fisher relation, which uses an observed relationship between the rotational velocity of a spiral galaxy and its luminosity. By measuring how fast a galaxy rotates (astronomers use the 21 cm hydrogen line described earlier), we can calculate its luminosity with remarkable accuracy. Once we know the luminosity of a galaxy, we can measure its apparent brightness and easily calculate its distance out to several hundred million parsecs.

Star Words

A **standard candle** is any object whose luminosity is well-known. Its measured brightness can then be used to determine how far away the object is. The brightest standard candles can be seen from the greatest distances.

The other tool is more general and involves identifying various objects whose luminosity is known. Such objects are referred to as *standard candles*. If we truly know the brightness of a source, we can measure its apparent brightness and determine how distant it is. A 100-watt light bulb is an example of a standard candle. It will look fainter and fainter as it recedes from us, and in fact, we can determine how far it is by measuring its apparent brightness versus its known luminosity.

One very interesting standard candle is a Type I supernova. It is interesting because its peak luminosity is very regular—and enormous (10 billion or 10^{10} solar luminosities!). We have discussed the Type II supernovae, the core collapse of a massive star. Type I supernovae occur when a white dwarf accretes enough material from its binary companion to exceed 1.4 solar masses. When this happens, the white dwarf begins to collapse, and the core ignites in a violent burst of fusion. The energies are sufficient to blow the star apart. These events are so luminous that we can see them *billions* of light-years away.

The Local Group and Other Galaxy Clusters

Armed with an ability to measure very great distances, we can begin looking at the relationships among galaxies.

Within 1 million parsecs (3 million light-years) of the Milky Way lie about 20 galaxies, the most prominent of which is Andromeda (M31). This galactic grouping, called the *Local Group,* is bound together by gravitational forces.

Star Words

The **Local Group** is a **galaxy cluster,** a gravitationally bound group of galaxies, which includes the Milky Way, Andromeda, and other galaxies.

The generic name for our Local Group is a *galaxy cluster,* of which thousands have been identified. Some clusters contain fewer than the 20 or so galaxies of the Local Group, while some contain many more. The Virgo Cluster, an example of a rich cluster, is about 15 million parsecs from the Milky Way and contains thousands of galaxies, all bound by their mutual gravitational attraction. Giant elliptical galaxies are often found at the centers of rich clusters where there are very few spirals.

From the velocities and positions of galaxies in clusters, one thing is very clear. We cannot directly observe at least 90 (perhaps as much as 99) percent of the mass that must be there. Galaxy clusters, like the outer reaches of spiral galaxies, must be teeming with dark matter.

Galaxy Cluster Abell 2218 HST • WFPC2
NASA, A. Fruchter and the ERO Team (STScI, ST ECF) • STScI-PRC00-08

The galaxy cluster called Abell 2218 is about 2 billion light years from the earth. When these photons left the cluster, life on earth was very simple indeed. The effect of the "unseen" mass in this cluster can be seen as the arcs of light that are "lensed" light from background sources.

(Image from NASA, A. Fruchter, and the ERO team)

Superclusters

Galaxy clusters themselves are grouped together into what are called *superclusters*. The Local Supercluster, which includes the Local Group, the Virgo Cluster, and other galaxy clusters, encompasses some 100 million parsecs.

On the very largest scales that we can measure (hundreds of millions of light-years across) the universe has a bubbly appearance, with superclusters concentrated on the edges of large empty regions or voids. In the final chapters of this book we will explore the possible explanations for this large-scale structure.

Star Words

A **supercluster** is a group of galaxy clusters. The Local Supercluster contains some 10^{15} solar masses.

Where Does It All Go?

The galaxies within a cluster do not move in an orderly manner. Like the stars in the galactic halo that envelops a spiral galaxy, the galaxies appear to orbit the cluster center in randomly oriented trajectories. But on larger scales, beyond the confines of a single cluster, there does appear to be orderly motion. This motion is the echo of a cataclysmic event: the event that brought the universe into being.

Hubble's Law and Hubble's Constant

It has been known since the early twentieth century that every spiral galaxy observed (sufficiently distant) exhibits a redshifted spectrum—they are all moving away from us. Recall from Chapter 17 the Doppler effect: Wavelengths grow longer (redshift) as an object recedes from the viewer. The conclusion is inescapable: All galaxies partake in a *universal recession*. This is the apparent general movement of all galaxies away from us. This observation does not mean that we are at the center of the expansion. Any observer located anywhere in the universe should see the same redshift.

In 1931, Edwin Hubble and Milton Humason first plotted the distance of a given galaxy against the velocity with which it receded. The resulting plot was dramatic. The rate at which a galaxy is observed to recede is directly proportional to its distance from us; that is, the farther away a galaxy is from us, the faster it travels away from us. This relationship is called *Hubble's Law*. This law relates the velocity of galactic recession to its distance from us. Simply stated, Hubble's Law says that the recessional velocity is directly proportional to the distance.

Astronomer's Notebook

Hubble's Law turns on Hubble's constant (H_0), the constant of proportionality between the velocity of recession and the distance from us. The value of H_0 is expressed in kilometers per second per megaparsec (a megaparsec [Mpc] is 1 million parsecs) and is found to be 60–65 km/s/Mpc. Sort of. Actually, the precise value of H_0 is the subject of dispute. Why should its value be so hotly debated? Because it can be used to determine nothing less momentous than the age of the universe. A different Hubble constant gives the universe a different age.

Picture a bunch of dots on the surface of a toy balloon. Our Galaxy is one of the dots, and all of the other galaxies in the universe are the other dots. As you inflate the balloon, the surface of the balloon stretches, and, from the point of view of *every* dot (galaxy), all the other dots (galaxies) are moving away. The farther away the dot, the more balloon there is to stretch, so the faster the dot will appear to recede.

One benefit of Hubble's Law is that it can be used to extend our cosmic distance scale to extraordinary distances. While the Tully-Fisher technique will get us out to about 200 million parsecs, and Type I supernovae to a bit less than 1 billion parsecs, Hubble's Law will go farther. In fact, Hubble's Law gives us the distance to any galaxy for which we can measure a spectrum (in order to get the Doppler shift). With the velocity from that spectrum, and the correct value of the Hubble constant, we arrive quite simply at the distance.

But there is a twist. We have assumed that the universal expansion happens at a constant rate. Is this a good assumption?

Recent observations suggest that the universe has not always been expanding at the same rate. Two magnificent forces are in opposition. There is the expansion of the universe (as we will see, set into play by the Big Bang), and there is the force of gravity, pulling every two particles with mass together, diminished by distance squared. If there were nothing pushing things apart, the universe would, due to gravity, collapse to a point. This would be a problem. And Einstein recognized it as such. When he was working on general relativity, he needed to add a term—the cosmological constant—to his equations to balance the force of gravity. This term was needed to keep the universe static, as it was then thought to be.

But then Edwin Hubble came along and showed that the universe was hardly static, but that, in fact, everything was rushing away from everything else, and while gravity might eventually win out, the universe wasn't going to collapse to a point any time soon. Einstein later called the introduction of the constant into his equations a "blunder."

As it turns out, however, Einstein's "blunder" may have been a useful one. Recent observations made by scientists of very distant ("high-redshift," in astronomer parlance) Type Ia supernovae show something very surprising. These "standard candles" can be seen out to 7 billion light years, and the distances to the galaxies that they are in show that the universe appears to have been expanding more slowly in the past than now. So not only is everything rushing away from everything else, but it's rushing away more quickly these days. In other words, the expansion of the universe is accelerating.

What is causing it to accelerate? Well, it might be the "vacuum energy" that Einstein thought he needed to support the universe from collapsing on itself. Interestingly, this force increases with distance (as opposed to gravity, which weakens with distance).

But we come back to the debated value of the Hubble constant. Our distances determined in this way are only as good as our value for the Hubble constant.

The Big Picture

Expansion implies a beginning in time. And we will explore where and how the expansion might have begun in Chapter 25, "What About the Big Bang?". In Chapter 26, "(How Will It End?", we'll explore the details of the expansion itself. But before we move on to some of these big questions, we turn our attention in the following chapter to some of the most energetic and unusual members of the galactic family: the quasars, black holes (again), and galactic jets, all of which keep the universe quite active, thank you very much.

The Least You Need to Know

➤ Galaxies were first classified by Edwin Hubble into three broad types: spiral, elliptical, and irregular.

➤ The majority of the mass (90 percent) of most galaxies and clusters (99 percent) is made of material we cannot observe, dark matter. We see it only by its gravitational effect.

➤ Distances on intergalactic scales are measured by observation of Cepheid variable stars, the Tully-Fisher relation, standard candles such as Type I supernovae, and Hubble's Law.

➤ Many galaxies are grouped in galactic clusters, which, in turn, are grouped into superclusters. Superclusters are found together on the edges of huge voids.

➤ Edwin Hubble first observed that galaxies recede from us at a rate proportional to their distance from us. The constant of proportionality, called the Hubble constant, tells us the age of the universe.

Chapter 23

Moving Out of Town

In This Chapter

➤ Distant galaxies, strange galaxies

➤ Active galaxies: Seyfert and radio galaxies

➤ What drives an active galaxy?

➤ Galactic jets

➤ What's in a quasar?

Remember Grote Reber? He's the man who, back in 1936, built a radio telescope in his backyard, and by the 1940s had discovered the three brightest radio sources in the sky. He didn't know it at the time, but two of them—the Galactic center, Sagittarius A; and the shrapnel of a supernova explosion, Cassiopeia A—are sources in our own Galaxy. But the third radio source, called Cygnus A, turned out to be much, much farther away. And far more strange.

In 1951, Walter Baade and Rudolph Minkowski located a dim optical source at the position of Cygnus A and, from its spectrum, measured a redshift (or its recessional velocity) of some 12,400 miles per second (almost 20,000 km/s). Whatever it was, Cygnus A was moving away from us—fast! For a while, astronomers tried to figure out how an object in our galaxy could be moving so fast. It turns out it wasn't in our galaxy at all, but very far away. Later, astronomers would discover that this was something never before seen: a distant inferno churning out the energy of a hundred normal galaxies.

In this chapter, we will examine objects like Cygnus A and the other monsters that hide deep in the hearts of distant galaxies. They are some of the most energetic and bizarre objects in the universe.

A Long Time Ago in a Galaxy Far, Far Away ...

Because of their great distances—hundreds of millions of parsecs away—the farthest "normal" galaxies are very faint. (We'll talk about what *normal* means in just a moment.) At such extreme distances, it becomes difficult to even *see* such galaxies, much less study their shapes or how they are distributed in space. But as far as we can see, there is little difference between *normal* distant galaxies and *normal* galaxies closer to home.

The operative word here is *normal.* Out in the farthest reaches of space, we see some objects that are not normal—at least they are not what we're accustomed to seeing in our cosmic neighborhood.

What does this tell us? Remember that the universe has a speed limit, the speed of light, and that information can travel no faster than this speed. Thus the farther away a certain star or galaxy might be, the longer it has taken its light to reach us.

So not only are we seeing far into space, we are in a very real sense seeing *back into time.* The strange objects that lurk in the distant universe existed in its earliest times. To study a quasar is to see into the origin of galaxies.

Star Words

Quasar is short for "quasi-stellar radio source." The first quasars were detected at radio frequencies, though most quasars do not emit large amounts of radio energy. Quasars are bright, distant, tiny objects, which produce the luminosity of 100 to 1,000 galaxies within a region the size of a solar system.

Quasars: Looks Can Be Deceiving

After seeing the spectacular optical images of emission nebulae (see Chapter 18, "Stellar Careers,") and galaxies such as Andromeda (see Chapter 21, "The Milky Way: Much More Than a Candy Bar,") a *quasar* can make a disappointing first impression. While the optical counterparts of the brightest of them can be seen with an amateur telescope on a dark night—3C 273 is 12th magnitude—they are undistinguished. In fact, they look so much like stars that astronomers at first thought they were simply peculiar stars (quasi-stellar objects).

But keep in mind how incredibly distant these objects are. The closest quasars are some 700 million light-years away! Though their apparent brightness might be small, their luminosities (the amount of energy that they put out each second) are astounding.

Quasars appear optically faint for two reasons: (1) much of their energy is emitted in the nonvisible part of the spectrum, and (2) quasars are very distant objects.

The great distance to quasars truly became apparent in the 1960s. The first quasars were discovered at radio frequencies, and optical searches at these locations showed objects that looked like stars. But the spectra of these "stars" told a different story. In the early 1960s, the astronomer Maarten Schmidt made a stunning proposal. The four bright spectral lines that distinguish hydrogen from the rest of the elements in the universe were seen to be shifting to longer wavelengths—much longer wavelengths. These *redshifted* lines, along with Hubble's Law, indicated that quasars were *very* distant—billions of light-years away, in fact.

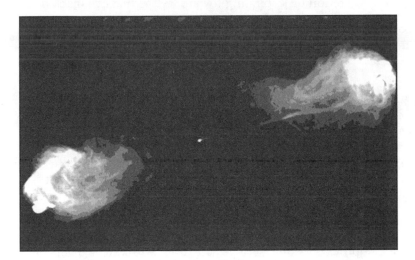

This image from the Very Large Array shows the detailed structure of Cygnus A, located in the constellation Cygnus, the swan. One of the brightest radio sources in the sky, Cygnus A is one of a large number of radio galaxies that has been discovered, with "jets" that arise from the region around a central black hole.

(Image from NRAO)

Quasar 3C 273 (the 3C stands for the Third Cambridge Catalogue) has its spectral lines redshifted in velocity by 16 percent of the speed of light! As we know, light from a source is redshifted when the object is moving away from us. In the context of Hubble's Law, such dramatic redshifts mean that quasars are receding at tremendous speeds and are very far away. The quasar 3C 273 travels at some 30,000 miles per second (48,000 km/s) and is some 2 billion light-years (640 million parsecs) distant from us.

Small and Bright...

It is an awe-inspiring thing to contemplate an object so powerful and so distant. The most distant quasars are some of the most distant objects we can see.

Quasars are distant and bright, but also small. How do we know they are small? Many quasars flicker, varying in brightness, rapidly—on scales of days. We know that light must be able to travel across the size of an object for us to see it vary in brightness. Why? Because for a region to appear to brighten, you must detect photons from the

Astronomer's Notebook

Quasars, among the most luminous objects in the universe, have luminosities in the range of 10^{38} watts to 10^{42} watts. These numbers average out to the equivalent of 1,000 Milky Way Galaxies.

far side as well as from the near side of the object that is brightening. If an object is one light-year across, the "brighter" photons from the far side of the source will not reach you until a year after the photons on the near side. So this source could only flicker on scales of a year.

Any object that flickers on scales of mere days must be small—light *days* across, to be exact. This characteristic flickering reveals that the source of energy is perhaps the size of our solar system—presumably a gaseous accretion disk spiraling toward a supermassive black hole.

If a supermassive black hole is the source of a quasar's power, then about 10 suns per year falling into the black hole could produce its enormous luminosity.

Quasars and the Evolution of Galaxies

Quasars may be more than strange, distant powerhouses of the early universe. In fact, they may be part of the family tree of every galaxy, including our own. Perhaps all galaxies (as they form) start out as quasars, which become less luminous, less energetic, as the early fuel supply of the galaxy is exhausted. Certainly quasars cannot burn fuel at the prodigious rates that they do forever.

But what could fuel a quasar? A generation of stars that forms near the center of the galaxy early in its life could, through the natural mass loss that occurs as stars age, provide the needed mass. Another possibility is a galactic train wreck. If two galaxies collide, one could provide fuel for the other's dormant black hole. As the fuel arrives, the black hole once again lights up. Bingo! A quasar.

When galaxies collide, star formation is often the result, as seen in the large ring of star formation that surrounds the galaxy to the left. This Hubble Space Telescope *image shows the famous "cartwheel" galaxy, and the escaping culprit from the recent hit and run. One of the two smaller galaxies to the right is the one to blame, and the ring surrounding the cartwheel galaxy is teeming with young massive stars.*

(Image from NASA)

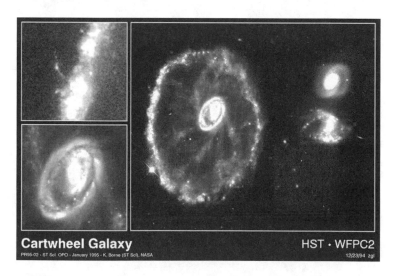

Cartwheel Galaxy

PR95-02 · ST ScI OPO · January 1995 · K. Borne (ST ScI), NASA

HST · WFPC2

12/23/94 zgl

A Piece of the Action

From what we have seen so far of galaxies, there aren't many slackers. They all seem to get a lot done on an average day. That is, they all seem quite active.

But active galaxy has a specific meaning to an astronomer. Indeed, what astronomers call active galaxies may well be an intermediate evolutionary stage between quasars and normal (or should we say mature?) galaxies.

Between the great distances to quasars and the more moderate distances to our local galactic neighbors are a vast number of normal galaxies, and a few galaxies that are more luminous (particularly in their central or nuclear region) than average. These latter objects are called *active galaxies*.

The excess luminosity tends to be concentrated in the nucleus of the galaxy, and the centers of these galaxies are referred to as active galactic nuclei. As a class, active galaxies have bright emission lines (implying that their centers are hot), or are variable on short time scales. Some of them also have jets of radio emission emanating from their centers, stretching hundreds of thousands of light-years into intergalactic space.

The Violent Galaxies of Seyfert

In 1944, Carl K. Seyfert, an American astronomer, first described a subset of spiral galaxies characterized by a bright central region containing strong, broad emission lines. These *Seyfert galaxies* have luminosities that vary, and some show evidence of violent activity in their cores.

Here are a few distinguishing characteristics of Seyfert galaxies:

➤ Spectra emitted by Seyfert nuclei have broad emission lines, which indicate the presence of very hot gas or gases that are rotating at extreme velocities.

➤ The radiation emitted from Seyfert galaxies is most intense at infrared and radio wavelengths and must be, therefore, nonstellar in origin.

➤ The energy emitted by Seyfert galaxies fluctuates significantly, over relatively short periods of time.

At the heart of all Seyfert galaxies is something relatively small but extremely massive, massive enough to (periodically) create the tremendous

Star Words

Active galaxies are galaxies that have more luminous centers than normal galaxies.

Astro Byte

Only one percent of all spiral galaxies are classified as **Seyfert galaxies.** Another way to think about this: Perhaps all spiral galaxies exhibit Seyfert properties one percent of the time.

335

activity evident at the Seyfert galaxy nucleus. One likely possibility? The core of a Seyfert galaxy may contain a massive black hole.

Cores, Jets, and Lobes: Radio Galaxy Anatomy

Another kind of active galaxy is called a *radio galaxy*. While Seyferts are an active subclass of spiral galaxies, what we call radio galaxies are an active subclass of elliptical galaxies.

Star Words

Radio galaxies are an active galaxy subclass of elliptical galaxies. They are characterized by strong radio emission, and in some cases, narrow jets and wispy lobes of emission located hundreds of thousands of light-years from the nucleus.

There are many types of radio galaxies, often classified by their shapes. Some radio sources have emission only in their nucleus. In others, two narrow streams—or jets—of oppositely directed radio emission emerge from the galaxy nucleus. The jets in these so-called double radio sources often end in wispy, complex puffs of radio emission much larger than the central elliptical galaxy. These are called radio lobes. In some radio galaxies the lobe emission dominates, and in others the jet emission dominates.

One remarkable aspect of the jets of radio emission is that they are observed over such a huge range of scales. The jets in some galaxies are linear for hundreds of thousands of light-years, and yet are observable down to the smallest scales that we can see—a few light-years. Radio jets are thought to be beams of ionized material that have been ejected from near the galactic center. The jets eventually become unstable and disperse into lobes.

The radio galaxy 3C31 is shown superimposed on an optical image of the same region. In the inset to the right, the VLA image (jets) is superimposed on a Hubble Space Telescope *image of the region. World-class telescopes need to improve their resolutions in sync so that overlays like these are possible.*

(Image from NASA)

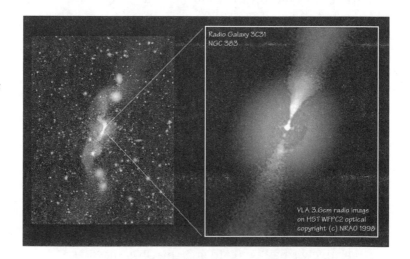

Close Encounter

Many differences in astronomy come down to a question of perspective. Astronomers are not entirely certain that *core-halo radio galaxies* (with a bright center and a diffuse envelope of emission) and *lobe radio galaxies* (with a distinct bright center and diffuse lobes) are unique objects. That is, a core-halo galaxy may be nothing more than a foreshortened view of a lobe radio galaxy. If the galaxy happens to be oriented so that we view it through the end of one of its lobes, it will look like a core-halo galaxy. If we happen to see the galaxy from its side, two widely spaced radio lobes will be detected on either side of a central core.

Material in radio jets is being accelerated to enormous velocity. In some radio jets, bright blobs of radio emission appear to be moving faster than light. And that's not allowed!

What's going on? The apparent *superluminal motion* (speed faster than light) results from jets that are moving toward us. It does not defy any laws of physics.

What is this mysterious *radio emission* we have been discussing? Well, radio emission comes in two basic flavors. One is rather bland, and the other a bit more spicy. The bland radio emission is *thermal emission,* which arises most commonly in regions of hot, ionized gas—like the HII ("H-two") regions around young, massive stars. This type of radio emission comes from free electrons that are zipping around in the hot gas.

The spicy variety is *synchrotron emission,* sometimes called *non-thermal emission*. This type of radiation arises from charged particles that are being accelerated by strong magnetic fields. (James Clark Maxwell first discussed the effects of magnetic fields on charged particles.) The intensity of synchrotron radiation is not tied to the temperature of the source, but to the strength of the magnetic fields that are there. Radio jets, filled with charged particles and laced with strong magnetic fields, are intense sources of *synchrotron radiation*.

Star Words

Superluminal motion is a term for the apparent "faster than light" motion of the blobs in some radio jets. This effect results from radio-emitting blobs that are moving at high velocity toward us.

Where It All Starts

Quasars and active galaxies are sources of tremendous energy. We have seen that stars cannot account for their energy output, and that most of their energy arises from a small region at the center of the galaxy. How can we account for the known properties of active galaxies and quasars, including their staggering luminosities?

Generating Energy

We saw in the last chapter that the rotation curves of many galaxies show evidence for large mass accumulations in the central regions. In fact, the masses contained are so large and in such a small volume that astronomers have in many cases concluded that a black hole must be present at the galaxy's center.

Star Words

Synchrotron radiation arises when charged particles are accelerated by strong magnetic fields. The emission from radio galaxies is mostly synchrotron.

Astronomer's Notebook

Remember the fluctuations in brightness seen in Seyfert galaxies and quasars? These fluctuations may be explained by brightness fluctuations in the accretion disk—the swirling disk of gas spiraling toward the black hole.

If a black hole were present in the center of active galaxies (and the quasars that they might have evolved from), many of the observed properties of these strange galaxy types might be explained. For one, the tremendous *nonstellar* energy output originating in a compact area points to the gravitational field and accretion disk of a black hole.

But not just any black hole.

These black holes dwarf the stellar remnant black holes that we discussed in Chapter 19, "Black Holes: One-Way Tickets to Eternity," and are called supermassive black holes.

The center of our own Galaxy, the Milky Way, may contain a black hole with a million (or more) solar masses. To account for the much greater luminosity of an active galaxy, the mass must be higher—perhaps a billion solar masses.

If black holes are the engines, we can explain the luminosity of active galaxies and quasars as the result of gas that spirals toward the black hole at great velocity, becoming heated in the process and producing energy in the form of electromagnetic radiation, x-ray photons are often produced and these are visible with the new Chandra X-ray Observatory.

In the black hole model, the radio jet arises when the hot gas streams away from the accretion disk in the direction with the least resistance—perpendicular to the accretion disk. These jets, then, can stream away from the disk in two directions, giving rise to the oppositely directed jets.

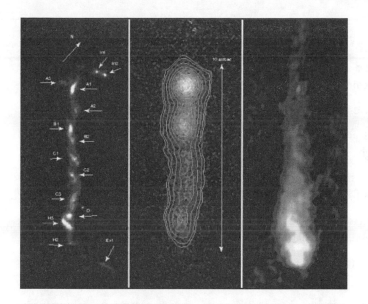

This image shows the jet associated with the quasar 3C273. On the left is the optical image (from the Hubble Space Telescope*). In the middle is the Chandra X-ray Observatory image, and on the right is the radio image of the jet as seen by the* Multi-Element Radio Linked Interferometer Network (MERLIN). *Notice that the different wavelengths emphasize different parts of the jet.*

(Image from NASA, Chandra and Merlin)

If black holes exist at the centers of all or most galaxies, then the differences between quasars, radio galaxies, Seyfert galaxies, and normal galaxies is not the engine, but the fuel. When fuel is plentiful (as it is early in the universe), the cores of all galaxies might burn bright as a quasar. Then they would move through a quieter, yet still active phase as Seyfert or radio galaxies. Finally, at our time in the universe, with fuel less plentiful, most black holes lie dormant. Models of galaxy evolution are still hotly debated (at least by astronomers), but the existence of central black holes may answer many questions.

A closing thought, then. Let's say you take the trouble to find the quasar 3C 273 for yourself. *You* see a quasar—but the light that you see left the galaxy some 2 billion years ago. The quasar you see may now (*now* as it is experienced at the distant location of the 3C 273) be a more mature galaxy, not unlike the Milky Way. Perhaps on some planet orbiting an average star somewhere in the distant universe, an amateur astronomer is pointing a telescope at our Milky Way, which appears to him or her as a faint, bluish blob of light: a quasar.

In the following chapters, we ask some of the big questions, the first of which is "Are We Alone?" And we will see just how likely it is that someone else might be watching us.

The Least You Need to Know

➤ Quasars were first discovered at radio wavelengths and are some of the most distant astronomical objects visible.

➤ Quasars may be the ancestors of all galaxies, the violent beginnings of us all.

➤ Active galaxies are any galaxies that more luminous than what we call normal galaxies, with bright star-like cores and broad, strong emission lines.

➤ Seyfert galaxies are the active subset of spiral galaxies, and radio galaxies are the active subset of elliptical galaxies.

➤ Radio jets originate at the cores of active elliptical galaxies and terminate in wispy patches of emission called lobes.

➤ Supermassive black holes are the most likely source of energy for quasars and active galaxies.

Part 6

The Big Questions

The questions themselves may be expressed simply enough: In this universe, are we alone? Is there life—other than on Earth—within the solar system? The Galaxy? Beyond? If so, is it intelligent? And if intelligent, will it communicate with us? Then on to questions about the origin and fate of the universe. Is the universe eternal? If not, how did it begin? And how will it end? Is the universe infinite?

The questions may be put simply enough. The answers ... well, read on.

Table for One?

Throughout this book we have been exploring distant worlds, places that are incompatible with any life we can imagine. We have traveled from the burning and frigid surfaces of the other planets in our solar system to the cores of stars generating the tremendous temperatures of nuclear fusion, to the neighborhoods of black holes that spit jets of charged particles into intergalactic space. These are not exactly life-friendly environments. But, undeniably, *we* are here. Our planet is literally crawling with all variety of life. Is our planet really so unique in the universe?

Probably not.

We have learned that there is nothing particularly special about our position in the cosmos: We inhabit the third planet orbiting an average star in a solar system tens of thousands of light-years from the center of a normal spiral galaxy. And there are a few hundred billion other stars in the Galaxy.

So if we're not so special, why has no one contacted us to let us know? Do we smell bad? Is it our hair? Or is our planet truly the only repository of life in the universe? (Hollywood seems to think not.) In this chapter we ask a big question, "Are we alone?" And if we aren't alone, where *is* everybody?

What Do You Mean by "Alone"?

In 1845, the naturalist and philosopher Henry David Thoreau set up housekeeping in a cabin on the shores of Walden Pond, near Concord, Massachusetts. Ten years later he wrote *Walden*, an account of his intellectual and spiritual experiences there, and became famous to generations of readers of American literature and philosophy as the man who went to live alone in the woods.

But did he live alone?

It is true that he had no close human neighbors, but he certainly had plenty of animals and plants around him. For that reason, Thoreau did not feel himself ever to be alone.

Maybe you feel the same way. The mere presence of life, in any form, is sufficient to keep you company. Then again, maybe it's not enough for you. Maybe without the proximity of *intelligent* life, you feel very much alone, despite the flora and fauna of your back-yard.

Astronomer's Notebook

Where did the early complex molecules on the earth's surface come from? Did the violent environment of the early Earth, complete with volcanic activity and energetic bombardment from space, provide the energy? Or did complex molecules rain down on the planet, having been created on dust grains in interstellar space? Complex organic molecules have been detected on Halley's Comet as well as on Comet Hale-Bopp. It is indeed possible that the stuff of life was imported from the stars. It is also possible that the basis of life on the earth is a combination of domestic and imported raw materials. We may be "Assembled on Earth from Domestic and Imported Components."

... If You Call This Living

In thinking about the prospects for life beyond our planet, most of us probably contemplate two distinct subjects: intelligent life, capable of creating civilizations, and, well, just plain life.

How plain can life get and still be called life?

You know how shrink-wrapped boxed software has "minimum requirements" printed on their sides? How many megs of RAM you need, how much hard drive space, and so on. Well, life seems to have some minimum requirements as well. Most biologists would agree that the requirements for something to be classified as a living organism would include:

➤ The ability to grow and reproduce, obtaining nourishment from the environment

➤ The ability to react to the environment

➤ The ability to evolve or change as the environment changes

Biologist Steven J. Gould has called the current age on Earth the Age of Bacteria, and he makes the case that, in terms of diversity and adaptability, bacteria rule. Kurt Vonnegut Jr. once suggested that the purpose of people is to bring viruses to other planets. Bacteria and viruses may be plentiful, but what about more macroscopic beings? Are these found—or even common—on other planets?

Is Earth Rare?

The requirements just listed are minimal and, for example, say nothing about movement, specific senses, or even consciousness (more specific attributes such as these narrow and refine the definition of life). We know that life on the earth ranges from tiny viruses (which hardly seem alive at all when inactive), to simple one-celled organisms, to the whole range of more complex plant and animal life, to ourselves.

Life at its most basic seems to require very little: the presence of a few chemical elements, the absence of certain harmful substances, and the availability of tolerable environmental conditions. While some scientists believe that the existence of these basic requirements on the earth is a sort of cosmic fluke—a rare, perhaps unique, lottery jackpot win—most assume that the earth, while special to us, is mediocre. The belief that there is nothing uncommon, let alone unique, about the conditions on the earth that support life is sometimes called the *assumption of mediocrity*.

As far as anyone can tell, the earth does not possess any special elements or conditions that aren't easily available elsewhere in the Galaxy and the universe. Not only, then, is there nothing to have prevented life from evolving elsewhere in the universe, but with hundreds of billions of stars in a single galaxy, there is every reason to believe that life *has* indeed sprung forth elsewhere. There are even recent suggestions that life may have originated on Mars, and have been brought to the earth on the back of asteroids. Researchers have been able to generate structures that might be the precursors to cell walls in conditions that simulate the harsh environments of interstellar space.

Star Words

The **assumption of mediocrity** basically states that we are not so special. The conditions that have allowed life to arise and evolve on the earth likely exist many other places in the Galaxy and universe.

The Chemistry of Life

The universe is about 15 billion years old. Earth, like other planetary bodies of the solar system, is about 4.6 billion years old, but the fossil record shows that Earth was devoid of life for some hundreds of millions of years after it coalesced from the solar nebula. The earliest fossils are of very simple organisms, bacteria and blue-green algae, dated at 3.1 billion years old. How did they get there?

Some time before these first life forms arose, during the earth's first several hundred million years, the simple molecules present (nitrogen, oxygen, carbon dioxide gas, ammonia, and methane) somehow formed into the more complex amino acids that are the chemical building blocks of life. Most scientists believe that the turbulent youth of the earth provided the energy that caused the transformation of simple elements and compounds into the building blocks of life. Also remember that, in its infancy, the earth was orbited by a much closer moon, so that once liquid water was present on its surface, the tides would have been enormous—perhaps 1,000 feet up and down as the planet rotated once every 6 (not 24) hours! This vigorous mixing of the primordial soup (shaken, not stirred) may have had a profoundly creative effect.

Astro Byte

Volcanic activity on the early Earth sent many gases coursing into the atmosphere. Its early carbon dioxide atmosphere literally arose from within. The earth had sufficient mass to hold onto the heavier molecules in its atmosphere, but not the lighter hydrogen and helium.

Once amino acids and nucleotide bases were available, the next step up in complexity would have been the synthesis of proteins and genetic material. The DNA molecule is made up of what are called nucleotide bases, or genes. Strung together, the genes tell our cells what to do when they reproduce, and make us different from, say, earthworms. The DNA molecule is the most durable, portable, and compact information storage device we know of. We are just beginning to unravel its mysteries.

There may be plenty of planets similar to the earth—that is hard to deny. But it is extremely difficult to estimate the likelihood that the molecules on any given planet will combine to form amino acids and nucleotides, let alone proteins and DNA (which are made from amino acids and nucleotide bases, respectively). Here is where the questions, and the debates, begin.

All life on Earth is carbon-based. That is, its constituent chemical compounds are built on carbon combinations and, furthermore, developed in a liquid water environment. Even creatures such as ourselves, who do not live in water, consist mostly of water. If life on the earth developed first in the oceans, we, billions of years later, still carry those oceans within us.

Close Encounter

In 1953, scientists Stanley Miller and Harold Urey decided to see if they could duplicate, experimentally, the chemical and atmospheric conditions that produced life on the earth. In a 5-liter flask, they replicated what is thought to be the earth's primordial atmosphere: methane, carbon dioxide, ammonia, and water. They wired the flask for an electric discharge, the spark intended to simulate a source of ultraviolet photons (lightning) or other form of energy (such as a meteor impact shock wave). The Miller-Urey Experiment didn't produce life—no Frankenstein's monster—but, remarkably, it did create a collection of amino acids, sugars, and other organic compounds. Thus the Miller-Urey Experiment implied an extension of the assumption of mediocrity. Not only could we reasonably conclude that there are other similar planets in the Galaxy, but also that they must have the chemical elements necessary for life, and energy alone could trigger the synthesis of the building blocks of life.

The Odds for Life on Mars

Science fiction has long portrayed Mars as home to intelligent life, and, as we saw in Chapter 13, "So Close and Yet So Far: The Inner Planets," Percival Lowell, early in the twentieth century, created a great stir with his theory of Martian canals. Actor-director-playwright Orson Welles triggered a nationwide panic with his 1938 radio dramatization of H. G. Wells's *War of the Worlds,* about a Martian invasion of Earth. People were ready to believe.

Given our fascination with the red planet and its proximity, it is no wonder that Mars has been the target of a number of unmanned probes (see Chapter 9, "Space Race: From *Sputnik* to the International Space Station"). In the mid- to late 1970s, the *Viking* probes performed robotic experiments on the Martian surface, including tests designed to detect the presence of simple life forms such as microbes. These tests yielded positive results, which, however, were subsequently reinterpreted as false positives resulting from chemical reactions with the Martian soil. However, on August 7, 1996, scientists announced that, based on its chemical composition, a meteorite recovered in Antarctica had originated on Mars and possibly contained fossilized traces of molecules that, on Earth, can be produced only by bacteria.

Most recently, the *Mars Pathfinder* (1997) and *Mars Global Surveyor* (1998) missions have added detailed panoramic views from the Martian surface and high-resolution satellite images of the surface. Both of these missions suggest the presence at some time in the past of liquid water on the Martian surface. Barry E. Digregorio, Gilbert V. Levin, and Patricia Ann Straat, authors of *Mars: The Living Planet* (North Atlantic Books, 1997), present evidence of possible subsurface water, though most evidence suggests that the Martian surface has been dry for a very long time.

All of these results are intriguing, but, so far, none of the Martian probes has found clear evidence of life (past or present) on Mars. In the past, the Martian atmosphere, now very thin, was thicker and, as a consequence, the planet's surface warmer and wetter. There is a possibility that microbial life once existed on Mars. Under the planet's current cold, dry, and generally harsh conditions; however, the presence of life is highly unlikely.

The Mars Orbiter Camera continues to transmit high-resolution images of the Martian surface that suggest at least the periodic presence of surface water.

(Image from NASA)

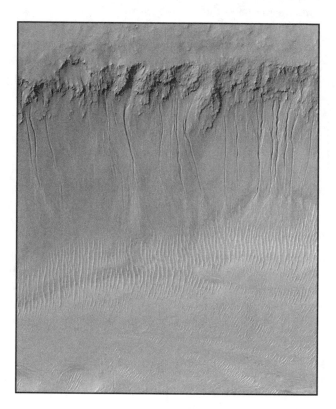

The Face on Mars

In 1976, *Viking 1,* surveying Martian landing spots for *Viking 2,* took a picture of the Martian surface that revealed what some of the observers interpreted as a giant monument, about a third of a mile (0.5 km) across, in the form of a human (or humanoid?) face. The Face on Mars has provoked much discussion in the popular press and (more recently) on the Internet. With the canals on Mars long since revealed as nonexistent, the Face on Mars provided new hope, as some saw it, that intelligent life either exists or once existed on Mars. Some commentators even suggested that objects resembling pyramids and other structures could be seen in the vicinity of the face.

Close Encounter

Europa (see Chapter 15, "The Far End of the Block,") a moon of Jupiter, has a frozen surface and may well have liquid water beneath it. Titan (see Chapter 15), orbiting Saturn, has an atmosphere rich in nitrogen, methane, and ammonia. It may also have liquid methane on its surface. So far, these planetary moons, rather than any planet, offer the best prospects for other life within the solar system. However, the prospects are dim, as temperatures on both of these moons are extremely low.

On April 5, 1998, the Mars Orbiter Camera on the *Mars Global Surveyor* made high-resolution images of the face, showing it to be nothing more than a natural formation. The angle of sunlight and the comparatively low resolution of the 1976 image, combined with human imagination, had temporarily transformed this geological feature—really somewhat like a mesa on Earth—into a giant face.

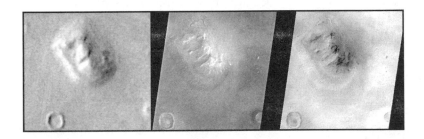

The Face on Mars—as photographed by Viking 1 *on July 25, 1976 (left), and as photographed on April 5, 1998, by the Mars Orbiter Camera on the* Mars Global Surveyor *(two images on the right). (Images from NASA)*

Hello! Is *Anybody* Out There?

Life beyond Earth but within the solar system seems unlikely. But that's okay. There's still the rest of the universe.

With hundreds of billions of stars in our own Galaxy, there are probably many other planetary systems. Virtually all stars that last long enough have planetary systems. Odds are that life does exist somewhere out there. What are the odds, exactly?

You Just Love the Drake Equation

With a few hundred billion stars in the Galaxy, the likelihood of life somewhere in the Milky Way other than on the earth seems great. But is there any way to get a handle on just how great? Not every star has a fair shake, however. Some are so massive that they don't last long enough, and others are not hot enough to warm any planets that might be there. Perhaps 15 percent of the stars in the Galaxy have the proper mass to be luminous enough (but not too luminous) to support habitable planets. In the 1960s, one astronomer attempted to roughly quantify the odds that intelligent life capable of communication exists elsewhere in the Galaxy.

In 1961, at a conference on the Search for Extra-Terrestrial Intelligence (SETI), held at the National Radio Astronomy Observatory in Green Bank, West Virginia, Professor Frank Drake proposed an equation to estimate the possible number of civilizations in the Milky Way. The *Drake Equation* contains a number of highly uncertain terms—variables that must simply be guessed at—however, it is a very useful way of breaking down a complex question into more readily digestible portions. In addition, some of the variables that were highly uncertain in 1961—the number of stars with planets, for example—are becoming more certain as further research is conducted.

Here's what the Drake Equation, in its basic form, looks like:

$$N = R_* \times f_p \times N_p \times f_e \times f_l \times f_i \times f_c \times L$$

And here's what the terms of the Drake Equation mean:

➤ N (the left-hand side of the equation) is the number of civilizations in our Galaxy with which we should be able to communicate via radio signals.

➤ R_* is the rate at which our Galaxy produces stars—its productivity in solar masses per year.

➤ f_p is the fraction of stars with planetary systems.

➤ N_p is the average number of planets per star.

➤ f_e is the fraction of Earth-like planets (planets suitable for life).

➤ f_l is the fraction of f_e planets on which life actually develops.

Star Words

The **Drake Equation** proposes a number of terms that help us make a rough estimate of the number of civilizations in our Galaxy, the Milky Way.

Astronomer's Notebook

The terms of Drake's Equation are highly uncertain. However, a number of the terms that were the most uncertain when the equation was first proposed in 1961 (including the star formation rate in the Galaxy and the number of stars with planetary systems) are now on much firmer footing as the result of recent astronomical research. Other terms, such as the likelihood that life will arise given the right conditions, are still hotly debated.

➤ f_i is the fraction of f_l planets on which intelligent civilizations arise.

➤ f_c is the fraction of f_i planets on which technological civilizations arise.

➤ L is the lifetime of a civilization in years.

A Closer Look at the Equation

Perhaps the most important term in the equation is L, the lifetime of a civilization. If civilizations last for many millions of years, then our own Galaxy could be teeming not only with life, but also with civilizations. If L is rather small (hundreds or thousands of years), then the Galaxy may be a lonely place indeed. Civilizations, like candles, may be lit frequently, but if they do not burn for very long, only a few others will be burning with us.

Let's look at some of the other key terms.

Close Encounter

Astronomers have two basic ways to look for planets around other stars—extrasolar planets. Both involve an understanding of gravity and mass. When we talked about the solar system, we said that the planets orbit the sun. That is approximately true. In detail, all of the masses in the solar system (the Sun, Jupiter, and all of the planets) orbit around their common center of mass. As a result, the sun is not exactly fixed, but wobbles back and forth in space every 11 years as Jupiter (most of the mass other than the Sun) orbits.

This wobble can either be detected directly (with a technique called astrometry) by watching a star wobble around in the plane of the sky, or indirectly. The indirect method (so far) has been the more successful, and involves the use of the Doppler shift. As a star wobbles back and forth (alternately coming at us and moving away), its spectral lines shift from shorter to longer wavelengths and back again. Of course, the planets discovered by either method must be massive; if not, the star wouldn't wobble enough.

For example, a planet 2.4 times the mass of Jupiter has been discovered at a distance of more than 2 A.U. from the star 47 Ursa Majoris (47Uma).

Want to get the latest on the search for extrasolar planets? Point your Web browser to www.obspm.fr/encycl/encycl.html.

Galaxy Productivity

Some astronomers figure that a minimum of 100 billion stars populate the Milky Way; others place this estimate as high as 300 billion. If the Galaxy is about 10 billion years old, it can be safely assumed that 10 solar-mass stars form each year.

Do They All Have Planets?

When the Drake Equation was first written down, the only planetary system we knew of was our own. But astronomers have expended great effort in the past three decades to uncover other planetary systems. It's not an easy task, as it amounts to detecting a firefly (the planet) sitting on the edge of a searchlight (the star). The firefly may well be there, but it's lost in the glow of the searchlight.

Astronomer's Notebook

There are three planets in the habitable zone of the Sun—Venus, Earth and Mars. So apparently just being habitable is not enough to have life!

Welcome to the Habitable Zone

Estimates based on numerical modeling of the typical number of planets per star range from 5 to 20. Our own solar system has 9 planets, so this number may be a good estimate.

There is a doughnut-shaped region around all stars that astronomers call the *habitable zone*. This zone is the space between some inner and outer distance from the star where a star with a given luminosity (rate of energy production) produces a temperature that's "just right." Just right for what? Well, just right for liquid water to exist without freezing or boiling away. We often estimate that about 1 in 10 of the planets around a star will fall within the habitable zone.

Let There Be Life

Although astronomers such as Carl Sagan (1934–1997) have assumed that life develops on virtually every planet capable of supporting it, estimates vary widely. This number, f_l, is expressed as a decimal, with the number 1 indicating that there is a 100 percent chance that life will develop on a habitable planet. Estimates range from Sagan's optimistic 1 down to 0.000001—that is, perhaps only 1 out of every million habitable planets actually develops life. This number is very uncertain.

Who Are You Calling Intelligent?

The terms in the Drake Equation are now becoming very uncertain indeed. Some scientists believe that it is possible that biological evolution is a universal phenomenon; that given time, life anywhere it exists will tend to evolve toward greater and greater complexity. If this is the case, intelligence will emerge (given sufficient time) wherever there is life. Others, however, believe that intelligence is not indispensable to survival and that, therefore, it is not the inevitable end of evolution. Are bacteria intelligent? They've done pretty well for themselves here.

Whatever this fraction actually may be, whether 1 or a number much, much smaller, we may need to reduce further by estimating what fraction of this fraction goes on to develop civilization and technology—f_c in our equation.

Once again, estimates show wide variation, with some scientists believing that, given time, technology and civilization are inevitable developments of intelligence. Others argue that the connection is not inevitable and put this figure much lower. But we have now entered into the realm of social science and philosophy—unfamiliar turf for astronomers! Other complicating factors involve the correlation between intelligence and aggression. Are intelligent life forms destined to destroy themselves? Or do civilizations enter a time of peace, a *pax planetaria?* Also, will all civilizations have the *desire* to communicate?

Astro Byte

Kurt Vonnegut once said that the ultimate purpose of humans might be to bring viruses to other planets.

The Life Span of a Civilization

The final term is perhaps the most interesting and speculative of them all: How long do civilizations capable of communication last? To estimate how many civilizations there are in the Galaxy at any given time, we must know how long they last. Once lit, how long does the candle of civilization (on average) burn? Consider our own civilization, one with a recorded history of no more than 5,500 years and that has been highly technological for less than a century.

Planets and stars have long life spans, but we have only a little over 5,000 years of experience with civilization. And what is most disquieting is that we have reached a point of technological development that allows us, at the touch of a few thermonuclear buttons, to destroy our civilization here and now.

What is the average life span of a civilization? A thousand, five thousand, ten thousand, a million years?

Even if, over time, there have been many, many civilizations in our Galaxy, how many are there now?

Traveling to the location of another civilization in the Galaxy is a highly impractical proposition. In the best-case scenario, civilizations in the Galaxy are hundreds of light-years apart, and getting even to the nearest star would take us (using current technology) tens of thousands of years. The civilization might be long gone by the time we got there—to say nothing of the mother of all jet lags incurred by the voyagers.

Close Encounter

If we approach any variable in the Drake Equation pessimistically, assigning it a low value, we end up with the possibility that there are very few technological civilizations in the Galaxy. But if we plug into the equation the most optimistic estimates for all the variables (such as life will always develop, intelligence will always evolve, technology will always arise), we end up essentially equating the number of technological civilizations in the Galaxy with the average lifetime of such a civilization.

If the average technological civilization survives 10,000 years, we can expect to find 10,000 civilizations in the Galaxy at any time. If 5,000 years is the norm, 5,000 civilizations may be expected. If a civilization usually lasts millions of years, then, assuming high values throughout the rest of the equation, we might expect to find a million civilizations in the Milky Way.

Where Are the Little Green Men?

The original goal of the Drake Equation was to justify that the search for life in the Galaxy was not a hopeless venture, that the Galaxy might indeed be teeming with life, and that if we just looked, we might well find something or someone looking for *us*.

What We Look For

It is unlikely that we'll be traveling to distant civilizations any time soon. Yet, for optimists, the urge to make contact persists, and small groups of researchers have trained radio antennas skyward in the hope of receiving broadcasts from extraterrestrial sources. Researchers are searching for any radio signals that are clearly unnatural—that is, artificially generated.

Later, on Oprah ...

The project of monitoring the heavens for radio broadcasts assumes that a technological civilization would have invented something that produces intelligible signals in the radio wavelengths. There is no need to assume, however, that the distant civilization actually *wants* to communicate and is intentionally broadcasting to other worlds. After all, we on the earth have been broadcasting since the early twentieth century and have been doing so intensively for more than 60 years.

Close Encounter

Monitoring the vast heavens for artificial extraterrestrial broadcasts is a daunting task—akin to looking for a needle in a haystack. Where should we look? What frequencies should we monitor? How strong will the signal be? Will it be continuous or intermittent? Will it drift? Will it change frequency? Will it even be recognizable to us?

At the very least, radio searches are time- and equipment-intensive. Frank Drake conducted the first searches in 1960, using the 85-foot antenna at the National Radio Astronomy Observatory in Green Bank, West Virginia. He called his endeavor Project Ozma, after the queen of the land of Oz. This search ultimately developed into Project SETI (Search for Extra-Terrestrial Intelligence), which, despite losing its Congressional funding, remains the most important sponsor of search efforts.

The SETI Institute is a private, non-profit group based in Mountainview, California. When NASA search funding was cut in 1993, SETI consolidated much of its effort into Project Phoenix (risen, like the mythical bird, from the ashes of the funding cut), a program that began in 1995 and that monitors 28 million channels simultaneously. SETI hopes eventually to monitor two billion channels for each of some 1,000 nearby stars. Computer software alerts astronomers to any unusual, repeating signals. So far, no positive findings have been obtained.

Want to do your bit for the Search for Extra-Terrestrial Intelligence (SETI)? Download a screensaver that just might find ET. You never know. It uses cycles on your computer (when not in use) to process one of the largest datasets mankind has ever assembled. (www.seti-inst.edu/science/setiathome.html.)

And if you just want to check the current status of SETI, go to www.seti-inst.edu. The SETI Institute recently received a large financial boost from Microsoft cofounders Paul Allen and Nathan Myhrvold, who have committed $25 million to the construction of a radio telescope array dedicated to the SETI search. Up to this point, searches have depended on cadging time on existing telescopes. The proposed Allen Telescope Array will be dedicated to the SETI project around the clock.

While radio signals at longer wavelengths do not penetrate beyond our atmosphere, those in the higher frequencies, FM radio and television, are emitted into space. We don't *intend* this to happen, but it is an inevitable by-product of our technological

Astro Byte

Radio signals? Why not an optical laser sent out into space? Because the dust in the plane of our Galaxy absorbs optical photons very well. So an optical-frequency beacon would not get very far. Radio waves, on the other hand, pass through the dust as if it weren't there. We (and our neighbors) cast a wider net in radio waves.

Star Words

The **water hole** is the span of the radio spectrum from 18 cm to 21 cm, which many researchers believe is the most likely wavelength on which extraterrestrial broadcasts would be made. The name is a little astronomical joke—the hydrogen (H) and hydroxyl (OH) lines are both located in a quiet region of the radio spectrum, a region where there isn't a lot of background noise. Since H and OH add up to H_2O (water), this dip in the spectrum is called the "water hole."

civilization. Perhaps some other civilization is producing a similar by-product. Our first glimpse of another civilization might be the equivalent of its sitcoms, advertising, and televangelists. Yet we keep on listening.

Down at the Old Water Hole

But SETI researchers also guess that civilizations might choose intentionally to broadcast their presence to their neighbors, sending some sort of radio-frequency beacon into space. What portion of the spectrum might they choose?

Researchers chiefly monitor a small portion of the radio wavelengths between 18 cm and 21 cm, called the *water hole*. The rationale for monitoring this slice of the radio spectrum is twofold. First, the most basic substance in the universe, hydrogen, radiates at a wavelength of 21 cm. Hydroxyl, the simple molecular combination of hydrogen and oxygen, radiates at 18 cm. Combine hydrogen and hydroxyl, and you get water.

If the symbolism of the water hole is not sufficiently persuasive to prompt extraterrestrial broadcasters to use these wavelengths, there is also the likelihood that this slice of the spectrum will be recognized as inviting on a more practical level. It is an especially quiet part of the radio spectrum. There is little interference here, a very low level of Galactic noise. Researchers reason that, at the very least, intentional broadcasters would see the water hole (which looks the same no matter where in the Galaxy you are located) as a most opportune broadcast channel.

Should We Reach Out?

Hollywood has done a pretty fair job of reflecting the range of opinion on extraterrestrials. The 1950s and even 1960s saw a number of movies about malevolent alien invaders; but then Steven Spielberg's 1977 *Close Encounters of the Third Kind* and *E.T.—the Extraterrestrial* (1982) suggested that contact with emissaries from other worlds might be thrilling, beneficial, and even heartwarming. The close of the 1990s pitted

Independence Day—in which aliens attempt to take over the earth—against *Contact,* in which communion with extraterrestrials is portrayed as a profoundly spiritual (if rather teary-eyed) experience. And then *X-Files: Fight the Future* proposed that our entire planet is being prepared for massive colonization by extraterrestrials.

The point is this: Not everyone is persuaded that reaching out is such a hot idea. After all, even the most sociable among us lock up our houses at night.

But, some argue, the point is moot. For the radio cat has been out of the bag for more than six decades. There is an ever-expanding sphere of our radio broadcasts moving out at the speed of light in all directions from the earth. The earliest television broadcast (which, unfortunately, includes images of Adolf Hitler at the 1936 Olympics) is now some 65 light-years from the earth. It's a bit too late to get shy now.

Astro Byte

The *Voyager 1* and *2* spacecraft (which have now left the solar system) each have on board a golden plaque and phonograph record. The plaque shows a man and woman, the trajectory of the spacecraft in the solar system, and a representation of the hydrogen atom, among other vital information. The phonograph record contains recordings of human voices and music.

The Arecibo message was broadcast from the Arecibo radio telescope in Puerto Rico in 1974. The message is 1679 bits long, the product of two prime numbers, 23 and 73. When the zeros and ones are arranged in a rectangle with sides 23 and 73 units long, they make this picture, which shows among other things a human form, our number system, and the double helix of DNA.

(Image from NASA)

357

On November 16, 1974, the giant radio dish of the Arecibo Observatory (see Chapter 8, "Seeing in the Dark") was used as a transmitter to broadcast a binary-coded message containing a compact treasure trove of information about humans. Reassembled, the message generates an image that shows the numbers 1 through 10, the atomic numbers for hydrogen, carbon, nitrogen, oxygen, and phosphorous, a representation of the double-helix of DNA, an iconic image of a human being, the population of the earth, the position of the earth in the solar system, and a schematic representation of the Arecibo telescope itself. No reply has yet been received. That should come as no surprise. Traveling at the speed of light, the message will take 25,000 years to reach its intended target, a globular cluster in the constellation Hercules.

Some may view the Arecibo Message as the work of latter-day Quixotes jousting with cosmic windmills. Others may view it as an interstellar message cast adrift in a digital bottle, containing the very human hope of contact. Let's just hope as well that we haven't told them too much.

The Least You Need to Know

➤ The chemicals and conditions on the earth that support life are probably not unique or even exceptional, but common throughout the Galaxy and the universe.

➤ The Harold–Urey Experiment gave experimental support to the theory that the basic chemicals of life can be readily transformed into the building blocks of DNA and proteins (nucleotide bases and amino acids).

➤ While the possibility of life beyond the earth within our solar system is very small, the existence of simple life, elsewhere in our Galaxy (let alone the entire universe) is highly likely.

➤ The Drake Equation is a rough way to calculate in a reasoned manner the number of civilizations that might exist in the Milky Way. It includes a number of factors, some scientific, and others highly speculative.

➤ Small but dedicated groups of researchers (including those of the SETI project) monitor the heavens for artificial radio signals of extraterrestrial origin. The SETI project is now on a very firm financial footing and is planning to build a dedicated SETI telescope.

Chapter 25

What About the Big Bang?

In This Chapter

➤ What a cosmologist does (it doesn't involve makeup)

➤ Understanding the cosmological principle

➤ Is the universe infinite? Eternal?

➤ The Big Bang: what, when, and where

➤ Critical density and the expanding universe

➤ Matter from energy and energy from matter

Cosmology. The word has an archaic ring to it. No wonder, because it's been in the English language since at least the 1600s, and because it describes the study of some of the oldest and most profound issues humankind has ever addressed: the nature, structure, origin, and end of the universe. Those are big, unanswered questions.

For centuries, cosmology has been the province chiefly of priests and philosophers. This chapter shows what happened when astronomers took up the subject in the twentieth century.

The Work of the Cosmologist

Most of us grow up believing that the universe is both infinite in extent and eternal in duration. The planets may move in the sky, and, farther away, stars may orbit the centers of galaxies. But on the largest scales, certainly the universe is changeless. It has always been and it will always be. Right? How, after all, could it be otherwise?

Star Words

Cosmology *is the study of the origin, structure, and fate of the universe.*

Astronomer's Notebook

It may seem odd that the answer to how long it has taken for any given galaxy to reach its present distance from us should be 13 billion years for every galaxy—until we recall that Hubble's Law shows that the rate of recession is directly proportional to the distance from us; that is, if galaxy A is three times farther from us than galaxy B, it is also moving three times faster than galaxy B.

Actually, long before *cosmologists* came on the scene to study the origins of the universe, there were creation stories. But these were mythological and religious narratives. In the Judeo-Christian tradition, for example, God said, "Let there be light." And there was.

The human mind naturally looks for origins, beginnings, grand openings. In the twentieth century, we stumbled across two bits of evidence of the biggest grand opening ever.

Cosmologists have put together a creation story of their own. It is the result of centuries of observation, modeling, and testing. The model is so well established that it is generally referred to as the *standard model*. As Steven Weinberg points out in his account of the start of it all, *The First Three Minutes* (Basic Books, 1993), the study of the origins of the universe was not always a scientifically respectable pursuit. It is only in relatively recent scientific history that we have had the tools to rigorously test theories of how the universe began.

I'll Give You Two Clues

The twentieth century has thrown some major curve balls at the human psyche, from the carnage of two world wars to the invention of the hydrogen bomb, capable of releasing the fury of a stellar interior on the surface of our planet. And while the seventeenth century saw the earth pushed from the center of the universe, it was twentieth-century astronomers who told us that we were located in the suburbs of an average galaxy, hurtling away from all the other galaxies that we can see.

Why are we all flying away from one another? Isn't it obvious? The universe exploded 15 billion years ago!

Redshifting Away

Recall from Chapter 22, "A Galaxy of Galaxies," our discussion of Hubble's Law and Hubble's constant. We know from observing the redshift in the spectral lines of galaxies that all the galaxies in the universe are flying away from us, and that the galaxies farthest away are moving the fastest. And an observer in any galaxy would see exactly what we do.

Armed with Hubble's Law and Hubble's constant, it is quite easy to calculate how long it has taken any given galaxy to reach its present distance from us. In a universe that is expanding at a constant rate, we simply divide the distance by the velocity. The answer for all the galaxies is 15 billion years.

So if we were able to rewind the expanding universe to see how long it would take for all the galaxies that are flying apart to come together, the answer would be about 15 billion years.

Pigeon Droppings and the Big Bang

The Russian-born American nuclear physicist and cosmologist George Gamow (1904–1968) was the first to propose (in the 1940s) that the recession of galaxies implied that the universe began in a spectacular explosion. Not an explosion *in* space, but an explosion *of all* space. The entire universe was filled with the explosion that started the universe.

British mathematician and astronomer Sir Fred Hoyle (born 1915) coined the originally derisive term *Big Bang* for this event, and he did not mean it as a compliment. With astronomers Herman Bondi and Thomas Gold, he proposed an alternative *steady state theory* of the universe, proposing that the universe is eternal, because it had no beginning and will have no end.

Quite soon after Gamow proposed the Big Bang theory, theorists realized that the early universe must have been incredibly hot, and that the electromagnetic echoes of this colossal explosion should be detectable as long-wavelength radiation that would fill all space. This radiation was called the *cosmic microwave background.*

You may recall from Chapter 8, "Seeing in the Dark," that the science of radio astronomy was the accidental by-product of a telephone engineer's search for sources of annoying radio interference. Another radio-frequency accident resulted in an even more profound discovery. From 1964 to 1965, two Bell Telephone Laboratories scientists, Arno Penzias and Robert Wilson, were working on a project to relay telephone calls via early satellites. Wherever they directed their antenna on the sky, however, they detected the same faint background noise. The noise was not associated with any particular point on the sky, but was the same everywhere. In an effort to remove all possible sources of interference, they went so far as to scrape off the pigeon droppings they found on the antenna because they were a possible source of heat. Their antenna was clean as a whistle, but the noise persisted.

Star Words

The **Big Bang,** a term coined by Sir Fred Hoyle in the 1950s, refers to the origin (in space and time) of the universe. Cosmologists propose that the universe started its life in an incredibly dense and hot state.

The "noise" that they had stumbled across was the very cosmic microwave background that had been predicted, the highly redshifted photons left behind by the Big Bang. The theory goes like this: Immediately after the Big Bang, the energy liberated was most intense in the shortest, highest-energy wavelengths—gamma rays. As the universe expanded and cooled over millions, then billions of years, the waves expanded as well, reaching us today as long-wavelength photons in the microwave part of the electromagnetic spectrum.

This microwave radiation carries information from the early moments of our universe. It tells us that there is a remnant echo of the Big Bang found permeating the universe, evidence that in past times the universe was a much hotter place. The Cosmic Background Explorer (COBE) satellite was launched in November 1989. In the first few minutes of its operation, it measured the temperature of the background radiation to be 2.735 K, by fitting a blackbody curve like the ones we discussed in Chapter 7 to the measured data points. This satellite has improved vastly on the early, crude mappings of the cosmic microwave background. In addition, from COBE, we have learned that the solar system has a small motion relative to the cosmic microwave background. Finally, in 1992, COBE data revealed an exciting discovery: There are tiny fluctuations on the sky in this background radiation—it is not exactly homogeneous, or smooth. These temperature differences are the result of density fluctuations in the early universe, and they may give clues as to how matter first started to arrange itself.

Star Words

The **cosmic microwave background** consists of the highly redshifted photons left behind by the Big Bang and detectable today throughout all space as radio-wavelength radiation.

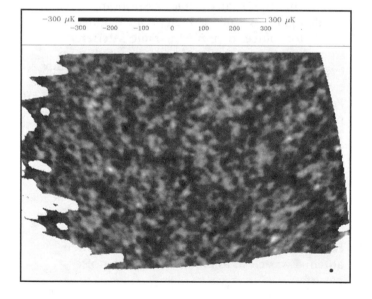

This image shows the cosmic microwave background as viewed the BOOMERANG mission. Notice that the BOOMERANG data is much higher resolution and shows the small scale of the variations. To give a sense of the portion of the southern sky seen in the BOOMERANG image to the right, the diameter of the Moon is indicated by the black circle in the lower right.

(Image from NASA/NSF/ University of California, Santa Barbara)

Same Old Same Old

Hubble made two very important discoveries in his studies of galaxy types and distributions. He found that the universe appeared to be both isotropic (the same in all directions), and homogeneous (one volume of space is much like any other volume of space).

His conclusion was the result of painstaking study of the distribution and type of a large number of galaxies at various redshifts, or distances from the earth.

The Cosmological Principle

Together, the homogeneity and isotropy of the universe make up what we call the *cosmological principle:* a cornerstone assumption in modern cosmology. If we could not make this assumption (based on observation), then our cosmology might only apply to a very local part of the universe. But the cosmological principle allows us to extrapolate our conclusions drawn from our local viewpoint to the whole universe. And consider these implications: A homogeneous universe can have no border or edge (since it is the same in any volume), nor can it have a center (since it should look the same in all directions from any viewpoint).

Perhaps the best model of the expanding universe is a toy balloon, on which we randomly draw some dots with a magic marker. But a warning first: If the universe is represented by the surface of a balloon, then we are talking about a two-dimensional universe. There is no inside of the balloon if you can draw only on its surface. Magic-marker dots on the balloon's surface represent galaxies.

As you look out into the universe, that is, out into the balloon's surface, you see no center and no edge—just like our balloon model. And if the balloon is inflated, the dots move away from one another. Regardless of their point of view, all the dots "see" all the others as moving away.

Star Words

The **cosmological principle** is a cornerstone assumption about the nature of the universe. It holds that the universe exhibits two key properties: homogeneity (sameness of structure on the largest scales) and isotropy (it looks the same in all directions).

So What Was the Big Bang?

What caused the Big Bang?

Who knows?

That is a question mostly outside the bounds of astronomy and physics. As scientists, we can only hope to understand things that are of this universe, and whatever caused the Big Bang is inherently unobservable.

Recall also from Chapter 22 that universal recession in time means the universe is expanding and that the concept of expansion implies an origin. Astronomers believe that at its origin, the universe was unimaginably dense. The universe was a point, with nothing outside the point. Ten to fifteen billion years ago, the entire universe exploded in the Big Bang, and the universe has been expanding ever since, cooling and coalescing into ever more organized states of matter. One very highly organized state of matter (you) is reading this book. Hubble's Law describes the rate of the expansion.

Big Bang Overview

What we are able to discuss is what happened very soon *after* the Big Bang. And we do mean *very* soon. From about $\frac{1}{100}$ of a second after the Big Bang onward, we can outline the major steps in the universe's evolution. It is basically a story of cooling and expanding. For a thorough review of modern cosmology, see Timothy Ferris's excellent and humorous *The Whole Shebang* (Touchstone, 1998).

At the earliest times that we can track, the universe was incredibly hot (10^{11} K) and filled with elementary particles, the building blocks of atoms: electrons, positrons, neutrinos, and photons of light. And for every billion or so electrons present at this time, there was one heavy particle (a neutron or proton). In this soup—an entire universe hotter than the core of a star—energy and matter were going back and forth as electrons and positrons *annihilated* to produce energy, and more were born from the energetic photons that filled the universe.

As the universe cooled, there was soon insufficient energy to create the electrons and positrons, so most of them annihilated, without new ones taking their place. After about the first three minutes of its existence, the universe had cooled to about 1 billion degrees, sufficiently cool for the nuclei of atoms to hold together, and later hydrogen, helium, and small amounts of lithium and beryllium were formed.

The universe then consisted of photons of light, neutrinos and antineutrinos, and a relatively small amount of nuclear material. Of course, the universe was still far too hot for electrons to come together with nuclei to form stable atoms. Not until about 300,000 years later (when the entire universe was at the temperature of a stellar photosphere) would nuclei be able to hold on to electrons.

It wasn't until after this first 300,000 years that the universe became transparent to its own radiation. What does that mean? Before the electrons settled out into atoms, they got in the way of all the photons in the universe and kept them bumping around. With

Star Words

Particles and antiparticles **annihilate** when they meet, converting their mass into pure energy. Electrons and anti-electrons (positrons) were continually created and annihilated in the early universe.

the electrons out of the picture, the universe went from a *radiation-dominated* state to a *matter-dominated* state.

From this point on, the universe continued simply to cool and coalesce, eventually forming the stars and galaxies of the observable universe.

A Long Way from Nowhere

Now just *where* did this Big Bang occur?

If the Big Bang took place *somewhere,* that place is, by definition, the center of the universe. Yet the cosmological principle forbids any such center. The Big Bang may resolve some open questions; but, it would seem, it also torpedoes the cosmological principle.

Actually, it doesn't—as long as we conclude that the Big Bang took place *everywhere.* And the only way such a conclusion could be true is if the Big Bang was not an explosion within an empty universe—the way a movie explosion is—but was an explosion of the universe itself. At early times in the universe, its entire volume was hotter than a stellar interior. The universe was and is all that exists and has ever existed.

Looked at this way, it is a symptom of geocentric bias to say that the galaxies are receding from us. Rather, the universe itself is expanding, and the galaxies (ourselves included) are moving along with that expansion.

How Was the Universe Made?

We believe that radiation dominated matter in the early universe. While photons still outnumber hydrogen atoms by about a billion to one in the universe, matter now contains far more energy than radiation. A simple calculation, using Einstein's celebrated equation showing the equivalence of energy and mass, $E = mc^2$, shows that matter, not radiation, is dominant in the current universe. The energy contained in all the mass in the universe is greater than the energy contained in all the radiation.

As the universe expands, both radiation and matter become less concentrated; they travel with the expansion. The energy of the radiation is diminished more rapidly than the density of matter, however, because the photons are redshifted, becoming less energetic as their wavelength lengthens. Thus, over time, matter has come to dominate the universe.

Astro Byte

At the moment of the Big Bang, the entire universe came into being. The elements hydrogen (73 percent) and helium (27 percent) were created in the primordial fireball, along with small amounts of lithium.

How Were Atoms Made?

When we looked at fusion in the core of stars (see Chapters 16, "Our Star," and 19, "Black Holes: One-Way Tickets to Eternity"), we noted that hydrogen served as the nuclear fuel in a process that produced helium. Yet there is so much helium in the universe—helium accounts for more than 23 percent of the mass of the universe—that stellar fusion cannot have produced it all. It turns out that most of the helium in the universe was created in the moments following the Big Bang—when the universe had cooled sufficiently for nuclei to hold together.

Hydrogen fusion in the early universe proceeded much as it does in the cores of stars. A proton and a neutron come together to form deuterium (an isotope of hydrogen),

Star Words

Primordial synthesis refers to the fusion reactions that occurred in the early universe at temperatures of about 10^{10} K. These fusion reactions produced mostly helium and small amounts of beryllium and lithium.

Astro Byte

When the entire universe was the temperature of a cool stellar photosphere (about 5,000 K), the electrons were able to join together with nuclei, forming hydrogen atoms and causing the universe to become transparent to its own radiation.

and two deuterium atoms combine to form a helium nucleus. Careful calculations show that between 100 and 120 seconds after the Big Bang, deuterium would be torn asunder by gamma rays as soon as it was formed. However, after about two minutes, the universe had sufficiently cooled to allow the deuterium to remain intact long enough to be converted into helium. As the universe continued to cool, temperatures fell below the critical temperature required for fusion, and *primordial synthesis* ended.

Only after about 300,000 years had passed would the universe expand and cool sufficiently to allow electrons and nuclei to combine into atoms of hydrogen and helium. It is at this point, with the temperature of the universe at some 5,000 K, that photons could first move freely through the universe. The cosmic microwave background we now measure consists of photons that became free to move into the universe at this early time.

The rest of the periodic table—the other elements of the universe—beyond beryllium would be filled out by fusion reactions in the cores of stars and supernova explosions.

Stretching the Waves

There is another interesting way to think about the observed redshift in galaxy spectra. The photons emitted by a receding galaxy are like elastic bands in the fabric of the universe, and expand with it. In this sense, the redshift we measure from distant galaxies is nothing less than a *direct* measurement of the expansion of the universe.

We have seen where the universe and all that is within it have come from. Now we turn finally to where it is all going, where it will end. Will the universe expand forever? Or will it turn back in on itself and end in a final collapse? As it turns out, it all depends on gravity.

Depending on how much mass there is in the universe, it will either expand forever or collapse back in on itself in the opposite of a Big Bang: a Big Crunch. Recent observations of high-redshift supernovae and the cosmic microwave background suggest that a Big Crunch will never happen. The universe appears doomed to expand forever.

The Least You Need to Know

➤ The redshift of galaxies and the presence of a cosmic microwave background are two clues that the universe had a beginning.

➤ The cosmological principle holds that the universe is homogeneous (uniform in structure on large scales) and isotropic (it looks the same in all directions). This principle allows us to generalize from local observations.

➤ The leading theory of the origin of the universe is the Big Bang, which holds that the universe began as an explosion that filled all space.

➤ For the first 300,000 years after the Big Bang, the universe was opaque due to its own radiation. Once the universe was cool enough for electrons and nuclei to combine, it became transparent.

➤ All matter in the universe was created at the moment of the Big Bang. Since that time, stellar fusion and supernovae have converted the hydrogen, helium, and lithium into all of the elements of the periodic table.

(How) Will It End?

In Woody Allen's 1977 movie *Annie Hall,* there is a flashback to the Brooklyn childhood of the main character, a stand-up comic named Alvy Singer. The family doctor has been summoned because young Alvy, perhaps 10 years old, is desperately depressed. He has just learned that the universe is expanding. He believes it will eventually fly apart. The physician (puffing on a cigarette) offers the comforting thought that universal calamity is many billions of years in the future. Alvy's mother is more strident: "We live in *Brooklyn.* Brooklyn is *not* expanding!"

In this chapter we strive to be more precise than either Alvy's doctor or his mother. We look toward the end of the universe—if there will be an end. Will it end with a bang or a whimper?

What the Redshift Means

Not that Alvy's mother was wrong. The truth is that Brooklyn *isn't* expanding. But the universe that revolves around it (we're taking a Brooklynite's perspective here) is expanding. Much to the chagrin of New Yorkers, Brooklyn is not somehow exempt from the laws of the universe. This borough—like the earth, like a star, like a galaxy, or like a human being—is held together by its own internal forces. The expansion of the universe that is the echo of the Big Bang is only visible on the largest scales, the distances between clusters and superclusters of galaxies.

Let's return to our picture of the universe as the surface of a rubber balloon. As the balloon is inflated, the distance between any two dots drawn on its surface is increasing. Another way to think of it is that the photons that leave another galaxy are part of the fabric of the universe, and as this fabric is stretched, the photon gets stretched along with it—to a longer wavelength. This type of wavelength increase is called a *cosmological redshift,* and it is a direct measure of universal expansion.

Star Words

Cosmological redshift is the lengthening of the wavelengths of electromagnetic radiation caused by the expansion of the universe.

Limited Options

Just because the universe is expanding now doesn't necessarily mean that it will go on expanding forever. The expansion started with the Big Bang is opposed by a relentless force that we have met many times before in this book—*gravity.* The universe has a couple of options, and it all depends on how much stuff (mass) there is in the universe. Perhaps young Alvy would have been even more worried if his mother had told him about the other possibilities.

There are basically three options: The universe will either expand forever (be unbound), or it will continually slow in its expansion but never stop (be marginally bound), or it will reach a certain size and then begin to contract (be bound).

Star Words

Critical density is the density of matter in the universe that represents the division between a universe that expands infinitely (unbound) and one that will ultimately collapse (bound). The density of the universe determines whether it will get colder forever, or end with a conflagration as dramatic as the Big Bang.

A Matter of Density

The reason for these limited options comes down to the density of the universe. If the universe contains a sufficient amount of matter—is sufficiently dense—then the force of gravity will be such that it will eventually succeed in halting expansion. The universe will gradually expand at a slower and slower rate, and then reverse itself and begin to collapse. If the universe (even counting all of the material that we can't see—the dark matter) does not contain sufficient mass—is too gruelly and thin—the expansion of the universe will continue forever.

Remember that in the Big Bang theory, there is no such thing as new or old matter. All the matter there is was created as a result of the Big Bang; therefore, the density of the universe—and its fate—was determined at the moment the Big Bang occurred.

There is a certain density of matter in the universe that is the sword edge between ultimate collapse and eternal expansion. This *critical density* has been calculated. If the universe is more dense than this, it will eventually collapse. If it is less dense than this, it will expand forever, and if it is exactly this dense, it will keep expanding, more and more slowly, forever.

Close Encounter

Some recent discoveries have a bearing on the question of the eventual fate of the universe. Two groups of astronomers using the *Hubble Space Telescope* (one at the Harvard-Smithsonian Center for Astrophysics, and the other at Berkeley Labs) have found evidence that the universe is likely to expand forever, that the universe does not have enough mass in it to cause the expansion to halt.

They have used what are called Type Ia supernovae (cataclysmic explosions that result when a white dwarf star borrows a bit too much material from its binary companion) as standard candles. Type Ia supernovae are so incredibly luminous that they allow us to see farther than just about any other phenomenon we know of. Using these supernovae, the astronomers have determined that the expansion of the universe (as described by Hubble's Law) doesn't seem to be slowing down at all. In fact, it may be accelerating.

A Surprising Boomerang

The BOOMERANG (Balloon Observations Of Millimetric Extragalactic Radiation And Geophysics) mission flew for 10 days in late 1998 and early 1999. The balloon-borne craft flew around Antartica, collecting high-resolution data at microwave frequencies sensitive to the part of the spectrum where the universe emits photons that are a remnant echo of the Big Bang. The results were nothing short of breathtaking. While the craft mapped only a small part of the sky, it did cover that small portion with much higher resolution than was available to the COBE mission. As a result, BOOMERANG detected the scale of the fluctuations in the cosmic microwave background. In a flat universe (one that has critical density) the scale of these fluctuations is expected to be about 1 degree. Fluctuations smaller or larger than this mean that the universe has a curvature.

Well, which is it?

The scale size of the detected fluctuations is about 1 degree. Now, a flat universe requires that the universe be at critical density, and we have seen that, even including dark matter, we are far from closing the universe with matter. All told, luminous and dark matter only account for about 30 percent of the required critical density. The remaining 70 percent may be contained in the energy of the vacuum. And if this other 70 percent is there, it might also explain our accelerating universe!

Close Encounter

We defined neutrinos in Chapter 18, "Stellar Careers," as subatomic particles without electric charge and with virtually no mass. As a result, these are sort of stealth particles, hard *little neutral ones* to detect. On June 5, 1998, U.S. and Japanese researchers in Takayama, Japan, announ-ced that, using an enormous detection device, they found evidence of mass in the neutrino. The device, called the Super-Kamiokande, is a tank filled with 12.5 million gallons of pure water and buried deep in an old zinc mine.

Because neutrinos lack an electric charge and rarely interact with other atoms, passing unnoticed through virtually any kind or thickness of matter, they have proven especially elusive. They are the greased pigs of the subatomic world, so plentiful that 100 billion of them (that's as many stars as there are in the Galaxy) pass through your body every second. The Super-Kamiokande contains so much water, however, that occasionally a neutrino collides with another particle and produces an instantaneous flash of light, which is recorded by a vast array of light-amplifying detectors.

Just how much mass does a neutrino have? This is not yet known, but it does not appear that neutrinos can contribute enough mass to the universe to affect critical density.

Run Away! Run Away!

To predict whether the universe will expand infinitely, it is necessary to calculate critical density and to see whether the density of our universe is above or below this critical figure. By measuring the average mass of galaxies within a known volume of space, astronomers derive a density of luminous matter well short of critical density, about 10^{-28}kg/m^3. That figure is about 1 percent of what is required. After factoring in the added contribution of the dark matter that has an apparent gravitational effect in galaxy clusters (see Chapter 21, "The Milky Way: Much More Than a Candy Bar"), we can account for up to 30 percent of the critical density.

It is possible, though, that the other 70 percent of the density of the universe is not in mass, but in energy. Remember that energy and mass can be interchanged through the process described by $E = mc^2$. Einstein's "blunder" (see Chapter 25, "What About the Big Bang?") may point the way. If the expansion of the universe is accelerating, and the universe has "critical density," these might both be explained by energy contained in the vacuum of space.

Why this obsession with critical density? Well, one argument goes that it is unlikely that the universe would have a density so *close* to being critical without actually being critical.

Astronomer's Notebook

How close are we to critical density? Well, not very. All of the stars and galaxies that we see account for at most 3 to 4 percent of the critical density. If we include the mass apparent in clusters of galaxies, we can account for about 20 to 30 percent of the critical density. Recent experiments on several fronts suggest that the other 70 percent of critical density might be contained not in mass, but in energy.

What Does It All Mean?

We see different possible fates for the universe. At present, the evidence is starting to point toward an open universe, one which will expand forever, growing ever colder and more distant with time. Eventually, all the stars in the universe, even the low-mass ones will have blinked off, and the universe will be a dark place indeed. But before we throw in the towel, let's consider what some of the possible fates of the universe might mean for us in the universe now.

What's the Point?

Remember your high school geometry class? You were told that a point is a fiction, represented by the mark of a very sharp pencil, perhaps, but actually without dimension: no length, no height, no width.

And we are to believe this *point* once contained the entire universe, and might again? But let's be careful with the analogy. The point on a page is not the universe *within* something else, some void.

The point we're speaking of is not a point in space, for it contains the entire universe. Everything that ever is or was in the universe is there.

The Universe: Closed, Open, or Flat?

So then the universe expanded from this point in the Big Bang. What was the result? What is the architecture of the universe?

To begin to visualize the shape of the universe, we have to stop thinking like Newton and start thinking like Einstein. Remember that Einstein thought of gravitation not as a force that objects exert upon one another, but as the result of the distortion in space that mass causes. Einstein explained that the presence of mass warps the space in its vicinity. The more mass (the greater the density of matter), the greater the warp (the more space curves).

Star Words

A **closed universe** is finite and without boundaries. A universe with density above the critical value is necessarily closed. An **open universe** will expand forever, because its density is insufficient to halt the expansion.

If the density of the universe is greater than the level of critical density, then the universe will warp (or curve) back on itself, closed off and finite. If the universe is closed, then our balloon analogy has been particularly accurate. In a *closed universe,* we think of the entire universe as represented by the surface of a sphere, finite in extent, but with no boundary or edge. If you shot out parallel beams of light in a closed universe, they would eventually cross paths.

If the density of the universe is below the critical level, an *open universe* will result. Whereas the spherical or closed universe is said to be positively curved, the open universe is negatively curved.

Saddle Up the Horses: Into the Wide-Open Universe

It is difficult to visualize this open universe of infinite extent. The closest two-dimensional approximation is the surface of a saddle, curved up in one direction and down in the other—and extending out to infinity. In an open universe, beams of light that were initially parallel would diverge. Our current estimate of the density of the universe and the recent results from distant supernovae suggest that this is the type of universe we inhabit.

Star Words

A **flat universe** results if density is precisely at the critical level. It is flat in the sense that its space is defined by the rules of ordinary Euclidean geometry—parallel lines never cross.

With precisely critical density, by the way, it becomes a *flat universe,* albeit still infinite in extent. In what we call flat space, parallel beams of light would remain parallel forever. For this reason, flat space is sometimes called Euclidean space.

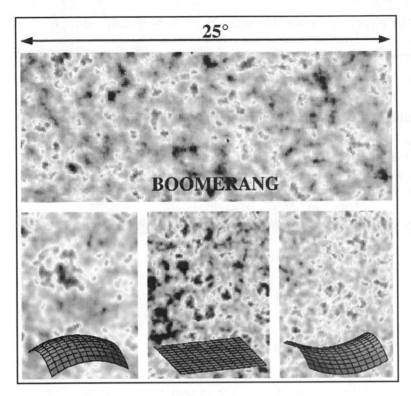

25°

BOOMERANG

*Images from the BOOM-
ERANG mission. The scale
size of the temperature
fluctuations in the microwave background can tell
us about the geometry of
the universe. A flat uni-
verse should have fluctuations with scale sizes of
about 1 degree. In a closed
universe the fluctuations
would be larger, and in an
open universe, they would
be smaller. Along with
other evidence, the
BOOMERANG data
strongly suggests that we
live in a universe with a
flat geometry.*

*(Image from NASA and
University of California,
Santa Barbara)*

We Have a Problem

Scientists are to their theories as overbearing parents are to their children. They have
for them only the very highest expectations. Contradictions are not allowed and
must be resolved. Even something less than an outright contradiction is intolerable. A
good theory should account for everything that is observed and make testable predictions. Any observations in contradiction to the theory are taken seriously and the
theory amended, if necessary.

In this regard, while it has been very successful, the Big Bang theory seems to fall short
in two areas: explaining the incredible sameness of the universe on its largest scales and
accounting for why the universe has a density that is apparently so close to critical.

First there is the so-called horizon problem, which deals with the incredible uniformity of the cosmic microwave background. No matter where we look, its intensity is
the same, even in parts of the universe that are far too distant to be in contact with
one another. Regions in the universe that never could have exchanged information
(because of the limiting speed of light) seem to know about each other.

The only current explanation we have is that the universe looks uniform now because
it always was uniform. Which is sort of like saying, it is so because it is so.

It is not that this uniformity contradicts anything in the Big Bang theory, it's just that
there is nothing in the theory that accounts for it. The theory provides no particular

reason why two widely separated regions should be the same—especially since regions very distant from one another would have never interacted.

The second issue is called the flatness problem. Its name comes from the fact that the critical density of the universe, once the mass of dark matter is figured into things, approaches 1, which means that the universe is almost flat.

So what's the problem?

Again, as with the horizon problem, the difficulty is not that flatness contradicts the Big Bang theory (it does not), but that the theory doesn't explain *why* the universe should have formed so close to critical density. As far as the theory goes, the universe might have been significantly more dense than 1 or significantly less. Scientists won't accept the luck of the draw. They want an explanation.

Close Encounter

There are four known forces in the universe: gravity, electromagnetism, the weak nuclear force, and the strong nuclear force. Gravity and the electromagnetic force act over large distances—the size of the universe—and govern the motions of planets and galaxies, molecules and atoms. The nuclear forces act over only small distances (within an atomic nucleus) and hold nuclei together. The nuclear forces are the strongest, but act over the shortest distances. Gravity is by far the weakest, but its long reach means that its pull will govern the fate of the universe.

In the early universe, it is believed that these forces started as a single force. As the universe expanded, each force established its own unique identity.

Down to Earth

Successful people in all fields—from astronomy to retail sales—learn to see problems not as obstacles, but as opportunities. The biggest, most basic issues astronomy deals with seem the farthest removed from our everyday experience; but the possible resolution of the horizon and flatness problems may help us to connect more intuitively with this strange thing called the universe.

Blow It Up

Within the first infinitesimal fractions of the first second following the Big Bang, astronomers have theorized, the three forces other than gravity (electromagnetic, weak nuclear forces, and strong nuclear forces) in the universe were united as a single force.

Between 10^{-35} seconds and 10^{-32} seconds after the Big Bang, gravity had already split out as a separate force, but the other forces were still one. For an unimaginably brief instant, gravity pushed the universe apart instead of pulling it together. Theorists call this moment *inflation,* for obvious reasons. Within this *inflationary epoch* (not to be confused with the late 1970s), the universe expanded 100 trillion trillion trillion trillion (10^{50}) times.

Of course, 10^{-35} to 10^{-32} seconds seems to us instantaneous; but there was time—however brief—before this, between the Big Bang and the onset of the epoch of inflation. During this period, all parts of the universe were in communication with one another. They had ample time to establish uniform physical properties *before* the epoch of inflation pushed them to opposite sides of the universe.

Star Words

The **inflationary epoch** refers to a time soon after the Big Bang when the universe was puffed up suddenly, increasing its scale by a factor of 10^{50} in an instant. This inflation could account for the incredible sameness, or uniformity, of the universe, even in regions that (without inflation) could never have been in contact with one another.

The early universe expanded faster than its constituent regions could communicate with each other. The universe, in effect, outran information, so that the most extreme regions have been out of communication with one another since 10^{-32} seconds after the Big Bang. Yet they share the properties they had at the very instant of creation. They continue to share these properties today. Thus, with the addition of inflation, the Big Bang can account for the horizon problem, and we've tied up one loose end of the theory.

Looks Flat to Me

That leaves us with the flatness problem. Why (in terms of the Big Bang theory) should the universe be at or near critical density?

If a mass of external evidence didn't exist, it would be almost impossible to convince anyone that the earth is round. After all, it *looks* flat.

Once we accept that the earth is curved, however, we understand that it looks flat because the radius of the globe is very large and, therefore, the arc of the curve it describes is extremely gradual. Take a small portion of any arc, and it will look, for all practical purposes, flat. If the earth were very small—or we were very large, or very far away—the curvature of the earth's surface would be apparent on a routine basis.

The universe may have been very curved at the instant of its creation. But in expanding 10^{50} times during the epoch of inflation, it became—*on the scale of the observable universe* (and even beyond this)—flat.

A flat universe is consistent with a universe whose density is exactly critical. If the universe were more dense, it would be positively curved; if less dense, negatively curved. If the universe is truly flat, its density must be exactly critical.

Coming Full Circle

We've come a long way since we considered our astronomical ancestors, looking up at the sky and wondering. We have come to the end of our story, a story that—as is obvious from the daily paper—is still being written. We live in an incredible age, when the discovery of accretion disks around black holes and distant Type Ia supernovae at the edge of the visible universe are reported on the evening news, along with doings of movie stars and musicians.

As we've seen in this chapter, it is possible to contemplate the universe on scales much larger than may be directly relevant to us—scales on which its curvature (positive or negative) becomes significant, and our comfortable Euclidean-Newtonian reality gets deliciously bent out of shape. And your imagination needs no telescope, nor even the night sky, to take this trip.

As exciting as these findings are, and as stunning as the NASA and other images are that you will find on the Internet, remember to go outside as often as you can and look at the night sky. With or without a telescope, the planets, stars, and galaxies that you see are indeed *out* there, real, and immeasurable in their beauty. Open, closed, or flat, the universe isn't going away any time soon.

The Least You Need to Know

➤ The future of the universe is limited to two basic possibilities: Either it will expand forever or eventually collapse.

➤ The density of matter in the universe determines whether gravity will be able to halt the expansion of the universe that started with the Big Bang.

➤ Depending on the observed density of the universe, we say that it is open, closed, or flat. An open universe will expand forever, a closed universe will collapse on itself, and a flat universe will expand at a slower and slower rate.

➤ Inflation, a sudden expansion in the scale of the universe that occurred early on, might explain why the universe looks flat and homogeneous.

➤ Recent observations suggest strongly that we live in a flat universe, and that there is not enough mass in the universe to pull it all back together in the endgame counterpart of the Big Bang—the Big Crunch.

Star Words Glossary

absolute magnitude   *See* **luminosity.**

accretion   The gradual accumulation of mass; usually refers to the build-up of larger masses from smaller ones through mutual gravitational attraction.

active galaxy   A galaxy that has a more luminous nucleus than most galaxies.

altazimuth coordinates   Altitude (angular distance above the horizon) and azimuth (compass direction expressed in angular measure from due north).

altitude   *See* **altazimuth coordinates.**

angular momentum   The rotating version of linear momentum. Depends on the mass distribution and rotational velocity of an object.

angstrom   Abbreviated A (or Å). A small size, equal to one one ten billionth of a meter, or 10^{-10} meter.

angular separation   Distance between objects on the sky expressed as an angle (such as degrees) rather than in distance units (such as feet or meters).

angular size   Size expressed as an angle (such as degrees) rather than in distance units (such as feet or meters).

annihilate   Used as an intransitive verb by astronomers. Particles and antiparticles annihilate when they meet, converting their mass into energetic photons. Electrons and anti-electrons (positrons) were continually created and annihilated in the early universe.

aperture   The diameter of the objective lens (that is, the main lens) or primary mirror of a telescope and the main lenses of binoculars.

apollo asteroids   Asteroids with sufficiently eccentric orbits to cross paths with the Earth (and other terrestrial planets).

apparent magnitude   A value that depends on the distance to an object. *See* **luminosity.**

arcminute   One sixtieth of an angular degree.

arcsecond   One sixtieth of an arcminute ($\frac{1}{3600}$ of a degree).

assumption of mediocrity   A scientific assumption that we on Earth are not so special. The conditions that have allowed life to arise and evolve on the Earth likely exist in many other places in the Galaxy and universe. Related to the more lofty **cosmological principle.**

asterism   An arbitrary grouping of stars within or associated with a constellation, perceived to have a recognizable shape (such as a Teapot or Orion's Belt) and, therefore, readily serve as celestial landmarks.

asteroid   One of thousands of small, rocky members of the solar system that orbit the Sun. The largest asteroids are sometimes called minor planets.

astronomical unit (A.U.)   A conventional unit of measurement equivalent to the average distance from the earth to the sun (149,603,500 kilometers, or 92,754,170 miles).

autumnal equinox   On or about September 21; on this date, the beginning of fall, day and night are of equal length because the Sun's apparent course against the background stars (the ecliptic) intersects the celestial equator.

azimuth   *See* **altazimuth coordinates.**

barred-spiral galaxy   A spiral galaxy that has a linear feature, or bar of stars, running through the galaxy's center. The bar lies in the plane of the spiral galaxy's disk, and the spiral arms typically start at the end of the bar.

Big Bang   The primordial explosion of a highly compact universe; the origin of the expansion of the universe.

binaries   Also called binary stars. Two-star systems in which the stars orbit a common center of mass. The way the companion stars move can tell astronomers much about the individual stars, including their masses.

black body   An idealized (theoretical) object that absorbs all radiation that falls on it and perfectly re-emits all radiation it absorbs. The spectrum (or intensity of light as a function of wavelength) that such an object emits is an idealized mathematical construct called a black-body curve, which can serve as an index to measure the temperature of a real object. Some astronomical sources (like stars) can be approximated as black bodies.

black dwarf   A burned-out star that has cooled from the white dwarf stage through yellow and red.

black hole A stellar-mass black hole is the end result of the core collapse of a high-mass (greater than 10 solar mass) star. It is an object from which no light can escape within a certain distance. Although space behaves strangely very close to a black hole, at astronomical distances, the black hole's only effect is gravitational.

Bode's Law Also called the Titius-Bode Law, Bode's Law is a numerical trick that gives the approximate interval between some of the planetary orbits in our solar system.

brightness The measured intensity of radiation from an object. The brightness of astronomical objects falls off with the square of the distance.

brown dwarf A failed star; that is, a star in which the forces of heat and gravity reached equilibrium before the core temperature rose sufficiently to trigger nuclear fusion.

calderas Craters produced not by meteoroid impact but by volcanic activity. *See also* **corona**.

cardinal points The directions of due north, south, east, and west.

Cassini division A dark gap between rings A and B of Saturn. It is named for its discoverer, Gian Domenico Cassini (1625–1712), namesake of the mission that will arrive at Saturn in 2004.

cataclysmic variable *See* **variable star.**

celestial equator An imaginary great circle dividing the Northern and Southern Hemispheres of the celestial sphere.

celestial sphere An imaginary sphere surrounding the Earth into which the stars are imagined to be fixed. For hundreds of years, people believed such a sphere (or bowl) really existed. Today, however, astronomers use the concept as a convenient way to describe the position of stars relative to one another.

chain reaction *See* **nuclear fission.**

circumpolar stars Stars near the celestial North Pole; from many locations on Earth, these stars never set.

closed universe A universe that is finite and without boundaries. A universe with density above the critical value is necessarily closed (*see* **critical density**). An open universe, in contrast, will expand forever because its density is insufficient to halt the expansion.

comet Also thought of as a "dirty snowball." Small celestial bodies composed mainly of ice and dust, comets complete highly eccentric orbits around the sun. As a comet approaches the Sun, some of its material is vaporized and ionized to create a gaseous head (coma) and two long tails (one made of dust, and one made of ions).

conjunction The apparent coming together of two celestial objects in the sky.

constellations   Arbitrary formations of stars that are perceived as figures or designs. There are 88 official constellations in the Northern and Southern Hemispheres. (*See* Appendix C)

convective motion   A gas-flow pattern created by the rising movement of warm gases (or liquids) and the sinking movement of cooler gases (or liquids).

core   The innermost region of a planet or star.

core-collapse supernova   The extraordinarily energetic explosion that results when the core of a high-mass star collapses under its own gravity.

core-halo galaxy   *See* **radio galaxy.**

core hydrogen burning   The principal nuclear fusion reaction process of a star. The hydrogen at the star's core is fused into helium, and the small amount of mass lost is used to produce enormous amounts of energy.

corona   In astronomy, a corona may be a luminous ring appearing to surround a celestial body, the luminous envelope of ionized gas outside the Sun's chromosphere, or a large upswelling in the mantle of the surface of a planet or the Moon that takes the form of concentric fissures and that is an effect of volcanic activity. *See also* **calderas.**

cosmic microwave background   The highly redshifted photons left behind by the Big Bang and detectable today throughout all space as radio-wavelength radiation, indicating a blackbody temperature for the universe of 2.73 K.

cosmological principle   A cornerstone assumption about the nature of the universe. It holds that the universe exhibits two key properties: homogeneity (sameness of structure on the largest scale) and isotropy (looking the same in all directions).

cosmological redshift   The lengthening of the wavelengths of electromagnetic radiation caused by the expansion of the universe.

cosmology   The study of the origin, structure, and evolution of the universe.

crater   The Latin word for "bowl," it refers to the shape of depressions in the Moon or other celestial objects created (mostly) by meteoroid impacts.

critical density   The density of matter in the universe that represents the division between a universe that expands infinitely (unbound) and one that will ultimately collapse (bound). The density of the universe determines whether it will expand forever or end with a conflagration as dramatic as the Big Bang.

crust   The surface layer of a planet.

dark halo   The region surrounding the Milky Way and other galaxies that contains **dark matter.** The shape of the dark halo can be probed by examining in detail the effects its mass has on the rotation of a galaxy.

dark matter   A catch-all phrase used to describe an apparently abundant substance in the universe of unknown composition. Dark matter is 100 times more abundant than luminous matter on the largest scales.

differential rotation   A property of anything that rotates and is not rigid. A spinning CD is a rigid rotator. A spinning piece of gelatin is not as rigid, and a spinning cloud of gas is even less so. For example, the atmospheres of the outer planets and of the Sun have equatorial regions that rotate at a different rate from the polar regions.

Drake Equation   Proposes a number of terms that help us make a rough estimate of the number of civilizations in our Galaxy, the Milky Way.

dust lanes   Dark areas sometimes visible within emission nebulae and galaxies; the term most frequently refers to interstellar absorption apparent in edge-on spiral galaxies.

eccentric   An ellipse (or elliptical orbit) is called eccentric when it is noncircular. An ellipse with an eccentricity of 0 is a circle, and an ellipse with an eccentricity of close to 1 would be very oblong.

eclipse   An astronomical event in which one body passes in front of another so that the light from the occulted (shadowed) body is blocked. When the Sun, Moon, and the Earth align, the Moon blocks the light of the Sun (solar eclipse).

eclipsing binaries   *See* **visual binaries.**

ecliptic   The ecliptic traces the apparent path of the Sun against the background stars of the celestial sphere. This great circle is inclined at 23½ degrees relative to the celestial equator, which is the projection of the Earth's equator onto the celestial sphere.

ejecta blanket   The debris displaced by a meteoroid impact on a planetary surface. Such debris are apparent on the lunar surface.

electromagnetic radiation   Energy in the form of rapidly fluctuating electric and magnetic fields and including visible light in addition to radio, infrared, ultraviolet, x-ray, and gamma-ray radiation. EM radiation often arises from moving charges in atoms and molecules, though high-energy radiation can arise in other processes.

electromagnetic spectrum   The complete range of electromagnetic radiation, from radio waves to gamma waves and everything in between.

ellipse   A flattened circle drawn around two foci instead of a single center point.

elliptical galaxy   A galaxy with no discernible disk or bulge and that looks like an oval or circle of stars on the sky. The true shapes of ellipticals vary from elongated ("footballs") to spherical ("baseballs") to flattened ("hamburger buns"). Elliptical galaxies consist of old stars and appear to have little or no gas in them.

emission lines   Narrow regions of the spectrum where a particular substance is observed to emit its energy. These lines result from basic processes occurring on the smallest scales in an atom (like electrons moving between energy levels).

emission nebulae   Glowing clouds of hot, ionized interstellar gas located near a young, massive star. (Singular is emission nebula.)

ephemerides   Special almanacs that give the daily positions of various celestial objects for periods of several years.

escape velocity   The velocity necessary for an object to escape the gravitational pull of another object.

event horizon   Coincides with the Schwarzschild radius and is an imaginary boundary surrounding a collapsing star or black hole. Within the event horizon, no information of the events occurring there can be communicated to the outside, as at this distance not even the speed of light provides sufficient velocity to escape.

fireball   A substantial meteoroid that can create a spectacular display when it burns up in the atmosphere.

flat universe   The universe that results if its density is precisely at the critical level. It is flat in the sense that its space is defined by the rules of ordinary Euclidean geometry—parallel lines never cross.

focal length   The distance from a mirror surface to the point where parallel rays of light are focused.

focus   The point at which a mirror concentrates parallel rays of light that strike its surface.

frequency   The number of wave crests that pass a given point per unit of time. By convention, this is measured in hertz (equivalent to one crest-to-crest cycle per second and named in honor of the nineteenth-century German physicist Heinrich Rudolf Hertz).

galactic bulge   Also called nuclear bulge. A swelling at the center of spiral galaxies. Bulges consist of old stars and extend out a few thousand light-years from the galactic centers.

galactic disk   The thinnest part of a spiral galaxy. The disk surrounds the nuclear bulge and contains a mixture of old and young stars, gas, and dust. In the case of the Milky Way, it extends out some 50,000 light-years from the Galactic center but is only about 1,000 light-years thick. It is the dust in the disk (a few hundred light-years thick) that creates the dark ribbon that runs the length of the Milky Way and limits the view of our own Galaxy (*see* **dust lanes**).

galactic halo   A large (50,000 light-year radius) sphere of old stars surrounding the Galaxy.

galactic nucleus   The core of a spiral galaxy. In the case of Seyfert (active) galaxies, the nucleus is extremely luminous in the radio wavelengths.

galaxy cluster   A gravitationally bound group of galaxies.

geocentric   Earth-centered; the geocentric model of the universe (or solar system) is one in which Earth is believed to be at the center of the universe (or solar system).

giant molecular clouds Huge collections of cold (10 K to 100 K) gas that contain many millions of solar masses of molecular hydrogen. These clouds also contain other molecules (like carbon dioxide) that can be imaged with radio telescopes. The cores of these clouds are often the sites of the most recent star formation.

gibbous A word from Middle English that means "bulging"—an apt description of the Moon's shape between its first and third quarter phases.

globular clusters Collections of a few hundred thousand stars, held together by their mutual gravitational attraction, that are found in highly eccentric orbits above and below the galactic disk.

gnomon Any object designed to project a shadow used as an indicator. The upright part of a sundial is a gnomon.

go-to computer controller A handheld "paddle" that stores location data on celestial objects. Select an object in the database or punch in right ascension and declination coordinates, and the controller will guide motors to point the telescope (if properly aligned) at the desired object or coordinates.

H-R diagram Short for Hertzsprung-Russell diagram. A graphical plot of luminosity versus temperature for a group of stars. Used to determine the age of clusters of stars.

heliocentric Sun-centered; describes our solar system, in which the planets and other bodies orbit the Sun.

helium flash A stellar explosion produced by the rapid increase in the temperature of a red giant's core driven by the fusion of helium.

homogeneity *See* **cosmological principle.**

Hubble's Law The linear relationship between the velocity of a galaxy's recession to the galaxy's distance from us. Simply stated, the law says that the recessional velocity is directly proportional to the distance. Used to determine the age of the universe.

inflationary epoch A time soon after the Big Bang when the universe was puffed up suddenly, increasing in size by a factor of 10^{50} in an instant. This inflation could account for the incredible sameness, or uniformity, of the universe, even in regions that (without inflation) could never have been in contact with one another.

interferometer A combination of telescopes linked together to create the equivalent (in terms of resolution) of a giant telescope. This method (while computing-intensive) greatly increases resolving power.

interstellar matter The material found between stars. Refers to the gas and dust thinly distributed throughout space, the matter from which the stars are formed. About 5 percent of our Galaxy's mass is contained in its gas and dust. The remaining 95 percent is in stars.

interstellar medium *See* **interstellar matter.**

intrinsic variable *See* **variable star.**

irregular galaxy   A galaxy type lacking obvious structure but containing lots of raw materials for the creation of new stars, and often, many young, hot stars.

isotropy   *See* **cosmological principle.**

jovian planets   The gaseous planets in our solar system farthest from the Sun (with the exception of Pluto, which is classed as neither terrestrial nor jovian): Jupiter, Saturn, Uranus, and Neptune.

Kelvin scale   The Kelvin (K) temperature scale is tied to the Celsius (C) temperature scale and is useful because there are no negative Kelvin temperatures. 0 C is the temperature at which water at atmospheric pressure freezes. 100 C is the temperature at which water boils. Absolute zero (0 K) is the coldest temperature that matter can attain. At this temperature, the atoms in matter would stop jiggling around all together. 0 K corresponds to approximately –273 C.

kiva   A Native American underground ceremonial chamber, partially open at the top. Associated with various rituals, many are clearly oriented to the sun.

leading face and **trailing face**   Moons that are tidally locked to their parent planet have a leading face and a trailing face; the leading face always looks in the direction of the orbit, the trailing face away from it.

libration   The slow oscillation of the Moon (or other satellite, natural or artificial) as it orbits a larger celestial body. Lunar libration gives us glimpses of a very small portion of the far side of our Moon.

light pollution   The effect of poorly planned lighting fixtures that allow light to be directed upward into the sky. Light pollution washes out the contrast between the night sky and stars.

light-year   The distance light travels in one year: approximately 5.88 trillion miles (9.46 trillion km). For interstellar measurements, astronomers use the light-year as a basic unit of distance.

(The) Local Group   A galaxy cluster, a gravitationally bound group of galaxies that includes the Milky Way, Andromeda, and other smaller galaxies.

luminosity   The total energy radiated by a star each second. Luminosity is a quality intrinsic to the star; magnitude may or may not be. Absolute magnitude is another name for luminosity, but apparent magnitude is the amount of energy emitted by a star and striking some surface or detection device (including our eyes). Apparent magnitude varies with distance.

magnetosphere   A zone of electrically charged particles trapped by a planet's magnetic field. The magnetosphere lies far above the planet's atmosphere.

magnitude   A system for classifying stars according to apparent brightness (*see* **luminosity**). The human eye can detect stars with magnitudes from 1 (the brightest) to 6 (the faintest). A 1st magnitude star is 100× brighter than a 6th magnitude star.

main sequence   When the temperature and luminosity of a large number of stars are plotted, the points tend to fall mostly in a diagonal region across the plot. The main sequence is this well-defined region of the Hertzsprung-Russell diagram (H-R diagram) in which stars spend most of their lifetime. The Sun will be a main sequence star for about 10 billion years.

mantle   The layer of a planet beneath its crust and surrounding its core.

maria (pronounced *MAH-ree-uh*)   The plural of "mare" (pronounced *MAR-ay*, with a long *a* at the end), this word is the Latin for "sea." Maria are dark-grayish plains on the lunar surface that resembled bodies of water to early observers.

meteor   The term for a bright streak across the night sky—a "shooting star." Meteoroid is the object itself, a rocky object that is typically a tiny fragment lost from a comet or an asteroid. A micrometeoroid is a very small meteoroid. The few meteoroids that are not consumed in the Earth's atmosphere reach the ground as meteorites.

meteor shower   When the Earth's orbit intersects the debris that litters the path of a comet, we see a meteor shower. These happen at regular times during the year.

meteorite   *See* **meteor.**

meteoroid   *See* **meteor.**

micrometeoroid   *See* **meteor.**

millisecond pulsar   A neutron star rotating at some 1,000 revolutions per second and emitting energy in extremely rapid pulses.

minor planet   *See* **asteroid.**

nebula   A term with several applications in astronomy but used most generally to describe any fuzzy patch seen in the sky. Nebulae are often (though not always) vast clouds of dust and gas.

neutron star   The superdense, compact remnant of a massive star, one possible survivor of a supernova explosion. It is supported by degenerate neutron pressure, not fusion. It is an entire star with the density of an atomic nucleus.

nova   A star that suddenly and very dramatically brightens, resulting from the triggering of nuclear fusion caused by the accretion of material from a binary companion star.

nuclear fission   A nuclear reaction in which an atomic unit splits into fragments, thereby releasing energy. In a fission reactor, the split-off fragments collide with other nuclei, causing them to fragment, until a chain reaction is under way.

nuclear fusion   A nuclear reaction that produces energy by joining atomic nuclei. While the mass of a nucleus produced by joining two nuclei is less than that of the sum of the original two nuclei, the mass is not lost; rather, it is converted into large amounts of energy.

objective lens   *See* **aperture.**

open universe   A universe whose density is below the critical value (*see* **critical density**). In contrast to a closed universe, an open universe will expand forever because its density is insufficient to halt the expansion.

optical window   An atmospheric property that allows visible light to reach us from space.

orbital period   The time required for an object to complete one full orbit around another object. The orbital period of the Earth around the Sun, for example, is a fraction over 365 days.

planetary nebula   The ejected gaseous envelope of a red giant star. This shell of gas is lit up by the ultraviolet photons that escape from the hot, white dwarf star that remains. (This term is sometimes confusing as it has nothing to do with planets.)

planetesimals   Embryonic planets in an early formative stage. Planetesimals, which are probably the size of small moons, develop into protoplanets, immature but full-scale planets. It is the protoplanets that go on to develop into mature planets as they cool.

precession   The slow change in the direction of the axis of a spinning object (such as the Earth), caused by an external influence or influences (such as the gravitational fields of the Sun and the Moon).

primary mirror   *See* **aperture.**

primordial synthesis   The fusion reactions that occurred in the early universe at temperatures of about 10^{10} K. These fusion reactions produced helium and a small amount of lithium.

proper motion   Motion of a star determined by measuring the angular displacement of a target star relative to more distant background stars. Measurements are taken over long periods of time, and the result is an angular velocity (measured, for example, in arcseconds/year). If the distance to the star is known, this angular displacement can be converted into a transverse velocity in km/s (*see* **transverse component**).

protoplanet   *See* **planetesimals.**

pulsar   A rapidly rotating neutron star whose magnetic field is oriented such that it sweeps across the Earth with a regular period.

pulsating variable   *See* **variable star.**

quasar   Short for "quasi-stellar radio source." Quasars are bright, distant, tiny objects that produce the luminosity of 100 to 1,000 galaxies within a region the size of a solar system.

radar   Short for "radio detection and ranging." In astronomy, radio signals are sometimes used to measure the distance of planets and other objects in the solar system.

radial component *See* **transverse component.**

radio galaxy A member of an active galaxy subclass of elliptical galaxies. Radio galaxies are characterized by strong radio emission and, in some cases, narrow jets and wispy lobes of emission located hundreds of thousands of light-years from the nucleus.

radio jets Narrow beams of ionized material that have been ejected at relativistic velocities from a galaxy's nucleus.

radio lobe The diffuse or wispy radio emission found at the end of a radio jet. In some radio galaxies, the lobe emission dominates; in others, the jet emission dominates.

radio telescope An instrument, usually a very large dish-type antenna connected to a receiver and recording and/or imaging equipment, used to observe radio-wavelength electromagnetic radiation emitted by stars and other celestial objects.

radio window A property of our atmosphere that allows some radio waves from space to reach Earth and that allows some radio waves broadcast from Earth to penetrate the atmosphere.

radioactive decay The natural process whereby a specific atom or isotope is converted into another specific atom or isotope at a constant and known rate. By measuring the relative abundance of parent-and-daughter nuclei in a given sample of material (such as a meteorite), it is possible to determine the age of the sample.

red giant A late stage in the career of stars about as massive as the Sun. More massive stars in their giant phase are referred to as supergiants. The relatively low surface temperature of this stage produces its red color.

redshift An increase in the detected wavelength of electromagnetic radiation emitted by a celestial object as the recessional velocity between it and the observer increases. The name derives from the fact that lengthening the wavelength of visible light tends to redden the light that is observed, but is used generally.

refracting telescope Also called a refractor. A telescope that creates its image by refracting (bending) light rays with lenses.

resolving power The ability of a telescope (optical or radio telescope) to render distinct, individual images of objects that are close together.

retrograde An orbit that is backward or contrary to the orbital direction of the planets.

Schwarzschild radius The radius of an object with a given mass at which the escape velocity equals the speed of light. As a rule of thumb, the Schwarzschild radius of a black hole (in km) is approximately three times its mass in solar masses, so a 5-solar mass black hole has a Schwarzschild radius of about $5 \times 3 = 15$ km.

seeing The degradation of optical telescopic images as a result of atmospheric turbulence. "Good seeing" denotes conditions relatively free from such atmospheric interference.

Seyfert galaxy   A type of active galaxy, resembling a spiral galaxy, whose strong radio-wavelength emissions and emission lines come from a small region at its core.

sidereal day   A day measured from star rise to star rise. The sidereal day is 3.9 minutes shorter than the solar day.

sidereal month   The period of 27.3 days that it takes the moon to orbit once around the Earth. *See also* **synodic month.**

sidereal year   The time it takes the Earth to complete one circuit around the Sun with respect to the stars.

singularity   The infinitely dense remnant of a massive core collapse.

solar day   A day measured from sunup to sunup (or noon to noon, or sunset to sunset). Slightly longer than a sidereal day. *See* **sidereal day.**

solar flares   Explosive events that occur in or near an active region on the Sun's surface.

solar nebula   The vast primordial cloud of gas and dust from which (it has been theorized) the Sun and solar system were formed.

solar wind   A continuous stream of radiation and matter that escapes from the Sun. Its effects may be seen in how it blows the tails of a comet approaching the Sun.

sounding rockets   Early suborbital rocket probes launched to study the upper atmosphere.

spectral lines   *See* **spectroscope.**

spectral classification   A system for classifying stars according to their surface temperature as measured by their spectra. The presence or absence of certain spectral lines is used to place stars in a spectral class.

spectrometer   *See* **spectroscope.**

spectroscope   An instrument that passes incoming light through a slit and prism, splitting it into its component colors. A spectrometer is an instrument capable of precisely measuring the spectrum thus produced. Substances produce characteristic spectral lines or emission lines, which act as the "fingerprint" of the substance, enabling identification of it.

spectroscopic binaries   *See* **visual binaries.**

spicules   Jets of matter expelled from the Sun's photosphere region into the chromosphere above it.

spiral arms   Structures found in spiral galaxies apparently caused by the action of spiral density waves.

spiral density waves   Waves of compression that move around the disk of a spiral galaxy. These waves are thought to trigger clouds of gas into collapse, thus forming hot, young stars. It is mostly the ionized gas around these young, massive stars that we observe as spiral arms.

spiral galaxy   A galaxy characterized by a distinctive structure consisting of a thin disk surrounding a galactic bulge. The disk is dominated by bright, curved arcs of emission known as spiral arms. The rotation curves of spiral galaxies indicate the presence of large amounts of dark matter.

standard candle   Any object whose luminosity is well known. Its measured brightness can then be used to determine how far away the object is. The brightest standard candles can be seen from the greatest distances.

standard solar model   Our current picture of the structure of the Sun. The model seeks to explain the observable properties of the Sun and also to describe properties of its mostly unobservable interior.

stellar occultation   An astronomical event that occurs when a planet passes in front of a star, dimming the star's light (as seen from the Earth). The exact way in which the light dims can reveal details, for example, in the planet's atmosphere.

summer solstice   On or about June 21; the longest day in the Northern Hemisphere, marking the beginning of summer.

sunspots   Irregularly shaped dark areas on the face of the Sun. They appear dark because they are cooler than the surrounding material. They are tied to the presence of magnetic fields at the Sun's surface.

Supercluster   A group of galaxy clusters. The Local Supercluster contains some 10^{15} solar masses.

superluminal motion   A term for the apparent "faster than light" motion of blobs of material in some radio jets. This effect results from radio-emitting blobs moving at high velocity toward the observer.

supernova   The explosion accompanying the death of a massive star as its core collapses.

synchronous orbit   A celestial object is in synchronous orbit when its period of rotation is equal to its average orbital period; the Moon, in synchronous orbit, presents only one face to the Earth.

synchrotron radiation   Synchrotron radiation arises when charged particles (e.g., electrons) are accelerated by strong magnetic fields. Some of the emission from radio galaxies is synchrotron.

synodic month   The period of 29.5 days that the Moon requires to cycle through its phases, from new Moon to new Moon. *See also* **sidereal month.**

telescope   A word from Greek roots meaning "far-seeing." Optical telescopes are arrangements of lenses and/or mirrors designed to gather visible light efficiently enough to enhance resolution and sensitivity. *See also* **radio telescope.**

terminator   The boundary separating light from dark, the daytime from nighttime hemispheres of the Moon (or another planetary or lunar body).

terrestrial planets   The planets in our solar system closest to the Sun: Mercury, Venus, Earth, and Mars.

thought experiment   A systematic hypothetical or imaginary simulation of reality, used as an alternative to actual experimentation when such experimentation is impractical or impossible.

tidal bulge   The deformation of one celestial body caused by the gravitational force of another extended celestial body. The Moon creates an elongation of the Earth's oceans—a tidal bulge.

time dilation   The apparent slowing of time (as perceived by an outside observer) as an object approaches the event horizon of a black hole, or moves at very high velocity.

trailing face   *See* **leading face.**

transverse component   Stellar movement across the sky, perpendicular to our line of sight. The radial component is motion toward or away from us. True space motion is calculated by combining the observed transverse and radial components.

triangulation   An indirect method of measuring distance derived by geometry or trigonometry using a known baseline and two angles from the baseline to the object.

tropical year   A year measured from equinox to equinox. *See* **sidereal year.**

Tully-Fisher relation   A correlation of the rotational velocity of any spiral galaxy with its luminosity.

universal recession   The apparent general movement of all galaxies away from us. This observation does not mean we are at the center of the expansion. Any observer located anywhere in the universe should see the same redshift.

Van Allen belts   Named for their discoverer, American physicist James A. Van Allen, these are vast doughnut-shaped zones of highly energetic, charged particles that are trapped in the magnetic field of the Earth. The zones were discovered in 1958.

variable star   A star that periodically changes in brightness. A cataclysmic variable is a star, such as a nova or supernova, that changes in brightness suddenly and dramatically as a result of interaction with a binary companion star, while an intrinsic variable changes brightness because of rapid changes in its diameter. Pulsating variables are intrinsic variables that vary in brightness in a fixed period or span of time and are useful distance indicators.

vernal equinox   On or about March 21; on this date, the beginning of spring, day and night are of equal duration because the sun's apparent course intersects the celestial equator at these times.

Very-Long Baseline Interferometry (VBLI)   A system combining signals from radio telescopes distant from one another in order to achieve very high degrees of image resolution.

visual binaries   Binary stars that can be resolved from the Earth. Spectroscopic binaries are too distant to be seen as distinct points of light, but they can be observed with a spectroscope. In this case, the presence of a binary system is detected by noting Doppler-shifting spectral lines as the stars orbit one another. If the orbit of one star in a binary system periodically eclipses its partner, it's possible to monitor the variations of light emitted from the system and thereby gather information about orbital motion, mass, and radii. These binaries are called eclipsing binaries.

water hole   The span of the radio spectrum from 18 cm to 21 cm, which many researchers believe is the most likely wavelength on which extraterrestrial broadcasts will be made. The name is a little astronomical joke—the hydrogen (H) and hydroxyl (OH) lines are both located in a quiet region of the radio spectrum, a region where there isn't a lot of background noise. Since H and OH add up to H_2O (water), this dip in the spectrum is called the water hole.

wavelength   The distance between two adjacent wave crests (high points) or troughs (low points). By convention, this distance is measured in meters or decimal fractions thereof.

white dwarf   The remnant core of a red giant after it has lost its outer layers as a planetary nebula. Since fusion has now halted, the carbon-oxygen core is supported against further collapse only by the pressure supplied by densely packed electrons.

winter solstice   On or about December 21; in the Northern Hemisphere, the shortest day and the start of winter.

zonal flow   The prevailing east-west wind pattern that is found on Jupiter.

Upcoming Eclipses

Total Solar Eclipses

Date	Duration of Totality (in Minutes)	Where Visible
2002 Dec. 4	2.1	South Africa, Australia
2003 Nov. 23	2.0	Antarctica
2005 Apr. 8	0.7	South Pacific Ocean
2006 Mar. 29	4.1	Africa, Asia Minor, Russia, and the C.I.S.
2008 Aug. 1	2.4	Arctic Ocean, Siberia, China
2009 July 22	6.6	India, China, South Pacific
2010 July 11	5.3	South Pacific
2012 Nov. 13	4.0	Northern Australia, South Pacific
2013 Nov. 3	1.7	Atlantic Ocean, Central Africa
2015 Mar. 20	4.1	North Atlantic, Arctic Ocean
2016 Mar. 9	4.5	Indonesia, Pacific Ocean
2017 Aug. 21	2.0	Pacific Ocean, United States, Atlantic Ocean

Total Lunar Eclipses

2003 May 16
2003 Nov. 9
2004 May 4
2004 Oct. 28
2007 Mar. 3
2007 Aug. 28
2008 Feb. 21

Appendix C

The Constellations

Name	Genitive	Abbreviation	Meaning	Approximate Position R.A. (hours)	Approximate Position DEC (degrees)
Andromeda	Andromedae	And	Chained Maiden	1	+40
Antlia	Antliae	Ant	Air Pump	10	–35
Apus	Apodis	Aps	Bird of Paradise	16	–75
Aquarius	Aquarii	Aqr	Water Bearer	23	–15
Aquila	Aquilae	Aql	Eagle	20	+5
Ara	Arae	Ara	Altar	17	–55
Aries	Arietis	Ari	Ram	3	+20
Auriga	Aurigae	Aur	Charioteer	6	+40
Boötes	Boötis	Boo	Herdsmen	15	+30
Caelum	Caeli	Cae	Chisel	5	–40
Camelopardalis	Camelopardalis	Cam	Giraffe	6	+70
Cancer	Cancri	Cnc	Crab	9	+20
Canes Venatici	Canum Venaticorum	CVn	Hunting Dogs	13	+40
Canis Major	Canis Majoris	CMa	Great Dog	7	–20
Canis Minor	Canis Minoris	CMi	Little Dog	8	+5
Capricornus	Capricorni	Cap	Sea-goat	21	–20
Carina	Carinae	Car	Keel	9	–60
Cassiopeia	Cassiopeiae	Cas	Queen	1	+60
Centaurus	Centauri	Cen	Centaur	13	–50
Cepheus	Cephei	Cep	King	22	+70

Name	Genitive	Abbreviation	Meaning	Approximate Position R.A. (hours)	DEC (degrees)
Cetus	Ceti	Cet	Whale	2	−10
Chamaeleon	Chamaeleontis	Cha	Chameleon	11	−80
Circinus	Circini	Cir	Compasses (art)	15	−60
Columba	Columbae	Col	Dove	6	−35
Coma Berenices	Comae Berenices	Com	Berenice's Hair	13	+20
Corona Australis	Coronae Australis	CrA	Southern Crown	19	−40
Corona Borealis	Coronae Borealis	CrB	Northern Crown	16	+30
Corvus	Corvi	Crv	Crow	12	−20
Crater	Crateris	Crt	Cup	11	−15
Crux	Crucis	Cru	Cross (southern)	12	−60
Cygnus	Cygni	Cyg	Swan	21	+40
Delphinus	Delphini	Del	Dolphin	21	+10
Dorado	Doradus	Dor	Swordfish	3	−20
Draco	Draconis	Dra	Dragon	17	+65
Equuleus	Equulei	Equ	Little Horse	21	+10
Eridanus	Eridani	Eri	River	3	−20
Fornax	Fornacis	For	Furnace	3	−30
Gemini	Geminorum	Gem	Twins	7	+20
Grus	Gruis	Gru	Crane (bird)	22	−45
Hercules	Herculis	Her	Hercules	17	+30
Horologium	Horologii	Hor	Clock	3	−60
Hydra	Hydrae	Hya	Water Snake (female)	10	−20
Hydrus	Hydri	Hyi	Water Snake (male)	2	−75
Indus	Indi	Ind	Indian	21	−55
Lacerta	Lacertae	Lac	Lizard	22	+45
Leo	Leonis	Leo	Lion	11	+15
Leo Minor	Leonis Minoris	LMi	Little Lion	10	+35
Lepus	Leporis	Lep	Hare	6	−20
Libra	Librae	Lib	Balance	15	−15
Lupus	Lupi	Lup	Wolf	15	−45
Lynx	Lyncis	Lyn	Lynx	8	+45
Lyra	Lyrae	Lyr	Lyre	19	+40
Mensa	Mensae	Men	Table Mountain	5	−80
Microscopium	Microscopii	Mic	Microscope	21	−35

Name	Genitive	Abbreviation	Meaning	Approximate Position R.A. (hours)	DEC (degrees)
Monoceros	Monocerotis	Mon	Unicorn	7	−5
Musca	Muscae	Mus	Fly (insect)	12	−70
Norma	Normae	Nor	Square (rule)	16	−50
Octans	Octantis	Oct	Octant	22	−85
Ophiuchus	Ophiuchi	Oph	Serpent Bearer	17	0
Orion	Orionis	Ori	Hunter	5	+5
Pavo	Pavonis	Pav	Peacock	20	−65
Pegasus	Pegasi	Peg	Flying Horse	22	+20
Perseus	Persei	Per	Hero	3	+45
Phoenix	Phoenicis	Phe	Phoenix	1	−50
Pictor	Pictoris	Pic	Easel	6	−55
Pisces	Piscium	Psc	Fishes	1	+15
Piscis Austrinius	Piscis Austrini	PsA	Southern Fish	22	−30
Puppis	Puppis	Pup	Stern (deck)	8	−40
Pyxis	Pyxidis	Pyx	Compass (sea)	9	−30
Reticulum	Reticuli	Ret	Net	4	−60
Sagitta	Sagittae	Sge	Arrow	20	+10
Sagittarius	Sagittarii	Sgr	Archer	19	−25
Scorpius	Scorpii	Sco	Scorpion	17	−40
Sculptor	Sculptoris	Scl	Sculptor's tools	0	−30
Scutum	Scuti	Sct	Shield	19	−10
Serpens	Serpentis	Ser	Serpent	17	0
Sextans	Sextantis	Sex	Sextant	10	0
Taurus	Tauri	Tau	Bull	4	+15
Telescopium	Telescopii	Tel	Telescope	19	−50
Triangulum	Trianguli	Tri	Triangle	2	+30
Triangulum Australe	Trianguli Australis Southern	TrA	Triangle	16	−65
Tucana	Tucanae	Tuc	Toucan	0	−65
Ursa Major	Ursae Majoris	UMa	Great Bear	11	+50
Ursa Minor	Ursae Minoris	UMi	Little Bear	15	+70
Vela	Velorum	Vel	Sail	9	−50
Virgo	Virginis	Vir	Maiden	13	0
Volans	Volantis	Vol	Flying Fish	8	−70
Vulpecula	Vulpeculae	Vul	Fox	20	+25

The Messier Catalog

Charles Messier (1730–1817) was a French astronomer who was the first to compile a systematic catalog of nebulae and star clusters. His purpose in this was to make comet hunting easier by taking careful note of permanent deep-sky objects that might be mistaken for comets. In Messier's time, the term *nebula* was applied to any blurry celestial object. The Messier Catalog that resulted is valued today by amateur astronomers as a checklist of deep sky objects easily observable with even a modest telescope. The 110 Messier objects are considered the most interesting deep-sky objects visible in the Northern Hemisphere. Some astronomy clubs award pins for successful observing the Messier list. Contact your local astronomy club for details.

For images of all the Messier objects, see www.seds.org/messier.

Legend

M# = Messier Catalog number

NGC# = *New General Catalog Number* (The *NGC* is a modern work, which, among professional astronomers, has superseded the Messier Catalog.)

Con = The constellation in which the object will be found. See Appendix C for the full names of the constellations abbreviated here.

RA = Right ascension in hours and minutes. See Chapter 1 for an explanation.

DEC = Declination in degrees and minutes. See Chapter 1 for an explanation.

vMag = Visual Magnitude of the object. See Chapter 17 for an explanation of magnitude.

Type = Type of object. The abbreviations are as follows:

DS	Double star
E (with a number)	Elliptical galaxy
EN	Emission nebula (glowing gas)
ERN	Emission and reflection nebula
GC	Globular cluster (stars)
Irr	Irregular galaxy
OC	Open cluster (stars)
PN	Planetary nebula
S	Spiral galaxy
SB	Barred spiral galaxy
SCL	Star cloud
SNR	Supernova remnant

Dia" = Angular diameter (where known) of the object. This number is in seconds of arc ($1/3600$ degree). See Chapter 1 for an explanation of angular measurement.

Name = The common name of the object

M#	NGC#	Con	RA	Dec	vMag	Type	Dia"	Name
1	1952	Tau	05h 34.5m	+22d 01'	8.4	SNR	300	Crab Nebula
2	7089	Aqr	21h 33.5m	−00d 49'	6.5	GC		
3	5272	CVn	13h 42.2m	+28d 23'	6.4	GC		
4	6121	Sco	16h 23.6m	−26d 32'	5.9	GC		
5	5904	Ser	15h 18.6m	+02d 05'	5.8	GC		
6	6405	Sco	17h 40.1m	−32d 13'	4.2	OC		Butterfly Cluster
7	6475	Sco	17h 53.9m	−34d 49'	3.3	OC		
8	6523	Sgr	18h 03.8m	−24d 23'	5.8	EN	2400	Lagoon Nebula
9	6333	Oph	17h 19.2m	−18d 31'	7.9	GC		
10	6254	Oph	16h 57.1m	−04d 06'	6.6	GC		
11	6705	Sct	18h 51.1m	−06d 16'	5.8	OC	750	Wild Duck Cluster
12	6218	Oph	16h 47.2m	−01d 57'	6.6	GC		
13	6205	Her	16h 41.7m	+36d 28'	5.9	GC		Hercules Cluster
14	6402	Oph	17h 37.6m	−03d 15'	7.6	GC		
15	7078	Peg	21h 30.0m	+12d 10'	6.4	GC		
16	6611	Ser	18h 18.8m	−13d 47'	6.0	EN		Eagle Nebula
17	6618	Sgr	18h 20.8m	−16d 11'	7.0	EN	1200	Swan Nebula (also known as Omega Nebula)
18	6613	Sgr	18h 19.9m	−17d 08'	6.9	OC		
19	6273	Oph	17h 02.6m	−26d 16'	7.2	GC		
20	6514	Sgr	18h 02.6m	−23d 02'	8.5	ERN	900	Trifid Nebula
21	6531	Sgr	18h 04.6m	−22d 30'	5.9	OC		
22	6656	Sgr	18h 36.4m	−23d 54'	5.1	GC		
23	6494	Sgr	17h 56.8m	−19d 01'	5.5	OC		
24		Sgr	18h 16.9m	−18d 29'	4.5	SCL		(star cloud)
25	4725	Sgr	18h 31.6m	−19d 15'	4.6	OC		
26	6694	Sct	18h 45.2m	−09d 24'	8.0	OC		
27	6853	Vul	19h 59.6m	+22d 43'	8.1	PN	420	Dumbbell Nebula
28	6626	Sgr	18h 24.5m	−24d 52'	6.9	GC		
29	6913	Cyg	20h 23.9m	+38d 32'	6.6	OC		
30	7099	Cap	21h 40.4m	−23d 11'	7.5	GC		
31	224	And	00h 42.7m	+41d 16'	3.4	SB	14400	Great Andromeda Sp.

continues

continued

M#	NGC#	Con	RA	Dec	vMag	Type	Dia"	Name
32	221	And	00h 42.7m	+40d 52'	8.2	E2		(M31 companion)
33	598	Tri	01h 33.9m	+30d 39'	5.7	S	3600	Triangulum Spiral
34	1039	Per	02h 42.0m	+42d 47'	5.2	OC		
35	2168	Gem	06h 08.9m	+24d 20'	5.1	OC		
36	1960	Aur	05h 36.1m	+34d 08'	6.0	OC		
37	2099	Aur	05h 52.4m	+32d 33'	5.6	OC		
38	1912	Aur	05h 28.7m	+35d 50'	6.4	OC		
39	7092	Cyg	21h 32.2m	+48d 26'	4.6	OC		
40		UMa	12h 22.4m	+58d 05'	8.0	DS		Winnecke 4
41	2287	CMa	06h 47.0m	−20d 44'	4.5	OC		
42	1976	Ori	05h 35.4m	−05d 27'	4.0	EN		Great Orion Nebula
43	1982	Ori	05h 35.6m	−05d 16'	9.0	EN		Orion Nebula (part)
44	2632	Cnc	08h 40.1m	+19d 59'	3.1	OC	5400	Beehive Cluster
45		Tau	03h 47.0m	+24d 07'	1.2	OC	7200	Pleiades
46	2437	Pup	07h 41.8m	−14d 49'	6.1	OC		
47	2422	Pup	07h 36.6m	−14d 30'	4.4	OC		
48	2548	Hya	08h 13.8m	−05d 48'	5.8	OC		
49	4472	Vir	12h 29.8m	+08d 00'	8.4	E4		
50	2323	Mon	07h 03.2m	−08d 20'	5.9	OC		
51	194&5	CVn	13h 29.9m	+47d 12'	8.1	S		Whirlpool galaxy
52	7654	Cas	23h 24.2m	+61d 35'	6.9	OC		
53	5024	Com	13h 12.9m	+18d 10'	7.7	GC		
54	6715	Sgr	18h 55.1m	−30d 29'	7.7	GC		
55	6809	Sgr	19h 40.0m	−30d 58'	7.0	GC		
56	6779	Lyr	19h 16.6m	+30d 11'	8.2	GC		
57	6720	Lyr	18h 53.6m	+33d 02'	9.0	PN		Ring Nebula
58	4579	Vir	12h 37.7m	+11d 49'	9.8	SB		
59	4621	Vir	12h 42.0m	+11d 39'	9.8	E3		
60	4649	Vir	12h 43.7m	+11d 33'	8.8	E1		
61	4303	Vir	12h 21.9m	+04d 28'	9.7	S		
62	6266	Oph	17h 01.2m	−30d 07'	6.6	GC		
63	5055	CVn	13h 15.8m	+42d 02'	8.6	SB		Sunflower galaxy
64	4826	Com	12h 56.7m	+21d 41'	8.5	SB		Black Eye galaxy

M#	NGC#	Con	RA	Dec	vMag	Type	Dia"	Name
65	3623	Leo	11h 18.9m	+13d 05'	9.3	SB		
66	3627	Leo	11h 20.2m	+12d 59'	9.0	SB		
67	2682	Cnc	08h 50.4m	+11d 49'	6.9	OC		
68	4590	Hya	12h 39.5m	−26d 45'	8.2	GC		
69	6637	Sgr	18h 31.4m	−32d 21'	7.7	GC		
70	6681	Sgr	18h 43.2m	−32d 18'	8.1	GC		
71	6838	Sge	19h 53.8m	+18d 47'	8.3	GC		
72	6981	Aqr	20h 53.5m	−12d 32'	9.4	GC		
73	6994	Aqr	20h 58.9m	−12d 38'	5.0	OC		
74	628	Psc	01h 36.7m	+15d 47'	9.2	S		
75	6864	Sgr	20h 06.1m	−21d 55'	8.6	GC		
76	650-1	Per	01h 42.4m	+51d 34'	11.5	PN		Little Dumbbell
77	1068	Cet	02h 42.7m	−00d 01'	8.8	SB		
78	2068	Ori	05h 46.7m	+00d 03'	8.0	PN		
79	1904	Lep	05h 24.5m	−24d 33'	8.0	GC		
80	6093	Sco	16h 17.0m	−22d 59'	7.2	GC		
81	3031	UMa	09h 55.6m	+69d 04'	6.8	SB		
82	3034	UMa	09h 55.8m	+69d 41'	8.4	Irr		"Exploding" galaxy
83	5236	Hya	13h 37.0m	−29d 52'	10.1	S		
84	4374	Vir	12h 25.1m	+12d 53'	9.3	E1		
85	4382	Com	12h 25.4m	+18d 11'	9.3	Ep		
86	4406	Vir	12h 26.2m	+12d 57'	9.2	E3		
87	4486	Vir	12h 30.8m	+12d 24'	8.6	E1		
88	4501	Com	12h 32.0m	+14d 25'	9.5	SB		
89	4552	Vir	12h 35.7m	+12d 33'	9.8	E0		
90	4569	Vir	12h 36.8m	+13d 10'	9.5	SB		
91	4548	Com	12h 35.4m	+14d 30'	10.2	SB		
92	6341	Her	17h 17.1m	+43d 08'	6.5	GC		
93	2447	Pup	07h 44.6m	−23d 52'	6.2	OC		
94	4736	CVn	12h 50.9m	+41d 07'	8.1	SB		
95	3351	Leo	10h 44.0m	+11d 42'	9.7	SB		
96	3368	Leo	10h 46.8m	+11d 49'	9.2	SB		
97	3587	UMa	11h 14.8m	+55d 01'	11.2	PN		Owl Nebula
98	4192	Com	12h 13.8m	+14d 54'	10.1	SB		
99	4254	Com	12h 18.8m	+14d 25'	9.8	S		
100	4321	Com	12h 22.9m	+15d 49'	9.4	S		

continues

continued

M#	NGC#	Con	RA	Dec	vMag	Type	Dia"	Name
101	5457	UMa	14h 03.2m	+54d 21'	7.7	S		Pinwheel galaxy
102	5866	Dra	15h 06.5m	+55d 46'	10.0	E6		
103	581	Cas	01h 33.2m	+60d 42'	7.4	OC		
104	4594	Vir	12h 40.0m	−11d 37'	8.3	SB		Sombrero galaxy
105	3379	Leo	10h 47.8m	+12d 35'	9.3	E1		
106	4258	CVn	12h 19.0m	+47d 18'	8.3	SB		
107	6171	Oph	16h 32.5m	−13d 03'	8.1	GC		
108	3556	UMa	11h 11.5m	+55d 40'	10.0	S		
109	3992	UMa	11h 57.6m	+53d 23'	9.8	SB		
110	205	And	00h 40.4m	+41d 41'	8.0	E6		(M31 companion)

Sources for Astronomers

Web Sites

A wealth of astronomy-related information and images are available on the World Wide Web. We've listed a few highlights for you to check out, but don't hesitate to use a good search engine (such as Google or Yahoo!) to look for additional or more specific information.

Institutions, Magazines, and Societies

Amateur Telescope Makers Association. Valuable for the do-it-yourselfer. **http://www.atmjournal.com**

Astronomy magazine. Includes links to other sites. **http://www.kalmbach.com/astro/astronomy.html**

Institute and Museum of History of Science, Florence, Italy. Includes a wealth of multimedia material (in English) devoted to the work of Galileo. **http://www.imss.fi.it**

Jet Propulsion Laboratory. NASA-related facility. A lot of information on solar system exploration projects. Many images. A must-see site. **http://www.jpl.nasa.gov/**

The Planetary Society. Founded by Carl Sagan to encourage the search for extraterrestrial life. **http://planetary.org**

SETI Institute. Dedicated to the search for extraterrestrial civilizations. **http://www.seti-inst.edu/**

Sky & Telescope magazine. An important source of information (dates, directions) on astronomical events. **http://www.skypub.com/**

Some Observatories

Anglo-Australian Observatory: **http://www.aao.gov.au**

Bradley Observatory (on-campus observatory at Agnes Scott College, built in 1950): **http://bradley.agnesscott.edu**

Keck Observatory: **http://astro.caltech.edu/mirror/keck/index.html**

Mount Wilson Observatory: **http://www.mtwilson.edu**

National Optical Astronomy Observatories: **http://www.noao.edu**

National Radio Astronomy Observatory: **http://www.nrao.edu**

Space Telescope Science Institute (source of *Hubble Space Telescope* images): **http://www.stsci.edu/public.html**

Space Telescope Science Institute Press Release Page: **http://oposite.stsci.edu/pubinfo/pr.html**

United States Naval Observatory (USNO). Information on timekeeping, sunrise and sunset times: **http://www.usno.navy.mil**

Yerkes Observatory, University of Chicago: **http://astro.uchicago.edu/yerkes/**

Planetary Positions, Solar System

Planetary position calculator: **http://imagiware.com/astro/planets.cgi**

Scale model solar system: **www.exploratorium.edu/ronh/solar_system**

Guides to Events (Eclipses, Meteor Showers, and So On)

Abrams Planetarium Sky Calendar: **http://www.pa.msu.edu/abrams/**

Mt. Wilson Online Stargazer Map: **http://www.mtwilson.edu/Services/Star Map**

For solar eclipse information, including link to the SOHO site, with frequently updated solar images at many wavelengths: **http://umbra.nascom.nasa.gov/sdac.html**

For comets and meteor showers: **http://comets.amsmeteors.org**

Images and Catalogs

Astronomy Pictures of the Day: **http://antwrp.gsfc.nasa.gov/apod/astropix.html**

The Messier Catalog (a visual catalog of Messier objects): **www.seds.org/messier**

Sky View. Truely a virtual observatory: **http://skyview.gsfc.nasa.gov/**

Links

The following sites serve as links to many others.

Astronomical Society of the Pacific: **http://www.aspsky.org/**

This site includes a link to *Mercury Magazine*.

American Astronomical Society (AAS): **http://www.aas.org**

Contains links to the online versions of the *Astronomical Journal* and the *Astrophysical Journal* in addition to information and statistics about the profession.

Astronomical World Wide Web Resources: **http://cws.stsci.edu/astroweb/net-www.html**

Astronomy Links: **http://www.open.hr/space/space/http.phtml**

Magazines

Astronomy (Kalmbach Publishing, PO Box 1612, Waukesha, WI 53187). A popular and well-written monthly journal.

Griffith Observer (Griffith Observatory, 2800 E. Observatory Rd., Los Angeles, CA 90027). Concentrates on the history of astronomy.

Mercury Magazine (390 Ashton Avenue, San Francisco, CA, 94112). A publication of the Astronomical Society of the Pacific (ASP); contains historical and scientific articles as well as excellent monthly columns.

Planetary Report (The Planetary Society, 65 N. Catalina Ave., Pasadena, CA 91106). Provides news on exploring the solar system and the search for extraterrestrial life.

Sky & Telescope (PO Box 9111, Belmont, MA 02178). Considered by some to be the standard for amateur astronomy magazines.

Books

As we told you, you're not alone in your fascination with astronomy. You'll find no shortage of books on the subject. The following are a few that are particularly useful.

Practical Guides

Berry, Richard. *Discover the Stars*. New York: Harmony Books, 1987.

Burnham, Robert, Jr. *Burnham's Celestial Handbook*. New York: Dover, 1978.

Carlson, Shawn, ed. *Amateur Astronomer*. New York: Wiley, 2000.

Charles, Jeffrey R. *Practical Astrophotography*. New York: Springer Verlag, 2000.

409

Covington, Michael A. *Astrophotography for the Amateur.* New York: Cambridge University Press, 1999.

Harrington, Philip S. *Star Ware: The Amateur Astronomer's Ultimate Guide to Choosing, Buying, and Using Telescopes and Accessories, 2nd ed.* New York: Wiley, 1998.

Levy, David H. *The Sky: A User's Guide.* Cambridge, England: Cambridge University Press, 1991.

Licher, David. *The Universe From Your Backyard.* Milwaukee: Kalmbach Publishing, 1988.

Mayall, R. Newton, et al. *The Sky Observer's Guide: A Handbook for Amateur Astronomers.* New York: Golden Books, 2000.

North, Gerald. *Advanced Amateur Astronomy.* New York: Cambridge University Press, 1997.

Tonkin, Stephen F., ed. *Amateur Telescope Making.* New York: Springer Verlag, 1999.

Guides to Events

Bishop, Roy, ed. *The Observer's Handbook.* Toronto: The Royal Astronomical Society of Canada, annual.

Westfall, John E., ed. *The ALPO Solar System Ephemeris.* San Francisco: Association of Lunar and Planetary Observers, annual.

Star and Lunar Atlases

Cook, Jeremy. *The Hatfield Photographic Lunar Atlas.* New York: Springer Verlag, 1999.

Dickinson, Terence, et al. *Mag 6 Star Atlas.* Barrington, NJ: Edmund Scientific, 1982.

Norton, Arthur P. *Norton's 2000.0, 18th ed.* Cambridge, MA: Sky Publishing Corporation, 1989.

Tirion, Wil. *Sky Atlas 2000.0.* Cambridge, MA: Sky Publishing Corporation and Cambridge University Press, 1981.

Introductory Textbooks and Popular Science Books

Chaisson, Eric, and Steve McMillan. *Astronomy: A Beginner's Guide to the Universe, 2nd ed.* Upper Saddle River, NJ: Prentice Hall, 1998.

Ferris, Timothy. *The Whole Shebang.* New York: Simon & Schuster, 1997.

Fraknoi, Andrew, et al. *Voyages Through the Universe.* Fort Worth, TX: Saunders College Publishing, 1997.

Hartman, William K. *Moons & Planets, 3d ed.* Belmont, CA: Wadsworth Publishers, 1993.

Kaufmann, William J., and Neil F. Comins. *Discovering the Universe, 4th ed.* New York: W.H. Freeman, 1996.

Life in the Universe

Davies, Paul. *The Fifth Miracle: The Search for the Origin of Life,* New York, NY: Simon & Schuster, 1999.

Goldsmith, Donald, and Tobias Owen. *The Search for Life in the Universe, 3rd ed.* Reading, MA: Addison-Wesley, 2001.

Index

Q-R

X-Y-Z